THE GLOBAL CASINO

THE GLOBAL CASINO

AN INTRODUCTION TO ENVIRONMENTAL ISSUES

Nick Middleton

Edward Arnold
A member of the Hodder Headline Group
LONDON NEW YORK SYDNEY AUCKLAND

First published in Great Britain 1995 by
Edward Arnold, a division of Hodder Headline PLC,
338 Euston Road, London NW1 3BH

Co-published in the USA by Halsted Press, an imprint of John
Wiley & Sons, Inc., 605 Third Avenue, New York, NY 10158–0012

British Library Cataloguing in Publication Data
A catalogue record for this book is available from the British
Library

Library of Congress Cataloging-in-Publication Data
A catalog record for this book is available from the Library of
Congress

ISBN 0 340 63210 0 (hb)
ISBN 0 340 59493 4 (pb)

1 2 3 4 5 95 96 97 98 99

Composition by Scribe Design, Gillingham, Kent UK
Printed by The Bath Press, Bath, Avon

CONTENTS

Preface   *vii*

Acknowledgements   *ix*

1   The Physical Environment   *1*

2   The Human Environment   *15*

3   Deforestation   *28*

4   Desertification   *43*

5   Food Production   *56*

6   Soil Erosion   *71*

7   Threatened Species   *85*

8   Climatic Change   *99*

9   Acid Rain   *114*

10   Oceans   *127*

11   Coastal Problems   *141*

12   Rivers, Lakes and Wetlands   *155*

13   Big Dams   *171*

14   Urban Environments   *184*

15   Transport   *198*

16   Waste Management   *212*

17   Energy Production   *224*

18   Mining   *237*

19   War   *251*

20   Natural Hazards   *265*

21   Conservation and Sustainable Development   *282*

Glossary   *298*

Bibliography   *302*

Index   *323*

PREFACE

This book is about environmental issues: concerns that have arisen as a result of the human impact on the environment and the ways in which the natural environment affects human society. The book deals both with the workings of the physical environment and the political, economic and social frameworks in which the issues occur. Using examples from all over the world, I have aimed to highlight the underlying causes behind environmental problems, the human actions which have made them issues, and the hopes for solutions.

Eighteen chapters on key issues follow the two initial chapters which outline the background contexts of the physical and human environments, and the book finishes with a chapter on conservation and sustainable development. The organisation of the book, therefore, allows it to be read in its entirety or dipped into for any particular topic, since each chapter stands on its own. Each chapter sets the issue in a historical context, outlines why the issue has arisen, highlights areas of controversy and uncertainty, and appraises how problems are being, and can be, resolved, both technically and in political and economic frameworks. Every chapter is followed by a critical guide to further reading on the subject, and the text is supported by a glossary of key terms.

I decided on the title *The Global Casino* because there are many parallels between the issues discussed here and the workings of a gambling joint. Money and economics underlie many of the eighteen issues covered here, which can be thought of as different games in the global casino, separate yet interrelated. Just like a casino, environmental issues involve winners and losers. The casino's chance element and the players' imperfect knowledge of the outcomes of their actions are relevant in that our understanding of how the Earth works is far from perfect. The casino metaphor also works on a socio-economic level, since some individuals and groups of individuals can afford to take part in the games while others are less able. Some groups are more responsible for certain issues than others, yet those who have little influence are still affected by the consequences. Different individuals and groups of people also choose to play the different games in different ways, reflecting their cultural, economic and political backgrounds and the information available to them.

The stakes are high: some observers believe that the global scale on which many of the issues occur represents humankind gambling with the very future of the planet itself. Everyone who reads this book has some part to play in the 'Global Casino'. I hope that the information presented here will allow those players to participate with a reasonable knowledge of how the games work, the consequences of losing, and the benefits that can be derived from winning.

ACKNOWLEDGEMENTS

I am indebted to many people who have helped in a variety of ways during the research and writing of this book. I would like to thank the following for their helpful comments on draft chapters. Colleagues at the School of Geography in Oxford: Tim Burt, Rachael McDonnell, Alisdair Rogers, Heather Viles and Robert Whittaker, and Andrew Goudie who commented on the entire manuscript; Brenda Boardman and Dave Favis-Mortlock at Oxford's Environmental Change Unit; Mark Carwardine; Helen Ellwood; David Thomas of the University of Sheffield Geography Department; Charles Toomer of Sedgwick Global, and Phillip Crowson, John Hughes and Jim Stevenson of RTZ.

I also thank a generation of undergraduates at the Oxford colleges of Oriel and St Anne's who have been exposed to my efforts at communicating the essence of environmental issues. These innumerable tutorials provided the basis for this book.

My thanks go to the technical staff at the School of Geography, University of Oxford for their patient and efficient help: Martin Barfoot for photographic work and Peter Hayward and Ailsa Allen who drew most of the figures. Paul Cole of the Cartographic Unit, University of Sheffield Geography Department drew the figures in Chapter 4. Thanks also to my editor at Edward Arnold, Laura McKelvie, for her encouragement and for taking me out to lunch a lot.

I am very grateful to the following for allowing me to reproduce their photographs: Mark Carwardine (figures 1.9, 7.5, 10.6 and 19.3), Charles Toomer (figures 1.5, 13.3 and 20.6), Stephen Stokes (figures 1.2 and 5.5), Australian Bureau of Meteorology (figure 6.5), US Department of Energy (figure 17.3) and the RTZ Corporation plc (figures 18.1, 18.4 and 18.5). All the other photographs are my own.

Thanks are due to the following for allowing me to reproduce figures: Academic Press Ltd (figure 18.3); American Association for the Advancement of Science (figures 7.3, 13.7 and 19.1); American Scientist (figure 10.4); Blackwell Publishers (figures 8.6 and 11.2); Blackwell Science Ltd (figure 16.3); Butterworth-Heinemann Ltd (figure 12.4); Cambridge University Press (figures 2.1 and 6.4); Chapman & Hall (figure 7.1); David Fulton Publishers Ltd (figure 13.1); Edward Arnold (figures 1.1, 2.3, 5.1 and 20.4); Elsevier Science (figures 2.2, 2.5, 2.8, 6.6 and 13.5); Helen Dwight Reid Educational Foundation (figure 20.5); HMSO (figure 16.1) – Crown copyright is reproduced with the permission of the Controller of HMSO; IOP Publishing Ltd (figure 12.2); Island Press (figure 17.1); John Wiley & Sons (figures 13.4, 13.6 and 11.3); Kluwer Academic Publishers (figure 15.5); McGill-Queen's University Press (figures 8.5 and 20.1); Oxford University Press (figures 3.1, 6.1 and 19.2); Woods Hole Oceanographic Institution (figure 11.4); Pan American Health Organisation (figure 14.8);

Penguin Books Ltd (figure 7.6); Routledge (figures 2.7, 12.1 and 20.3); Royal Swedish Academy of Sciences (figure 9.7); Scientific American (figure 19.5); sigma, Swiss Reinsurance Company (figure 20.2); Taylor & Francis Group Ltd (figure 11.6); The American Geographical Society (figure 5.6); The Swedish NGO Secretariat on Acid Rain (figures 9.3 and 9.4); The Swedish Society for Anthropology and Geography (figure 6.2); Tokyo Metropolitan Government (figure 14.2); United Nations Environment Programme (figures 4.1, 4.5, 8.3, 9.6, 10.5, 14.3, 14.4, 14.5, 14.6, 15.3, 16.2, 17.4 and 17.5); University of Chicago Press (figure 3.3); World Bank (figure 8.7); World Meteorological Organization – The Intergovernmental Panel on Climate Change – (figures 8.1 and 8.4).

Every effort has been made to trace the copyright holders of quoted material. If, however, there are inadvertent omissions these can be rectified in any future editions.

1

THE PHYSICAL ENVIRONMENT

The term environment is used in many ways. This book is about issues that arise from the physical environment which is made up of the living (biotic) and non-living (abiotic) things and conditions that characterise the world around us. While this is the central theme, the main reason for the topicality of the issues covered here is the way in which people interact with the physical environment. Hence, it is pertinent also to refer to the social, economic and political environments to describe those human conditions characteristic of certain places at particular times and to explain why conflict has arisen between human activity and the natural world. This chapter looks at some of the basic features of the physical environment, and Chapter 2 is concerned with the human factors that affect the ways in which the human race interacts with the physical world.

CLASSIFYING THE NATURAL WORLD

Geography, like other academic disciplines, classifies things in its attempt to understand how they work. The physical environment can be classified in numerous ways, but one of the most commonly used classifications is that which breaks it down into four interrelated spheres: the lithosphere, the atmosphere, the

biosphere and the hydrosphere. These four basic elements of the natural world can be further subdivided. The lithosphere, for example, is made up of rocks which are typically classified according to their modes of formation (igneous, metamorphic and sedimentary); these rock types are further subdivided according to the processes which formed them and other factors such as their chemical composition. Similarly, the workings of the atmosphere are manifested at the Earth's surface by a typical distribution of climates; the biosphere is made up of many types of flora and fauna, and the hydrosphere can be subdivided according to its chemical constituents (fresh water and saline, for example), or the condition or phase of the water: solid ice, liquid water or gaseous vapour.

These aspects of the natural world overlap and interact in many different ways. The nature of the soil in a particular place, for example, reflects the underlying rock type, the climatic conditions of the area, the plant and animal matter typical of the region and the quantity and quality of water available. Suites of characteristics are combined in particular areas called ecosystems. These ecosystems can also be classified in many ways, but one of the basic approaches uses the amount of organic matter or biomass produced per year – the net

TABLE 1.1 *Annual net primary production by major world ecosystem types*

Ecosystem type	Mean net primary productivity (g C/m²/year)	Total net primary production (billion tonnes C/year)
Tropical rain forest	900	15.3
Tropical seasonal forest	675	5.1
Temperate evergreen forest	585	2.9
Temperate deciduous forest	540	3.8
Boreal forest .	360	4.3
Woodland and shrubland	270	2.2
Savanna	315	4.7
Temperate grassland	225	2.0
Tundra and alpine	65	0.5
Desert scrub	32	0.6
Rock, ice and sand	1.5	0.04
Agricultural land	290	4.1
Swamp and marsh	1125	2.2
Lake and stream	225	0.6
Total land	324*	48.3
Open ocean	57	18.9
Upwelling zones	225	0.1
Continental shelf	162	4.3
Algal bed and reef	900	0.5
Estuaries	810	1.1
Total oceans	69*	24.9
Total for biosphere	144*	73.2

*The means for land, oceans and biosphere are weighted according to the areas covered by specific ecosystem types
Source: after Whittaker and Likens (1973)

production – which is simply the solar energy fixed in the biomass minus the energy used in producing it by respiration (see below). The annual net primary production of carbon, a basic component of all living organisms, by major world ecosystem types is shown in Table 1.1. Clear differences are immediately discernible between highly productive ecosystems such as forests, marshes, estuaries and reefs, and less-productive places such as deserts, tundras and the open ocean. All of the data are averaged and variability around the mean is perhaps greatest for agricultural ecosystems which, where intensively managed, can reach productivities as high as any natural ecosystem. One of the main reasons for agriculture's low average is the fact that fields are typically bare of vegetation for significant periods between harvest and sowing.

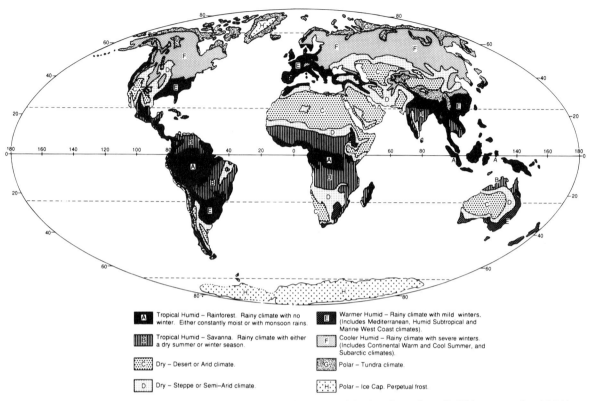

Tropical Humid – Rainforest. Rainy climate with no winter. Either constantly moist or with monsoon rains.

Tropical Humid – Savanna. Rainy climate with either a dry summer or winter season.

Dry – Desert or Arid climate.

Dry – Steppe or Semi–Arid climate.

Warmer Humid – Rainy climate with mild winters. (Includes Mediterranean, Humid Subtropical and Marine West Coast climates).

Cooler Humid – Rainy climate with severe winters. (Includes Continental Warm and Cool Summer, and Subarctic climates).

Polar – Tundra climate.

Polar – Ice Cap. Perpetual frost.

FIGURE 1.1 *Present-day morphoclimatic regions of the world's land surface (Williams* et al., *1993)*

One of the main factors determining productivity is the availability of nutrients, key substances for life on Earth: a lack of nutrients is often put forward to explain the low productivity in the open oceans, for example. Climate is another important factor. Warm, wet climates promote higher productivity than cold, dry ones. Differences in productivity may also go some way towards explaining the general trend of increasing diversity of plant and animal species from the poles to the equatorial regions. Despite many regional exceptions such as mountain tops and deserts, this latitudinal gradient of diversity is a striking characteristic of nature that fossil evidence suggests has been present in all geological epochs. The relationship with productivity is not straightforward, however, and many

other hypotheses have been advanced, such as the suggestion that minor disturbances promote diversity by preventing a few species from dominating and excluding others (Connell, 1978).

The relationships between climate and the biosphere are also reflected on the global scale in maps of vegetation and climate, the one reflecting the other. Figure 1.1 shows the world's morphoclimatic regions, which are a combination of both factors. Despite wide internal variations, immense continental areas clearly support distinctive forms of vegetation which are adapted to a broad climatic type. Such great living systems, which also support distinctive animals and to a lesser extent distinctive soils, are called biomes, a concept seldom applied to aquatic zones. Different ecologists produce various lists of biomes, and the following eight-fold classification may be considered conservative (Colinvaux, 1993):

- tundra
- coniferous forest (also known as boreal forest or taiga)
- temperate forest
- tropical rain forest
- tropical savanna
- temperate grassland
- desert
- maquis (also known as chaparral).

A notable aspect of the tundra biome is the absence of trees. Vegetation consists largely of grasses and other herbs, mosses, lichens and some small woody plants which are adapted to a short summer growing season. The cold climate ensures that the shallow soils are deeply frozen (permafrost) for all or much of the year, a condition which underlies about 20 per cent of the Earth's land surface. Many animals hibernate or migrate in the colder season, while others such as lemmings live beneath the snow. The main tundra region is located in the circumpolar lands north of the Arctic circle, which are bordered to the south by the evergreen, needle-leaved boreal or taiga forests. Here, winters are very cold, as in the tundra, but summers are longer. These forests are subject to periodic fires, and a burn–regeneration cycle is an important characteristic to which populations of deer, bears and insects, as well as the vegetation, are adapted. Much of the boreal forest is underlain by acid soils.

The temperate forests, by contrast, are typically deciduous. They are, however, like the boreal forests in that they are found almost exclusively in the Northern hemisphere. This biome is characteristic of northern Europe, eastern China, and eastern and midwest USA, with small stands in the Southern hemisphere in South America and New Zealand (Fig. 1.2). Tall broadleaf trees dominate, the climate is seasonal but water is

FIGURE 1.2 *Temperate forest dominated by beech trees on Stewart Island, off South Island, New Zealand (courtesy of Stephen Stokes)*

always abundant during the growing season, and this biome is less homogeneous than tundra or boreal forest. Amphibians, such as salamanders and frogs, are present, while they are almost totally absent from the higher-latitude biomes.

The tropical rain forest climate has copious rainfall and warm temperatures in all months of the year. The trees are always green, typically broad-leaved, and most are pollinated by animals (trees in temperate and boreal forests, by contrast, are largely pollinated by wind). Many kinds of vines (llianas) and epiphytes, such as ferns and orchids, are characteristic. Most of the nutrients are

FIGURE 1.3 *Temperate grassland in central Mongolia*

FIGURE 1.4 *This strange-looking plant, the welwitschia, is found only in the Namib Desert. Its adaptations to the dry conditions include long roots to take up any moisture in the gravelly soil and the ability to take in moisture from fog through its leaves. The welwitschia's exact position in the plant kingdom is controversial, but it is grouped with the pine trees*

stored in the biomass and the soils contain little organic matter. Above all, tropical rain forests are characterised by a large number of species of both plants and animals.

Savanna belts flank the tropical rain forests to the north and south in the African and South American tropics. The trees of tropical savannas are stunted and widely spaced, which allows grass to grow between them. Herds of grazing mammals typify the savanna landscape, along with large carnivores such as lions and other big cats, jackals and hyenas. These mammals, in turn, provide a food source for large scavengers such as vultures. The climate is warm all year, but has a dry season several months long when fires are a common feature.

The greatest expanses of the temperate grassland biome are located in Eurasia (where they are commonly known as steppe – Fig. 1.3), North America (prairie) and South America (pampa), with smaller expanses in South Africa (veldt). The climate is temperate, seasonal and dry. There are certain similarities with savannas in terms of fauna and the occurrence of fire, but unlike savannas, trees are absent. Temperate grassland soils tend to be deep and rich in organic matter.

In many parts of the world, where climates become drier, temperate grasslands fade into the desert biome. Hyper-arid desert supports virtually no plant life and is characterised by bare rock or sand dunes, but some species of flora and fauna are adapted to the high and variable temperatures and general lack of moisture, since some water is always available (Fig. 1.4). Sporadic rain promotes rapid growth of annual plants, and animals such as

FIGURE 1.5 *The US city of Boston, part of one of the world's most extensive areas of urban development. Only a few of the original temperate forest trees in the area survive in parks and gardens. Urban areas are now so widespread that they are often treated as a type of physical environment in their own right (courtesy of Charles Toomer)*

locusts, which otherwise lie dormant for several years as seeds or eggs.

A very distinctive form of vegetation is commonly associated with Mediterranean climates in which summers are hot and dry and winters are cool and moist. It is found around much of the Mediterranean Basin (where it is known as maquis), in California (chaparral), southern Australia (mallee), Chile and South Africa. Low evergreen trees (forming woodland) and shrubs (forming scrub) have thick bark and small, hard leaves. They are frequently exposed to fire during the arid summer period.

All these natural biomes have been affected to a greater or lesser extent by human action. Much of the maquis, for example, may represent a landscape where forests have been degraded by people, through cutting, grazing and the use of fire. The human use of fire may also be an important factor in maintaining, and possibly forming, savannas and temperate grasslands. Biomes considered by people to be harsh, such as the tundra and deserts, show less human impact, but the temperate forests have been severely altered over long histories with high population densities for farming and urban development (Fig. 1.5). The human influence is but one factor that promotes change in terrestrial as well as oceanic and freshwater ecosystems, because the interactions between the four global spheres have never been static. Better understanding of the dynamism of the natural world can be gained through a complementary way of studying the natural environment. Study of the processes which occur in natural cycles also takes us beyond description, to enable explanation.

NATURAL CYCLES

A good method for understanding the way the natural world works is through the recognition of cycles of matter in which molecules are formed and reformed by chemical and biological reactions, and are manifested as physical changes in the material concerned. The major stores and flows of water in the global hydrological cycle are shown in Fig. 1.6. Most of the Earth's water (about 97 per cent) is stored in liquid form in the oceans. Of the 3 per cent fresh water, most is locked as ice in the ice caps and glaciers, and as a liquid

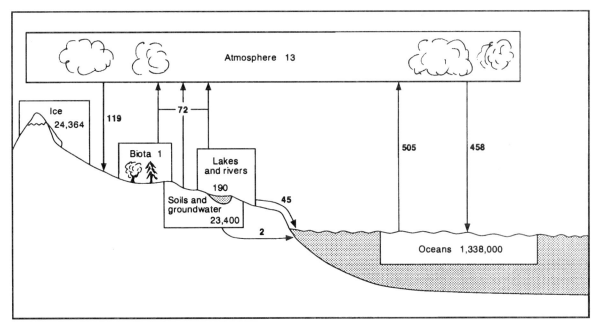

FIGURE 1.6 *Global hydrological cycle showing major stores and flows (data from Shiklomanov, 1993). The values in stores are in thousand km³, values of flows in thousand km³ per year*

in rocks as groundwater. Only a tiny fraction is present at any time in lakes and rivers. Water is continually exchanged between the Earth's surface and the atmosphere – where it can be present in gaseous, liquid or solid form – by evaporation, transpiration from plants and animals, and precipitation. The largest flows are directly between the ocean and the atmosphere. Smaller amounts are exchanged between the land and the atmosphere, with the difference accounted for by flows in rivers and groundwater to the oceans. Fresh water on the land is most directly useful to human society, since water is an essential prerequisite of life, but the oceans and ice caps play a key role in the workings of climate.

Similar cycles, commonly referred to as biogeochemical cycles, can also be identified for other forms of matter. Nutrients such as nitrogen, phosphorus and sulphur are similarly distributed among the four major environmental spheres and are continually cycled between them. Carbon is another key

element for life on Earth and the stores and flows of the carbon cycle are shown in Fig. 1.7. The oceans and rocks are the major stores of carbon with much smaller proportions present at any time in the atmosphere and biosphere. The length of time carbon spends in particular stores also varies widely. Under natural circumstances, fossil carbon locked in rocks, particularly carbonate sedimentary rocks such as limestones, and the hydrocarbons (coal, oil and natural gas) remain in these stores for millions of years. Carbon reaches these stores by the processes of sedimentation and evaporation and is released from rocks by weathering, vulcanism and sea-floor spreading. In recent times, however, the rate of flow of carbon from some of the lithospheric stores – the hydrocarbons or fossil fuels – has been greatly increased by human action: the burning of fossil fuels which liberates carbon by oxidation. Hence, a significant new flow of carbon between the lithosphere and the atmosphere has been introduced by human

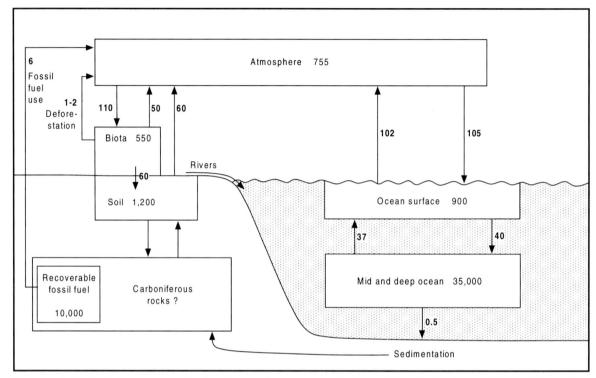

FIGURE 1.7 *Global carbon cycle showing major stores and flows (after Schlesinger, 1991). The values in stores are in units of Pg C, values of flows in Pg C per year. 1Pg C = 10^{15} g C = 1 billion tonnes of carbon as CO_2*

society, and the natural atmospheric carbon store is being increased as a consequence.

Carbon also reaches the atmosphere by the respiration of plants and animals, which in plants is part of the two-way process of photosynthesis. Photosynthesis is the chemical reaction by which green plants convert carbon from the atmosphere, with water, to produce complex sugar compounds (which are either stored as organic matter or used by the plant) and oxygen. The reaction is written as follows:

$$6CO_2 + 6H_2O \rightarrow C_6H_{12}O_6 + 6O_2$$

This equation shows that six molecules of carbon dioxide and six molecules of water yield one molecule of organic matter and six molecules of oxygen. The reaction requires an input of energy from the sun, some of which is stored in chemical form in the organic matter formed.

The process of respiration is written as the opposite of the equation for photosynthesis. It is the process by which the chemical energy in organic matter is liberated by combining it with oxygen to produce carbon dioxide and water. All living things respire to produce energy for growth and the other processes of life. The chemical reaction for respiration is, in fact, exactly the same as that for combustion. Humans, for example, derive energy for their life needs from organic matter by eating (just as other animals do) and also by burning plant matter in a number of forms, such as fuelwood and fossil fuels.

The flow of converted solar energy through living organisms can be traced up a hierarchy of life-forms known as a food chain. Figure 1.8

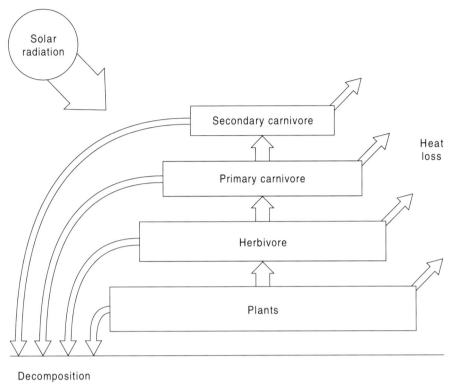

FIGURE 1.8 *Energy flow through a food chain*

shows a simple food chain in which solar energy is converted into chemical energy in plants (so-called producers) which are eaten by herbivores (so-called first-order consumers) which, in turn, are eaten by other consumers (primary carnivores) which are themselves eaten by secondary carnivores. An example of such a food chain on land is

$$\text{grass} \rightarrow \text{cricket} \rightarrow \text{frog} \rightarrow \text{heron}$$

Each stage in the chain is known as a trophic level. In practice, there are usually many, often interlinked, food chains which together form a food web, but the principles are the same. At each trophic level some energy is lost by respiration, through excretory products and when dead organisms decay, so that available energy declines along the food chain away from the plant. In general terms, animals also tend to be bigger at each sequential trophic level enabling them to eat their prey safely. This model helps us to explain the basic structures of natural communities: with each trophic level, less energy is available to successively larger individuals and thus the number of individuals decreases. Hence, while plants are very numerous because they receive their energy directly from the sun they can only support successively fewer larger animals. With the exception of humans, predators at the top of food chains are therefore always rare.

Food chains, the carbon cycle and the hydrological cycle are all examples of 'systems' in which the individual components are all related to each other. Most of the energy which drives these systems comes from the sun, although energy from the Earth also contributes. All the cycles of energy and matter referred to in this section are affected by human action, deliberately manipulating natural cycles to human advantage. One of the human impacts on the carbon cycle has been mentioned, but humans also affect other cycles. The cycle of minerals in the rock cycle is affected by the construction industry, for example. Human activity affects the hydrological cycle by diverting natural flows: the damming of rivers or extracting groundwater for human use. The nitrogen cycle is affected by concentrating nitrogen in particular places by spreading fertilisers on fields, for example. Food chains are widely affected: human populations manipulate plants and animals to produce food.

However, since all parts of these cycles are interrelated, human intervention in one part of a cycle also affects other parts of the same and other cycles. These knock-on effects are the source of many environmental changes which are undesirable from human society's viewpoint. A better appreciation of these changes can be gained by looking at the various scales of time and space through which they occur.

TIMESCALES

Changes in the natural environment occur on a wide range of timescales. Geologists believe that the Earth is about 4600 million years old, while fossil evidence suggests that modern humans (*Homo sapiens*) appeared between 100 000 and 200 000 years before present (BP), developing from the hominids whose earliest remains, found in Africa, date to around 3.75 million BP (Table 1.2). The very long timescales over which many changes in the natural world take place may seem at first to have little relevance for today's human society other than to have created the world we know today. It is difficult for us to appreciate the age

of the Earth, for example, and the thought that the present distribution of the continents dates from the break-up of the supercontinent Pangaea which began during the Cretaceous. Indeed, relative to the forces and changes due to tectonic movements, the human impact on the planet is very minor and short-lived. However, such Earth processes do have relevance on the timescale of a human lifetime. Tectonic movements cause volcanic eruptions which can affect human society as natural disasters at the time of the event. Volcanic eruptions also affect day-to-day human activities on slightly longer timescales, by injecting dust into the atmosphere which affects climate, for example, and by providing raw materials from which soils are formed. This example also illustrates the fact that the same event may be interpreted as bad from a human viewpoint on one timescale (a volcanic disaster) and good on another timescale (fertile volcanic soils).

It is important to realise that the timescale we adopt for the study of natural systems can affect our understanding as well as our perception of them. Many such systems are thought to be in 'dynamic equilibrium', meaning that the input and output of matter and energy is balanced, but recognition of dynamic equilibrium in natural systems depends upon the timescale over which the system is studied. To take the Earth as a whole, for example, the idea of dynamic equilibrium has been proposed to explain why the temperature of the Earth has remained relatively constant for the last 4 billion years, despite the fact that the sun's heat has increased by about 25 per cent over that period. The Gaia hypothesis suggests that life on the planet has played a key role in regulating the Earth's conditions to keep it amenable to life (Lovelock, 1989). The theory is not without its critics, but even if we accept it, the dynamic equilibrium only holds for the few billion years of the Earth's existence. Astronomers predict that eventually the sun

TABLE 1.2 *Geological timescale classifying the history of the Earth*

Era	Period		Start (million years BP)	Important events
		Holocene	0.01	Early civilisations
	Quaternary	Pleistocene	1.8	First humans
		Pliocene	5	First hominids
		Miocene	22.5	
		Oligocene	38	
		Eocene	54	
Cenozoic	Tertiary	Palaeocene	65	Extinction of dinosaurs
	Cretaceous		136	Main fragmentation of Pangaea
	Jurassic		190	
Mesozoic	Triassic		225	First birds
	Permian		280	Formation of Pangaea
		Pennsylvanian	315	
	Carboniferous	Mississipian	345	
	Devonian		395	
	Silurian		440	First land plants and animals
	Ordovician		500	First vertebrates
Palaeozoic	Cambrian		570	
Precambrian			4600	Formation of Earth

Source: after Goudie (1993a); Colinvaux (1993); Williams *et al.* (1993).

will destroy the Earth, so that over a longer timescale, dynamic equilibrium does not apply. This example applies over a very long timescale, and from our perspective the destruction of the Earth by the sun is not imminent, but the principle is relevant to all other natural systems. Adoption of different timescales of analysis dictates which aspects of a system we see and understand because the importance of different factors changes with different timescales. Indeed, even within the lifespan of the Earth, dramatic changes are known to have occurred, such as the progression of glacial and interglacial periods. The longer the time frame, the coarser the resolution, and vice versa. In a simple example, a human being who contracts a cold might feel miserable for a few days, but in terms of that person's lifetime career the cold is a very minor influence.

One of the key components of natural cycles and dynamic systems is the operation of feedbacks. Feedbacks may be negative, which tend to dampen down the original effect and thereby maintain dynamic equilibrium, or they may be positive, and hence tend to enhance the original effect. An example of negative feedback can be seen in the operation of the global climate system: more solar energy is received at the tropics than at the poles, but the movement of the atmosphere and oceans continually redistributes heat over the Earth's surface to redress the imbalance. Positive feedbacks can result in a change from one dynamic equilibrium to another: if a forest is cleared by human action, for example, the soil may be eroded to the extent that recolonisation by trees is impossible.

This last example illustrates another important aspect of natural cycles: the existence of thresholds. A change in a system may not occur until a threshold is reached: snow will remain on the ground, for example, until the air temperature rises above a threshold at which the snow melts. Crossing a threshold may be a function of the frequency or intensity of the force for change: a palm tree may be able to withstand, or be 'resilient' to, winds up to a certain speed, but will be blown out of the soil by a hurricane-force wind which is above the tree's threshold of resilience. Conversely, thresholds may be reached by the cumulative impacts of numerous small-scale events: regular rainfall inputs of moisture to a slope may build up to a point at which the slope fails, or in the erosion example, the gradual loss of soil reaches the point at which there is not enough soil left for trees to take root and grow. To complicate things further, there may be a lag in time between the onset of the force for change and the change itself: the response time of the system. An animal seldom dies immediately upon contracting a fatal disease, for example; its body ceases to function only after a period of time. Likewise in the erosion example, trees are unable to colonise only after a certain amount of soil has been lost. Consideration of feedbacks, thresholds and lags leads to some other characteristics of natural systems: their 'sensitivity' to forces for change, which dictates their 'stability'.

SPATIAL SCALES

Just as the choice of timescales is important to our understanding of the natural environment, so too is the spatial scale of analysis. Studies can be undertaken at scales which range from the microscopic – the effects of salt weathering on a sand grain, for example – through an erosion plot measured in square metres, to drainage basin studies which can reach subcontinental scales in the largest cases, to the globe itself. As with time, the resolution of analysis becomes coarser with increasing spatial scale. We draw a line on a world map to divide one biome from another, but on the ground there is usually no line, more a zone of transition, which may itself vary over different timescales.

Similarly, the types of influences which are important differ according to spatial scale. To use another example involving humans, a landslide which results in the loss of a farmer's field may have a significant impact on the farmer's ability to earn a living, but the same landslide has a minimal impact on national food production.

Thresholds and feedbacks also have relevance on the spatial scale. Certain areas may be more sensitive to change than others and if a threshold is crossed in these more sensitive areas, wider-scale changes may be triggered. Soil particles entrained by wind erosion from one small part of a field, for example, can initiate erosion over the whole field. On a larger scale, sensitive areas such as the Labrador–Ungava plateau of northern Canada appear to have played a key role in triggering global glaciations during the Quaternary because they were particularly susceptible to ice-sheet growth. A contemporary large scale example can be seen in the tundra biome (Fig. 1.9) which could release large amounts of methane locked in the permafrost if the global climate warms due to human-induced pollution of the atmosphere. Methane is a greenhouse gas so positive feedback could result, enhancing the warming effect worldwide.

The key factors influencing natural events also vary at different combined spatial and temporal scales. Individual waves breaking on a beach constantly modify the beach

FIGURE 1.9 *Tundra in Canada's northern Manitoba: part of a biome which may be particularly sensitive to a human-induced warming of global climate and which could create a positive feedback by releasing large amounts of methane, a greenhouse gas (see Chapter 8) (courtesy of Mark Carwardine)*

profile which is also affected by the daily pattern of tides dictating where on the beach the waves break. Individual storms alter the beach too, as do the types of weather associated with the seasons. However, all these influences are superimposed upon the effects of factors which operate over longer timescales and larger spatial scales, such as sediment supply and the sea level itself (Clayton, 1991).

The environmental issues in this book have arisen as a consequence of human activity conflicting with environmental systems. Resolution of such conflicts can only be based on an understanding of how natural systems work. For issues that stem from human impact upon the physical environment, as most do, we need to be able to rank the temporal and spatial scale of human impact in the natural hierarchy of influences on the natural system in question. Inevitably, we tend to focus on scales directly relevant to people, but we should not forget other scales which may have less direct but no less significant effects.

THE STATE OF OUR KNOWLEDGE

We already know a great deal about how the natural world works, but there remains a lot more to learn. We have some good ideas about the sorts of ways natural systems operate, but we remain ignorant of many of the details. Some of the difficulties involved in ascertaining these details include a lack of data and our own short period of residence on the Earth. Direct measurements are used in the contemporary era to monitor environmental processes. Historical archives, sometimes of direct measurements, otherwise of more anecdotal evidence, can extend these data back over decades and centuries. Good records of high and low water levels for the River Nile at Cairo extend from AD 641 to 1451, although they are intermittent thereafter until the nineteenth century, and continuous monthly mean temperature and precipitation records have been kept at several European weather stations since the eighteenth century. Other types of written historical evidence date back to ancient Chinese and Mesopotamian civilisations as early as 3000 BC. The further back in time we go, however, the patchier the records become, and in some parts of the world historical records begin only in the present century.

These data gaps for historical time, and for longer time periods of thousands, tens of thousands and millions of years, can be filled in using natural archives. The geological timescale given in Table 1.2 is based on fossil evidence. Such 'proxy' methods are based on our knowledge of the current interrelationships between the different environmental spheres. Particular plants and animals thrive in particular climatic zones, for example, so that fossils can indicate former climates. Other proxy 'palaeoenvironmental indicators' include pollen types found in cores of sediment taken from lake or ocean beds, and the rate of sediment accumulation in such cores can tell us something about erosion rates on the surrounding land. Landforms, too, become fossilised in landscapes to provide clues about past conditions. Examples include glacial and periglacial forms in central and northern Europe, indicating colder conditions during previous glaciations, and fossilised sand dunes in the Orinoco Basin of South America, also dating from periods of high-latitude glaciation, which indicate an environment much drier than that of today.

The understanding gained from all these lines of evidence can then be used to predict future environmental changes, incorporating any human impact, using models which simulate environmental processes. The human impact may still provide further complications, however, because in many instances through prehistory, history, and indeed in the present era, it can be difficult to distinguish between purely natural events and those which owe something to human activities. It is the interrelationships between human activities and natural functions which form the subject matter of this book.

FURTHER READING

Colinvaux, P. 1993 *Ecology 2*. New York, Wiley. A comprehensive textbook which covers all the basics of ecology and highlights many areas of ignorance and conflicting interpretations of the natural world.

Goudie, A.S. 1993 *The nature of the environment*, 3rd edn. Oxford, Blackwell. A good basic overview of physical geography, including the influence of geology and climatology, profiles of major world environments and the workings of landscapes and ecosystems.

2

THE HUMAN ENVIRONMENT

Knowledge of the physical environment only illuminates one-half of any environmental issue, since an appreciation of factors in the human environment is also required before an issue can be fully understood. Relationships between human activity and the natural world have changed greatly in the relatively short time that people have been present on the Earth. A very large increase in human population, along with widespread urbanisation associated with advances in technology and related developments of economic, political and social structures have all combined to make the interaction between humankind and nature very different from the situation just a few thousand years ago. This chapter is concerned with these aspects of humanity which together provide the human dimensions of environmental issues.

HUMAN PERSPECTIVES ON THE PHYSICAL ENVIRONMENT

Nobody knows what prehistoric human inhabitants thought about the natural world, but fossil and other archaeological evidence can be marshalled to give us some idea of how they interacted with it. The fact that humans have always interacted with the physical environment is obvious, since all living things do so by definition, but the ability of humans to conceptualise has allowed us to formalise our view of these interactions. One such framework, which must have been around in one form or another for as long as people have inhabited the planet, is the idea of resources. We call anything in the natural environment which may be useful to us a natural resource. Aspects of the human environment, such as people and institutions, can also be thought of as resources, while anything that we perceive as detrimental to us is sometimes called a negative resource. Since resources are simply a cultural appraisal of the material world, individual aspects of the environment can vary at different times and in different places between being resources and negative resources. A tiger, for example, can be viewed as a resource (for its skin) or a negative resource (as a dangerous animal). Similarly, different cultures recognise resources in different ways (the Aztecs, for example, were mystified by the Spaniards' insatiable demand for gold, thinking that perhaps they ate it or used it for medical purposes), and some groups see resources where others do not (a grub is a food resource to an Aboriginal Australian, but not to most Europeans). Hence, resources vary in character and some are also more difficult to manage than others.

TABLE 2.1 *A classification of resources*

PHYSICAL ENVIRONMENT

Continuous resources, which for all intents and purposes will never run out (e.g. solar energy, wind, tidal energy)

Renewable resources, which can naturally regenerate so long as their capacity for so doing is not irreversibly damaged, perhaps by natural catastrophe or human activity (e.g. plants, animals, clean water, soil)

Non-renewable resources, which are available in specific places and only in finite quantities because although they are renewable, the rate at which they are regenerated is extremely slow on the timescale of the human perspective (e.g. fossil fuels and other minerals, some groundwaters)

HUMAN ENVIRONMENT

Extrinsic resources, which include all aspects of the human species, all of which are renewable (e.g. people, their skills, abilities and institutions)

The classification shown in Table 2.1 gives some idea as to their characteristics and manageability.

All environmental issues can be seen as the result of a mismatch between extrinsic resources and natural resources: they stem from people deliberately or inadvertently misusing or abusing the natural environment. The reasons for such inappropriate uses are to be found within the nature of human activity, but before moving on to examine human activity it is appropriate to look briefly at the history of human interest in environmental issues.

INTEREST IN ENVIRONMENTAL ISSUES

Although environmental issues are a prominent concern of government and the general public in most countries of the world, today

more widely so than at any time in the past, worry over environmental matters is by no means a recent phenomenon. Observers of the natural world in classical Greece, imperial Rome and Mauryan India expressed reservations about the wisdom of human actions which promoted accelerated soil erosion and deforestation. In the present millenium, some of the earliest environmental legislation was introduced to protect the quality of natural resources. At the end of the thirteenth century, for example, laws were passed by the Mongolian state to protect trees and animals from overexploitation, and in England atmospheric pollution prompted a royal decree in 1306 forbidding the burning of coal in London. Warning voices were also raised at the damage caused by deforestation and plantation agriculture during the earliest periods of European colonial expansion, on the Canary Islands and Madeira from the early fourteenth century and in the Caribbean after 1560.

Indeed, the origins of today's global concern for the environment can be traced to some of these early European colonial experiences (Grove, 1990). While European colonies were often the sites of environmental degradation, they were also the cradles of modern scientific views on the issues. Colonial administrators and professional scientists working for trading companies in the eighteenth and nineteenth centuries developed Western conservationist ethics and environmentally sound management practices. Although such approaches in themselves were not new, since many traditional rural societies had employed similar practices locally for a long time before, this was the first time that a global perspective could be attained. Hence, international travel and observations by naturalist explorers such as Charles Darwin and Alexander von Humboldt allowed them to develop scientific theories which revolutionised Western thinking on the way nature was organised and worked.

TABLE 2.2 *Timetable of major political acceptance of some environmental issues and the dates when they were first identified as potential problems*

Environmental issue	Potential problem first identified	Decade of major political acceptance
Air and water pollution	Carson (1962)	1960s
Tropical deforestation	Marsh (1874)	
Acid rain	Smith (1852)	
Stratospheric ozone depletion	Molina and Rowland (1974)	1970s
Carbon dioxide-induced climatic change	Wilson (1858)	
Additional greenhouse gases	Wang et al. (1976)	1980s

Source: after Döös (1994)

Identification of many of today's global environmental concerns was first made more than 100 years ago but most have only gained major political acceptance in recent decades (Table 2.2). While some people have reservations over the domination of science in thinking on the environment and other areas of society (e.g. Feyerabend, 1978), that is the status quo. Some observers blame the delays in acceptance on scientists themselves, for being poor communicators to a wider audience and for being overcautious in their conclusions (Döös, 1994), but scientists need to gather data and test theories before they can recommend action. Their endeavours are also passed to wider audiences through a multitude of public and private organisations interested in environmental issues. Such organisations include government ministries and research bodies as well as many non-governmental organisations which operate at international, national and local scales. On occasions, political recognition is forced by civil agitation from environmental activists, but science usually plays a role somewhere in the process. The difficulties scientists face in understanding the natural environment and the human effects on it have been outlined in Chapter 1. This chapter will now look in more detail at the human factors behind environmental issues.

HUMAN FORCES BEHIND ENVIRONMENTAL ISSUES

The interactions between humankind and the physical environment result from our attempts to satisfy real and perceived needs and wants. The specific actions which cause environmental issues directly are well known. They include such things as modifying natural distributions of vegetation and animals, overusing soils, polluting water and air, and living in hazardous areas. At a deeper level, the forces which enable, encourage or compel people to act in inappropriate ways can be traced back to humankind's underlying behaviour, which is expressed as a series of driving forces and

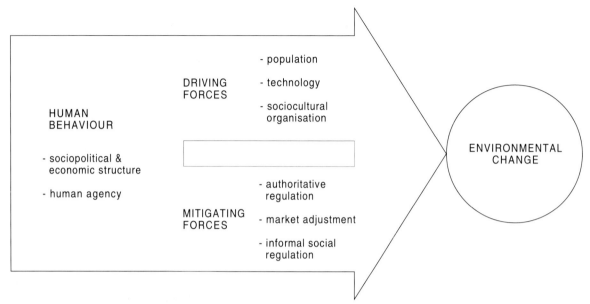

FIGURE 2.1 *Human forces of environmental change (after Kates, R.W., Turner, B.L., II, and Clark, W.C. 1990 in Turner, B.L., II, Clark, W.C., Kates, R.W., Richards, J.F., Mathews, J.T. and Meyer, W.B. (eds).* The earth as transformed by human action. *Reproduced by permission of Cambridge University Press).*

mitigating forces in Fig. 2.1. To delve as deep as the nature of human behaviour is beyond the scope of this book, but further investigation of the driving forces, which promote human impact on the environment, and the mitigating forces, which act as checks to the driving forces or their impacts, will comprise the remainder of this chapter.

Population

Growth in the global population of human beings is widely recognised as one of the most clear-cut driving forces behind increased human impact on the environment. The exponential rise in human population numbers is clearly indicated in Table 2.3. While in 1750 the world population of about 800 million had a doubling period of more than 1000 years, 250 years later as we near the year 2000, the world population of between 5 and 6 billion has a doubling period of less than 50 years. The logic is simple and undeniable: more people means a greater need for natural resources. As the population grows, more resources are used and more waste is produced.

The relationship between the human population, whose numbers can change, and natural resources, which are essentially fixed, has been an issue for a long time. At the end of the eighteenth century, the Englishman Thomas Malthus suggested that population growth would eventually outstrip food production and lead to famine, conflict and human misery for the poor as a consequence (Malthus, 1798). This Malthusian perspective continues to influence many interpretations of environmental issues. However, in many cases the relationship between population numbers and environmental degradation is complicated by numerous other factors. The rapid growth of population in an area can lead to overexploitation of the area's

TABLE 2.3 *World human population and growth rates*

Year	Total population (millions)	Annual growth rate (%)	Doubling period (years)
I million BC	A few thousand	—	—
8000 BC	8	0.0007	100 000
AD I	300	0.046	1500
1750	800	0.06	1200
1900	1650	0.48	150
1970	3678	1.9	36
2000	6199	1.7	41

Source: Otzen (1993)

resources, but rapid population decline can also lead to environmental degradation if some aspects of good management are discarded because there are not enough people to keep them up. Furthermore, high population densities do not necessarily lead to greater degradation of resources. A small number of people living in a particular area can cause greater damage than larger numbers in a similar area, for example. One important factor influencing the equation is the level of technology used. Political and economic forces also affect the ways in which people use or abuse their resources. In addition, many human impacts are not direct: people living in a city, for example, have influence on resource use far from their immediate surrounds. A walk down a supermarket aisle in any city will indicate the great distances some products have been transported before they reach the urban consumer.

Technology

Developments in technology have been closely associated with population growth. One view sees technological developments as a spur to growth, so that agricultural innovations provide more food per unit area and enable more people to be supported, for example. An opposite perspective sees technological change as a result of human inventiveness reacting to the needs created by more people (Boserüp, 1965). This latter view can be used to suggest that a growing human population is not necessarily bad from an environmental perspective. Any problems created by increasing population will be countered by new innovations which ease the burden on resources by using them more effectively.

However, many arguments can be presented to indicate that technological developments are responsible for much environmental degradation. Technology influences demand for resources by changing their accessibility and people's ability to afford them, as well as creating new resources (uranium, for example, was not considered a resource until its energy properties were recognised). The Industrial Revolution, for example, has been associated with high population growth and a greatly enhanced level of resource use and misuse; it has promoted urban development and the globalisation of the economy. Technology also has direct impact on resources, of course, and virtually every environmental issue can be interpreted as a consequence, either deliberate or inadvertent, of the impact of technology on the natural world. The numerous examples of inadvertent impacts are usually the result of ignorance: a new technology being tried and tested and found to have undesirable consequences. On the other hand, technology is not environmentally damaging by definition, since technological applications can be designed and used as a mitigating force, in many cases in response to a previous undesirable impact. When and where technology is used or abused is dictated by the nature of society and its organisation.

Sociocultural organisation

The influences of population and technology are intimately linked to the organisation of human society. Economic, political and social values, norms and structures are a diverse set of driving forces which influence environmental change. They also underpin the mitigating forces which have been developed by societies to offset some of the damaging aspects of environmental issues. Dramatic transformations in population and technology have been associated with the two most significant changes in the history of humankind, the agricultural revolution of the late Neolithic period and the Industrial Revolution of the past two centuries, but these changes have also been reflected in equally marked modifications to the ways that society is organised.

One of the most striking of these social changes is the rise of the city. Although cities have been a feature of human culture for about 5000 years, virtually the entire human species lived a rural existence just 300 years ago. Today, the proportion of the world's population living in cities is approaching 50 per cent, and individual urban areas have reached unprecedented sizes. In most cases, urban areas have much higher population densities than rural areas and the consumption of resources per person in urban areas is greater than that of their rural counterparts. The environmental impacts of cities within urban boundaries are obvious, since cities are a clear illustration of the human ability to transform all the natural spheres, but the influence of cities is also felt far beyond their immediate confines. The high level of resources used by city dwellers has been fuelled by extending their resource flows. Economic and political forces have developed to facilitate this extension, so that the world is now an integrated whole.

Although integrated, the world has also become a very unbalanced place in terms of human welfare and environmental quality. Much attention has been focused on how economic and political forces have produced global imbalances in the way human society interacts with the environment, not just between city and countryside, but between groups within society with different levels of access to power and influence (e.g. women and men, different political parties, ethnic groups.) On the global scale, there are clear imbalances between richer nations and poorer nations. In general terms, the wealthiest few are disproportionately responsible for environmental issues, but at the other end of the spectrum the poorest are also accused of a responsibility that is greater than their numbers warrant. The motivations for these disproportionate impacts are very different, however. The impact of the wealthy is driven by their intense resource use, many say their overconsumption of resources. The poor, on the other hand, may degrade the environment because they have no other option. Economic power is seen as a vital determinant: the wealthy have become wealthy because of their high-intensity resource use and can afford to continue overconsuming and to live away from the problems this creates. The poor cannot afford to do anything other than overuse and misuse the resources that are immediately available to them and as a consequence they are often the immediate victims of environmental issues. This difference in economic power is also manifested in political power: the wealthy generally have more influence over decisions that affect interaction with the environment than the poor, although when marginalised people are pushed to the edge of environmental destruction they may become active in forcing political changes (Broad, 1994). Some of the causes and manifestations of these global and regional imbalances are investigated further in the following sections.

HUMAN-INDUCED IMBALANCES

Human society has created a set of sociocultural imbalances that is superimposed on geographical patterns of environmental resources on all spatial scales from the global to the individual. These patterns of uneven distribution have been developed by, and are maintained by, the processes and structures of economics, politics and society. These imbalance theories, combined with the human driving force theories outlined above, together make up a diverse set of explanations which have been put forward to explain the human dimension behind environmental issues (Table 2.4).

Ownership and value

Two interrelated theories which explain the underlying causes of imbalance between human activities and the environment stem from differential ownership of certain resources and the values put on them. Some environmental resources are owned by individuals while others are under common ownership. One theory argues that resources under common ownership are prone to overuse and abuse for this very reason – the so-called 'tragedy of the commons' (Hardin, 1968). The example often given to illustrate

TABLE 2.4 *Some theories to explain why environmental issues occur*

Theory	Explanation
Neo-Malthusian	Demographic pressure leads to overuse and inappropriate use of resources
Ignorance	Ignorance of the workings of nature means that mistakes are made, leading to unintended consequences
Tragedy of the commons	Overuse or misuse of certain resources occurs because they are commonly owned
Poor valuation	Overuse or misuse of certain resources occurs because they are not properly valued in economic terms
Dependency	Inappropriate resource use by certain groups is encouraged or compelled by the influence of more powerful groups
Exploitation	Overuse and misuse of resources is pursued deliberately by a culture driven by consumerism
Human domination over nature	Environmental issues result from human misapprehension of being above rather than part of nature

Source: after Barrow (1991)

this principle is that of grazing grounds which are commonly owned in pastoral societies. It is in the interest of an individual to graze as many livestock as possible, but if too many individuals all have the same attitude the grazing lands may be overused and degraded: the rational use of a resource by an individual may not be rational from the viewpoint of a wider society. The principle can also be applied to explain the misuse of other commonly owned resources, such as the pollution of air and water or catching too many fish in the sea.

A related concept is the undervaluation of certain resources. Air is a good example. For all intents and purposes, air is a commonly owned continuous resource which, in practice, is not given an economic value. The owner of a windmill does not pay for the moving air which the windmill harnesses, nor does the owner of a factory who uses the air as a sink for the factory's wastes. Since air has no economic value it is prone to be overused. A simple economic argument suggests that if an appropriate economic value was put on the resource, the workings of the market would ensure that as the resource became scarce so the price would increase. As the value of the resource increased, theory suggests that it would be managed more carefully.

Exploitation and dependency on the global scale

A complex series of economic, political and social processes has resulted in patterns of exploitation and dependency which are associated with the misuse of resources. Inequalities exist between many different groups of people (see above) and can be identified on several different scales. Three levels are identified by Lonergan (1993), two spatial (international and national), and one temporal (intergenerational).

Some of the main structural inequalities of the global system are shown in Fig. 2.2. They

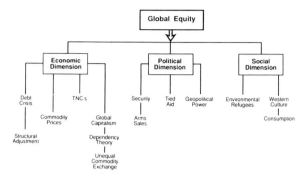

FIGURE 2.2 *Structural inequalities in the global system (reprinted with permission from Lonergan, S.C. 1993. Impoverishment, population and environmental degradation: the case for equity.* Environmental Conservation 20: 328–34. *Copyright © 1993 Elsevier Science)*

have evolved from colonial times to the point, today, where direct political control of empires has been superseded by more subtle control by wealthier countries over poorer ones. The economic dimension is particularly important, since it influences the rate of exploitation of natural resources in particular countries, the relative levels of economic development apparent in different countries, and the power of certain governments to control their own future.

The realities of global inequality in economic terms are stark. One billion people live in unprecedented luxury, while 1 billion live in destitution. American children even have more in pocket money (US$ 230 a year) than the half-billion poorest people alive (Durning, 1991). The gap between countries has also been widening in recent decades. In 1960, the richest 20 per cent of the world's population absorbed 70 per cent of global income, about thirty times more than the world's poorest 20 per cent, who took just 2.3 per cent of global income. By 1990, the share of the rich was more than sixty times that of the poorest, and disparities between people, as opposed to country averages, had grown even further (UNDP, 1993).

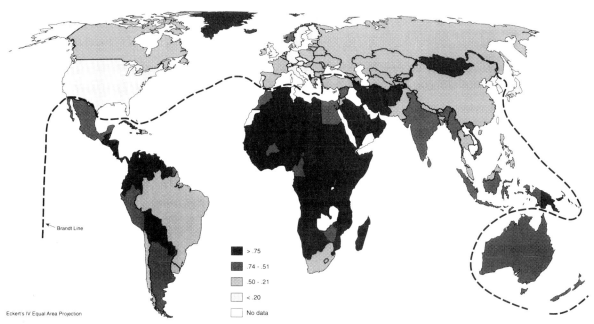

FIGURE 2.3 *Index of commodity concentration of exports (Knox and Agnew, 1994). The 'Brandt line' represents the division between countries of the more affluent 'North' and the poorer countries of the 'South'*

The structural aspects of this economic dimension have many facets, including the fact that many poorer countries are in debt to the countries and banks of the rich world (the so-called 'debt crisis') and many less-developed countries rely on a limited number of exports, which are usually primary products such as agricultural goods and minerals. Fig. 2.3 illustrates this picture globally with regard to exports. A low index of commodity concentration of exports reflects a diverse export base while a high index reflects a country's reliance on the export of a few agricultural or mineral resources. This pattern is maintained by terms of trade and prices for primary exports ('unequal commodity exchange') which are set largely by the countries of the North that represent the major markets for these exports. Overexploitation of resources in poorer countries often occurs in response to falling commodity prices that have been typical of recent decades, and the need to service debts.

Transnational corporations (TNCs), the very large majority of which have headquarters in the advanced capitalist regions, particularly Japan, North America and Western Europe, also play an important role in the workings of global economics. Comparison of the annual turnover of TNCs with the Gross National Product (GNP) of entire nation-states gives an indication of this role:

By this yardstick, all of the top 50 transnational corporations – including the likes of Exxon, General Motors, Ford Motor Company, Matsushita Electronics, IBM, Unilever, Philips, ICI, Union Carbide, ITT, Siemens and Hitachi – carry more economic clout than many of the world's smaller peripheral nation-states; while the very biggest transnationals are comparable in size with the national economies of semi-peripheral states like Greece, Ireland, Portugal, and New Zealand.

(Knox and Agnew, 1994: 38–39)

In development terms, investment by TNCs in developing countries has been portrayed as an engine of growth capable of eliminating economic inequality on the one hand, and as a major obstacle to development on the other. Some view such investment as a force capable of dramatically changing productive resources in the economically backward areas of the world, while others see it as a primary cause of underdevelopment because it acts as a major drain of surplus to the advanced capitalist countries (Jenkins, 1987).

The negative view has been illustrated by comparing the development paths of Japan and Java, which had a similar level of development in the 1830s. While Japan has subsequently developed much further, in spite of a poorer endowment of natural resources, Java has remained underdeveloped, despite a position on main trade routes and a good stock of natural resources, because few of the profits from resource use have been reinvested in the country (Geertz, 1963).

One aspect of these relationships which some consider to be detrimental to the environment in poorer countries is the movement of heavily polluting industries from richer countries that have developed high pollution-control standards, to poorer countries where such regulations are less stringent. Fear of an increase in this trend has been expressed during the negotiations for the Uruguay round of the General Agreement on Tariffs and Trade (GATT) and the establishment of the North American Free Trade Agreement (NAFTA) (Daly, 1993). Daly notes examples of 'maquiladoras', US factories which have located mainly in northern Mexico to take advantage of lower pollution-control standards and labour costs.

Economic power is closely related to political power, which can be seen in the sale of arms and the giving of aid by richer countries to poorer ones. The workings of global capitalism also have social dimensions, including the spread of Western cultural

FIGURE 2.4 *Domination of the world's poorer countries by their richer counterparts takes many forms, including what some term 'cultural imperialism', epitomised by this advertisement for Pepsi in Ecuador. Transfer of Western technology is often in the form of polluting industries, as can be seen behind the advertisement*

norms and practices (Fig. 2.4). The sale of Western products can also be seen to reinforce the role of developing nations as suppliers of raw materials, as Grossman (1992) shows for the increasing use of pesticides on Caribbean agricultural plantations where produce is destined for export markets, a trend influenced by foreign aid, among other criteria.

Exploitation and dependency on the national scale

Perhaps the most important underlying aspect of the global inequalities is the way in which they combine to maintain a situation where a minority of wealthy nations and most of their inhabitants are able to live an affluent life consuming large quantities of resources at the expense of many more poorer nations and their inhabitants. Nevertheless, some countries do move up the economic development scale, and the terms 'North' and 'South' disguise much internal variability (Fig. 2.3). Structural inequalities also exist at the national level, however, and these inequalities similarly have implications for the ways in

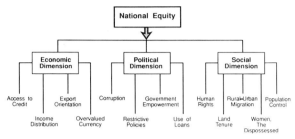

FIGURE 2.5 *Structural inequalities in national systems (reprinted with permission from Lonergan, S.C. 1993, Impoverishment, population and environmental degradation: the case for equity.* Environmental Conservation 20: *328–34. Copyright © 1993 Elsevier Science)*

which people interact with the environment within countries. Some of the aspects of inequality on the national scale, categorised according to their economic, political and social dimensions, are shown in Fig. 2.5.

As at the global level, the national scene is often characterised by a small élite group which has more economic, political and social power than the majority of people in more marginal groups. In general terms, this pattern is often manifested in a rural–urban divide: cities are centres of power, although they also typically contain a poorer 'under-class' in both rich and poor countries. Outside the city in the South, most rural people do not have access to economic power: in Africa, Asia and Latin America, for example, more than 85 per cent of farmers are estimated to lack access to credit (Holmberg, 1991). Corruption and centralised control are all too frequent characteristics of political and economic management in developing countries, which often means that local communities lack power over their own resources and how they are managed. These aspects are often tied up with social dimensions such as human rights. Another important social dimension is land tenure; poorer, marginalised groups tend to lack ownership of land and its associated resources, and even

where they do have ownership it is often of low-quality resources.

The ways in which national inequalities translate into environmental issues often mean that the poor and disadvantaged are both the victims and agents of environmental degradation. The poor, particularly poor women, tend to have access only to the more environmentally sensitive areas and resources. They therefore suffer greatly from productivity declines due to degradation such as deforestation or soil erosion. Their poverty also means that there may be little alternative but to extract what they can from the sparse resources available to them, with degradation often being the result. High fertility rates typical of poor households puts further strains on the stock of natural resources.

The impacts can be seen in numerous examples. Political power in the hands of a minority group can be used to force large concentrations of marginal groups into small parts of the national land area, causing environmental degradation through overpopulation. Examples include the homelands system of apartheid South Africa, and numerous similar inequalities exerted by more powerful ethnic groups in many other countries. The exploitation of resources by élite groups can also result in coerced migration of indigenous inhabitants. Such 'environmental refugees' often take refuge in neighbouring countries (Fig. 2.6). Political, economic and social forces also combine to make marginal groups more vulnerable to natural hazards, as the model developed by Blaikie *et al.* (1994) illustrates (Fig. 2.7).

Intergenerational inequalities

The spatial dimensions of exploitation and dependency are also closely related to the temporal dimension. As wealthy, powerful groups exploit available resources on the global and national scales, the inheritance of future generations is compromised.

FIGURE 2.6 *Refugees in northern Thailand who have fled their native lands in neighbouring Myanmar (Burma). A long-running armed struggle between indigenous peoples and the government has been exacerbated in recent years by government-sponsored clearance of tropical forests in north-east Myanmar*

FIGURE 2.7 *Model of pressures that create vulnerability and result in disasters (after Blaikie, P., Cannon, T., Davis, L. and Wisner, B. 1994,* At risk: natural hazards, people's vulnerability and disasters. *Reproduced by permission of Routledge)*

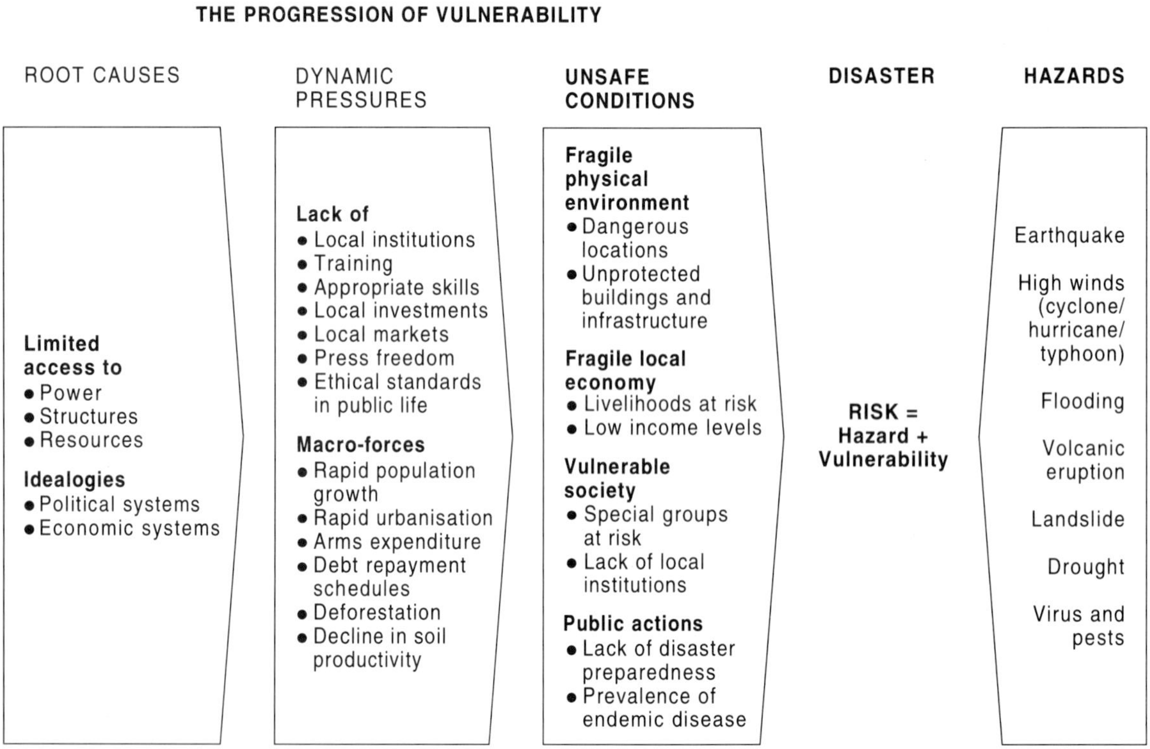

THE PROGRESSION OF VULNERABILITY

ROOT CAUSES	DYNAMIC PRESSURES	UNSAFE CONDITIONS	DISASTER	HAZARDS
Limited access to • Power • Structures • Resources **Idealogies** • Political systems • Economic systems	**Lack of** • Local institutions • Training • Appropriate skills • Local investments • Local markets • Press freedom • Ethical standards in public life **Macro-forces** • Rapid population growth • Rapid urbanisation • Arms expenditure • Debt repayment schedules • Deforestation • Decline in soil productivity	**Fragile physical environment** • Dangerous locations • Unprotected buildings and infrastructure **Fragile local economy** • Livelihoods at risk • Low income levels **Vulnerable society** • Special groups at risk • Lack of local institutions **Public actions** • Lack of disaster preparedness • Prevalence of endemic disease	**RISK =** **Hazard +** **Vulnerability**	Earthquake High winds (cyclone/ hurricane/ typhoon) Flooding Volcanic eruption Landslide Drought Virus and pests

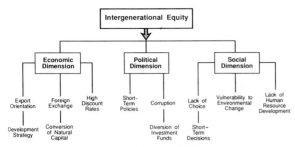

FIGURE 2.8 *Structural inequalities across generations (reprinted with permission from Lonergan, S.C. 1993, Impoverishment, population and environmental degradation: the case for equity.* Environmental Conservation *20: 328–34. Copyright © 1993 Elsevier Science)*

Marginalised groups who are compelled to overexploit today's resources are doing so at the expense of their own futures. 'Discounting the future' in this way will be offset to some extent, for example, by changes in technology which can change our perception of resources, as well as changes in political, economic and demographic factors. Various aspects of inter-generational inequalities are shown in Fig. 2.8. Resolving these inequalities across generations is the central aim of 'sustainable development', a concept to which we will return in Chapter 21, after consideration of a range of individual environmental issues.

FURTHER READING

Durning, A.B. 1989 *Poverty and the environment: reversing the downward spiral.* Worldwatch Paper 92. Worldwatch Institute, Washington, DC. This pamphlet looks at the many interactions between poverty and the environment and proposes ways in which situations can be improved.

Rees, J. 1990 *Natural resources: allocation, economics and policy,* 2nd edn. London, Routledge. This book reviews the spatial distribution of renewable and non-renewable resources, their development and consumption. It also considers the distribution of wealth derived from resources, and the roles of management, economics and politics.

Simmons, I.G. 1989 *Changing the face of the earth.* Oxford, Blackwell. The author traces the interactions between people and the natural environment from the earliest times to the present.

3

DEFORESTATION

Forests have been cleared by people for many centuries, to use both the trees themselves and the land on which they stand. Throughout the history of most cultures, deforestation has been one of the first steps away from a hunter–gathering and herding way of life towards sedentary farming and other types of economies. Forests were being cleared in Europe in Mesolithic and Neolithic times, but in central and western parts of the continent an intense phase occurred in the period 1050–1250. Later, when European settlers arrived in North America, for example, deforestation took place over similarly large areas but at a much faster rate: more woodland was cleared in North America in 200 years than in Europe in over 2000 years. Estimates suggest that since pre-agricultural times, about 1 billion hectares of the world's forests have been cleared, leaving a forest cover today of 4 billion hectares, or around one-third of the global land surface (Mather, 1990). Although most of this loss has taken place in the temperate latitudes of the Northern hemisphere, in recent decades the loss in this zone has been largely halted, and in many countries reversed by planting programmes. Meanwhile, in the tropics, rapidly increasing human populations and improved access to forests have combined in recent times to create an accelerating pace of deforestation which has become a source of considerable concern both at national and international levels, and it is on these latitudes that this chapter will concentrate.

DEFORESTATION RATES

Despite the high level of interest in deforestation, our knowledge of the rates at which it is currently occurring is far from satisfactory. In part, this is because of the lack of standard definitions of just what a forest is and what deforestation means. A distinction is often made between 'open' and 'closed' forest or woodland, which is sometimes defined according to the percentage of the land area covered by tree crowns. A 20 per cent crown cover is sometimes taken to be the cut-off point for closed forest, for example, while open woodland is defined as land with a crown cover of 5–20 per cent. Measurements of deforestation rates by the Food and Agriculture Organisation (FAO) define deforestation as the clearing of forest lands for all forms of agriculture and for other land uses such as settlements, other infrastructure and mining. In tropical forests this entails clearing to reduce tree crown cover to less than 10 per cent, and thus does not include some damaging activities such as selective logging which can seriously affect soils, wildlife and its habitat. In this case, the change in the status of the forest is referred to as degradation rather than deforestation.

The collection of data for all countries of the world and their compilation by different

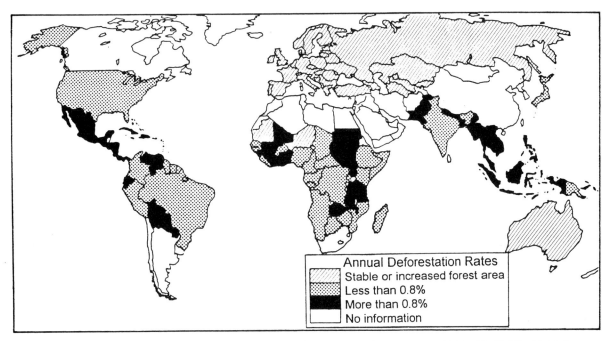

FIGURE 3.1 *Estimated annual rates of change in forest cover by country, 1981–90 (WRI, 1994 from FAO data)*

experts inevitably causes problems, since different countries and individuals often define the terms in different ways and the quality and availability of data varies between countries. One tool which is being used increasingly to avoid some of the problems of definition and poor data is satellite imagery, which can be evaluated using one set of criteria. But even these data can vary, depending on the satellite sensors used, differing methods of interpretation, and other difficulties such as that of distinguishing between primary and secondary forests. Estimates, from numerous sources, of the annual rate of deforestation of closed forests in the humid tropics (which include two main types: tropical rain forest and tropical moist deciduous forest) have varied from 11 to 15 million ha for the early 1970s, 6.1 to 7.5 mha for late 1970s, and 12.2 to 14.2 mha for the 1980s (Grainger, 1993b). This author suggests that the uncertainties are primarily the result

of lack of attention to remote sensing measurements and overconfidence in the use of expert judgement. One of the most recent assessments of the rates of change of forest cover by country, which has been prepared by the FAO, is shown in Fig. 3.1. This map shows clearly that while forest cover in the high latitudes is generally stable or increasing, virtually all tropical countries are experiencing a loss.

CAUSES OF DEFORESTATION

A concise summary of the causes of tropical deforestation is no easy task, since cutting down trees is the end result of a series of motivations and driving forces which are interlinked in numerous ways. On a worldwide basis, the people who actually cut down the trees can generally be agreed upon. They are:

TABLE 3.1 *Important factors influencing deforestation in the tropics by major world region*

Region	Main factors
Latin America	Cattle ranching, resettlement and spontaneous migration, agricultural expansion, road networks, population pressure, inequitable social structures
Africa	Fuelwood collection, logging, agricultural expansion, population pressure
South Asia	Population pressure, agricultural expansion, corruption, fodder collection, fuelwood collection
South-East Asia	Corruption, agricultural expansion, logging, population pressure

Source: after Kummer (1991)

- agriculturalists
- ranchers
- loggers
- fuelwood collectors.

However, a proper understanding of the deforestation process requires a deeper investigation of the driving forces behind these agents. Access to forests is an important aspect, for example, and this is usually provided by road networks. Hence, construction of a new road, whether it be by a logging company or as part of a national development scheme, is an integral part of deforestation. Another part of the equation is the role played by socio-economic factors which drive people to the forest frontier: poverty, low agricultural productivity and an unequal distribution of land are often important, while the rapid population growth rates which characterise many countries in the tropics also play a part. The role of national government is another factor in encouraging certain groups to use the forest resource, through tax incentives to loggers and ranchers, for example, or through large-scale resettlement schemes. On the global scale, international markets for some forest products, such as lumber and the produce from agricultural plantations, must also be considered.

The importance of these factors varies from country to country and from region to region and may change over time. Table 3.1 is an attempt to identify the prime factors which lie behind deforestation in the major forest regions of the tropics, based on a review of the large literature on the subject.

The scale of clearance that has occurred in some countries, and the influence of external factors, can be illustrated by the experience of Viet Nam, a country that in pre-agricultural times was almost entirely covered in forests but which has lost more than 80 per cent of its original forest area, much of it during the second half of this century (Fig. 3.2). Clearance of the coastal plains and valleys for agriculture took place over the last few centuries, and during the French colonial period when large areas in the south were cleared for banana, coffee and rubber plantations, but 45 per cent of the country was still forested in the 1940s. That proportion had fallen to about 17 per cent in the late 1980s as extensive zones were destroyed during the Viet Nam War (see Chapter 19) and still greater areas have since been destroyed by a rapidly growing population rebuilding after the war.

The forces of international economics which are integral to many areas of deforestation have played a central role in West Africa, where in most countries there is hardly any stretch of natural, unmodified vegetation left.

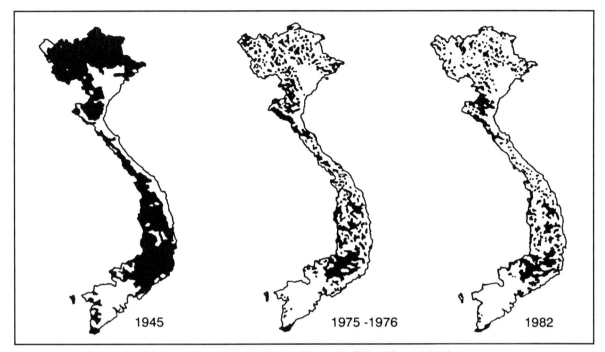

FIGURE 3.2 *Decline of forest cover in Viet Nam, 1945–82 (Viet Nam, 1985)*

A classic example is Côte d'Ivoire. Virtually every study of deforestation in the tropics has concluded that Côte d'Ivoire has experienced the most rapid forest clearance rates in the world, at 2800–3500 km² per year for the past 35 years.

Like most African governments, Côte d'Ivoire has viewed the nation's forests as a source of revenue and foreign exchange. However, given the country's high external debt, and declining international prices received for agricultural export commodities, the country has been faced with little alternative but to exploit its forests heavily. In 1973, logs and wood product exports provided 35 per cent of export earnings, but this figure had fallen to 11 per cent by the end of that decade due to the rapidly declining resource base. In the late 1970s, about 5.5 million m³ of industrial roundwood was extracted annually but this production had fallen below 3 million m³ by 1991. Nevertheless, tax incentives favour continued logging and further reduce govern-

ment income from the forestry sector (Repetto and Gillis, 1989). Deforestation has also been fuelled by a population which grew from 5 million in 1970 to almost 13 million in 1990, uncontrolled settlement by farmers, and clearance for new coffee and cacao plantations encouraged by government incentives.

Logging and agriculture have also been the primary agents of deforestation in the Philippines since the 1940s, and the series of factors which lie behind the loggers and agriculturalists, as interpreted by Kummer (1991), is shown graphically in Fig. 3.3. Forest cover in the Philippines declined from 70 per cent of the national land area to 50 per cent between 1900 and 1950, and declined further to below 25 per cent by the early 1990s. The most common pattern of events has been the conversion of primary forest to secondary forest by logging, followed by clearance of the secondary forest for the expansion of agriculture. Both activities are preceded by the construction of roads, built by provincial and

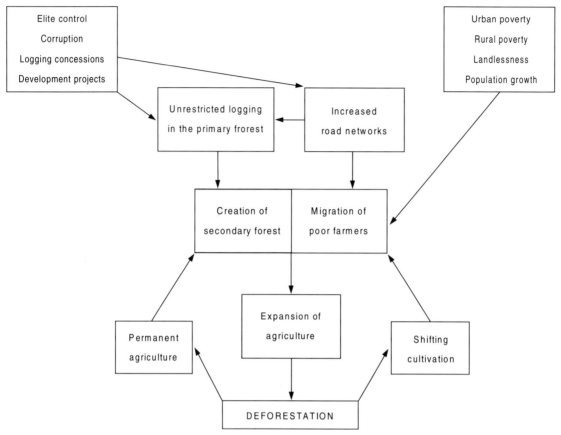

FIGURE 3.3 *Factors affecting deforestation in the Philippines (from* Deforestation in the postwar Philippines *by Kummer, D.M. Copyright © 1991 University of Chicago Press)*

national governments for general development purposes, or by loggers to access their concessions. Kummer suggests that the granting of logging concessions has occurred to foster development but also as political favours to Filipino élites and foreign-based transnationals. Much of the financial gain has also flowed to a small number of well-connected individuals. This concentration of the benefits from the Philippines' forest resource has been partly responsible for widespread and increasing poverty in the country, with no improvement in the living standards of the bottom 50–75 per cent of the population over the past 40 years. Most of the agriculturalists who clear the secondary

forests are subsistence farmers, spurred on by poverty, a rising population and a lack of land.

The role of government policy and practice has also been central to deforestation of the Amazon Basin in Brazil, which began on a large scale in the mid-1970s with a concerted effort to develop the country's tropical frontier. Agricultural expansion was the most important factor responsible for forest clearance at this time, both by smallholders and large-scale commercial agriculturalists, including ranchers producing beef for the domestic market. The Brazilian government's view of the Amazon as an empty land rich in resources spurred programmes of resettlement, particularly in the states of Rondônia and Pará under the slogan 'people without land in a land without people' (Moran, 1981), along with agricultural expansion programmes and plans to exploit mineral, biotic and hydroelectric resources.

The various agents of deforestation in the Brazilian Amazon have been summarised by Fearnside (1990). Cattle ranching has been encouraged by government subsidies and has been additionally attractive as a means of storing wealth, both in land and cattle, during periods of high inflation. Slash-and-burn agriculture practised by pioneer farmers arriving from outside the Amazon, as opposed to the shifting cultivation long practised by the indigenous peoples of the region, is another serious cause of deforestation, as it has long been in the Amazonian parts of Peru and Ecuador. Pioneers' slash-and-burn occurs at too great a human density of population, often leaves insufficient fallow periods and/or follows an initial crop with pasture planting, making it unsustainable. The construction of hydroelectric dams, with the inundation of reservoirs, has the potential for further major impacts to follow on from some of those existing, such as the Tucuruí and Balbina dams. The destruction caused by mining is also important, although usually on a local scale. In the case of the Grande Carajás Programme in the state of Pará, however, the proposed development of iron ore, bauxite, copper and manganese involves local smelting, a new railway and highway network, and a very large agricultural scheme. Lumbering is rapidly increasing in importance; agribusiness, until now a minor cause of deforestation in Brazil, could expand significantly, and the military bases being constructed along the Brazilian Amazon's northern frontier under the Calha Norte Programme pose another serious threat.

The broader set of dynamic circumstances underlying this policy of frontier development in Brazil has been highlighted by Turner *et al.* (1993) who trace them back to the OPEC (Organisation of Petroleum Exporting Countries) oil crisis of the 1970s which resulted in a large transfer of economic wealth from industrial countries to oil producers who, in turn, deposited these revenues in US and European banks. Large-scale lending of these monies to Brazil, amongst other developing countries, allowed development programmes to be financed. In Brazil, agricultural modernisation took much of this finance and was channelled into export crops such as wheat, soybeans and coffee. Soybean production was boosted particularly, because relative to coffee the international market for soybeans was much more dependable, and this expansion was concentrated in two states: Rio Grande do Sul and Paraná. However, widespread replacement of coffee cultivation, a labour-intensive crop, with soybean, which is more capital and energy intensive, has resulted in large-scale emigration, particularly from Paraná. Large numbers of migrants went to new opportunities on the forest frontier in Rondônia, practising small-scale slash-and-burn agriculture along the World Bank-financed Highway BR-364, as it was extended into the state.

In parts of the Third World inside and outside the tropics, a primary and increasingly serious cause of deforestation is the shortage of fuelwood which will affect 3 billion people by the year 2000 according to one estimate (FAO, 1983). The majority of these people live in Asia, although dryland regions of Africa (see Table 4.4) and the Andean plateau of South America are also particularly affected. Fuelwood collection is a serious cause of deforestation of intertidal mangrove forests in West Africa and Central America, for example (see Chapter 11). In India, by contrast, the overall forest area is not currently being depleted, but fuelwood collection continues to degrade and remove forests around many urban areas. In one study using satellite imagery, Bowonder *et al.* (1988) found considerable loss of closed forests in the urban

TABLE 3.2 *Loss of closed forest cover around selected major cities in India*

City (state)	Closed forest cover (km²)		Loss (%)
	1972–75	1980–82	1972–82
Bangalore (Karnataka)	3853	2762	28
Bhopal (Madhya Pradesh)	3031	1417	53
Bombay (Maharashtra)	5649	3672	35
Delhi (Delhi)	254	101	60
Hyderabad (Andhra Pradesh)	35	40	26
Jaipur (Rajasthan)	1534	786	49
Madras (Tamil Nadu)	918	568	38

Source: after Bowonder et al. (1988)

hinterlands within a 100-km radius around thirty-three major Indian cities, a selection of which are shown in Table 3.2. India's urban poor have been forced to turn to fuelwood for cooking by sharp increases in the price of kerosene, coal and charcoal over the last 15 years. The increased demand has pushed up the price of fuelwood, encouraging rural people to collect and sell it in the cities. Although they receive a good income for their work, they seldom replant the trees they cut. The fuelwood price per tonne in the southern city of Bangalore rose from 47 rupees in 1960 to 657 rupees in 1986, and had reached 1135 rupees by 1992.

CONSEQUENCES OF DEFORESTATION

Much of the concern over the issue of tropical deforestation stems from the perturbation it represents to the forests' role as part of human life-support systems, by regulating local climates, water flow, and nutrient cycles, as well as their role as reservoirs of biodiversity and habitats for species. The numerous deleterious environmental consequences resulting from the loss of tropical forests range in their scale of impact from the local to the global. Human clearance is by no means the only form of disturbance that forests experience, however. A tropical cyclone, for example, can cause tremendous damage over large areas, and through geological time the global extent of tropical forests has varied with climate and is thought to have been much smaller than its current extent during periods of glacial maxima (e.g. Endler, 1982).

The human impact on tropical forests also has a long history. In practice it can be difficult to distinguish in the field between so-called 'primary forest' and 'secondary forest' which is the result of human impact, sometimes dating back over long periods. For instance, large areas of tropical forest in the Yucatan Peninsula in Mexico, long believed to be primary forest, are now considered to be secondary forest managed by the Mayan people more than 1000 years ago (Gómez-Pompa et al., 1987). To the casual observer, secondary forests more than 60–80 years old are often indistinguishable from undisturbed primary forests and are, in fact, treated as primary forests in the FAO assessments of tropical forests. Secondary forests less than 60–80 years of age are extensive in the tropics, making up about 40 per cent of the total forest area (Brown and Lugo, 1990). However, although humans have registered an impact on tropical forests over many centuries, which

TABLE 3.3 *Some of the major constraints to agricultural development of two dominant Amazon Basin soils*

Constraint	Oxisols	Ultisols
Low plant nutrient levels	X	X
Nitrogen losses through leaching	X	X
High levels of exchangeable aluminium	X	X
Low cation exchange capacity	X	
Weak retention of bases	X	
Strong fixation or deficiency of phosphate	X	
Soil acidity	X	
Very low calcium content	X	
Essential to continue to maintain soil organic-matter levels		X
Weak structure in surface layers		X

Source: after Nortcliff and Gregory (1992)

cumulatively has affected great areas, the intensity, speed and relative permanence of clearance by human populations in the modern era puts this form of disturbance into a different category.

Soil degradation – including erosion, landslides, compaction and laterisation – is a common problem associated with deforestation in any biome, but is often particularly severe in the tropics. Tropical rainfall is characterised by large annual totals, it is continuous throughout the year and highly erosive. Hence, clearance of protective vegetation cover quickly leads to accelerated soil loss, although the rate will vary according to what land use replaces the forest cover (see Chapter 6). Soil erosion can also be exacerbated by compaction caused by heavy machinery, trampling from cattle where pasture replaces forest, and exposure to the sun and rain. A compacted soil surface yields greater runoff and often more soil loss as a consequence. Laterisation, the formation of a hard, impermeable surface, was once feared to be a very widespread consequence of deforestation in the tropics, but is probably a danger on just 2 per cent of the humid tropical land area (Grainger, 1993a).

The poor nutrient content of many tropical forest soils is one factor that tends to limit the potential of many of the land uses for which forest is cleared. In parts of the Amazon Basin, for example, the available phosphorus and other nutrients in the soil initially peaks after the burning of cleared vegetation, but declines thereafter due to leaching, and is insufficient to maintain pasture growth after 10 years. Soil nutrient depletion, compaction and invasion by inedible weeds quickly depletes the land's usefulness as pasture. Similarly, unsustainable shifting cultivation quickly depletes soil nutrients, and, in both cases, rapidly diminishing yields from inappropriate land uses encourage settlers to move on and clear further forest areas in a process of positive feedback. Some of the other major constraints for two of the most dominant soil orders found in the Amazon Basin are shown in Table 3.3.

The effects of widespread deforestation on hydrology have been widely discussed, particularly the concerns stemming from recognition of the forests' role in regulating the flows of rivers and streams from upland catchments. The feared type and scale of regional impact on natural systems consequent upon deforestation is well illustrated in the Himalayas, where the theory that widespread deforestation since 1950 has been responsible for increased flooding of the Ganges and Brahmaputra plains has been widely accepted. The theory has not been without its critics, however, who, upon looking at the evidence have shown that there is, in fact, little reliable data with which to support the proposed link between deforestation and increased flooding (Ives and Messerli, 1989). Although there is clearly a need to conserve forest cover on catchments, the role of deforestation in flooding appears to have been exaggerated in many cases.

A less-controversial effect is that on sediment loads in rivers due to enhanced erosion and runoff. High sediment loads have been the cause of problems, particularly in reservoirs – shortening their useful lives as suppliers of energy and irrigation water, for example, and requiring costly remediation (see Chapter 13).

The possible climatic impacts of large-scale forest clearance, resulting from effects on the hydrological and carbon cycles, may occur on regional and even global scales. In a large region like Amazonia, where about half of the rainfall is returned to the atmosphere via evapotranspiration, widespread removal of forest could have a serious impact on the hydrological cycle, with consequent negative effects on rainfall and, hence, continued forest survival and agriculture. Such theories have been tested by simulating future conditions with general circulation models, but the results to date have been inconclusive. At the global level, large-scale deforestation releases carbon dioxide into the atmosphere, either by burning or decomposition, possibly contributing to enhanced global warming through the greenhouse effect. Estimates of just how much additional carbon dioxide reaches the atmosphere due to tropical deforestation are uncertain, due both to our imprecise knowledge of deforestation rates

and of the amount of carbon locked into a unit area of forest. Although deforestation's contribution to rising atmospheric carbon dioxide concentrations is less than that of fossil fuels, it could contribute up to a third as much again. The possible consequences of climatic change due to the resulting greenhouse warming are discussed in Chapter 8.

One of the most widely discussed consequences of deforestation in the tropics is the loss of plant and animal species. The diversity of life in tropical rain forests relative to other biomes is legendary. Study of a single tree in Peru, for example, yielded no less than forty-three ant species belonging to twenty-six genera, roughly equivalent to the ant diversity of the entire British Isles (Wilson, 1989). Examples of this level of diversity are numerous: on just 1 ha of forest in Amazonian Ecuador, researchers have recorded no less than 473 species, 187 genera and fifty-four families of tree (Valencia et al., 1994), the highest number of tree species found on a tropical rain forest plot of this size, which is more than half of the total number of tree species native to North America. Although our knowledge of the total number of species on the planet is scant (see Chapter 7), most estimates agree that more than half of them live in the tropical moist forests which cover only 6 per cent of the world's land surface. Many species are highly localised in their distribution and, hence, can be destroyed without clearing a very large area. Overall, there is widespread agreement in the scientific community that deforestation in the tropics is a severe threat to the species and genetic diversity of the planet (although not *all* authorities are so pessimistic; see, for example, Lugo, 1988). From a purely practical viewpoint, the loss of species, most of which have not been documented, has been likened to destroying a library of potentially useful information having only flicked through the odd book. The types of goods which could be derived from tropical forests include new pharmaceutical products, food and industrial crops, while a diversity of genetic material is of importance to improve existing crops as they become prone to new strains of diseases and pests.

Forest structure and diversity are not just affected by clear-cutting, but also by fragmentation (see Chapter 7) and lesser forms of disturbance. The 1982/83 drought fires which seriously affected 950 000 hectares of dryland forest in Sabah and 2.7 million hectares of swamp and dryland forest in East Kalimantan, for example, raged mainly in selectively logged areas where dry combustible material from tree extraction littered the forest floor (Wirawan, 1993). While the small-scale, low-intensity, low-frequency fires commonly associated with traditional shifting cultivation systems are considered to be a moderate-level disturbance which is likely to stimulate the maintenance of high biodiversity in tropical rain forest, such large-scale fires represent a much greater disturbance which it will take a very long time to recover from.

Another all too common casualty of deforestation in the tropics is indigenous peoples, for whom death or loss of cultural identity usually follows the arrival of external cultures and the frontier of forest clearance (Fig. 3.4). Whether it is deliberate annexation of their territory by the dominant society, or a less-formalised erosion of their way of life, the result is often the same. The indigenous Indian population of Brazil, for example, is thought to have numbered 1 million in 1900, but has been reduced to about 200 000 today, largely due to the effects of deforestation. To extend the library metaphor with regard to the potential wider uses of biodiversity, the impacts of deforestation on tribal peoples is equivalent to killing off the librarians who know a great deal about what the library contains. At a more fundamental level, the plight of indigenous peoples is a simple question of human rights.

FIGURE 3.4 *The outsiders' perception of indigenous tropical rain forest inhabitants as savages, as this engraving from a seventeenth century book of travels in Brazil aptly illustrates, still tends to prevail. The all too common consequences for indigenous inhabitants include loss of cultural identity, and suffering due to the introduction of new diseases*

TROPICAL FOREST MANAGEMENT

The answers to the problems caused by exploitation of tropical forests are by no means straightforward to formulate or easy to implement. One difficulty which should be appreciated is that most remaining tropical forest areas are remote from national seats of government and that the rule of law in these frontier zones may be less than total. Therefore passing legislation which supposedly limits the agents of deforestation is often meaningless in practice. Besides, since deforestation is in many parts of the world a symptom of the underlying causes that have been noted above, it is these causes which must be addressed if development is to be directed towards a more sensible use of the forest resource. Many of the structural changes needed to strike at the root causes of deforestation will require years of endeavour. Nevertheless, as Fearnside (1990) suggests for the Brazilian Amazon, a start can be made now.

Among Fearnside's proposals for Brazil are some deep-seated national economic policies which need to be addressed, such as slowing population growth and improving employment opportunities both in labour-intensive agriculture and in urban areas outside the Amazon. Among his suggestions for the Brazilian government's immediate action are:

- stop highway construction in Amazonia,
- abolish subsidies for cattle pastures,
- end energy subsidies in Amazonia,
- ensure proper protection for natural forest reserves and stop reneging on previous commitments to reserves whenever the land is wanted for another purpose.

While these proposals might seem sensible from one perspective, the needs of national economic development cannot be denied. For Brazil to cease development of its largest sovereign natural resource is clearly neither desirable nor feasible. In practical terms, too, the fact that such suggestions often come from developed countries which have previously all but destroyed their own forests in the course of their development can quickly be interpreted as 'ecoimperialism'. There are many parallels between today's developing countries and the past experience of those which have developed further. In Britain, for example, almost the entire land area was covered by forest about 5000 years ago, about the time that human activity began to have an impact on the landscape. Today, the UK is one of the least-wooded countries in Europe. Woodland covers just 8 per cent of England and Wales, but most of this is recent plantings. The area covered by ancient woodland is about 2.6 per cent (Spencer and Kirby, 1992); most of the lost forests have been converted into farmland, which occupies some 72 per cent of the total land area today, and many wildlife species have become extinct in Britain as a result of this destruction of habitat that made them more susceptible to hunters (see Table 7.2).

Outsiders can make a positive contribution to controlling the poor use of forest resources, although direct pressure on governments from well-intentioned individuals abroad can backfire. Improvements in the environmental performance of such players as the World Bank, commercial banks and other investors can be called for, and moves made to tighten their financial approach to reduce the temptation of borrowing countries to run up heavy debts and promote short-sighted land use policies to pay them back, as in the case of investment in soybean production in southern Brazil. Debt-for-nature swaps have proved to be a successful approach in several countries, making a positive contribution to conservation and developing countries' debt burdens (see Chapter 21).

A pragmatic approach to the problems of deforestation cannot seriously obstruct all extension of cultivated lands or all clearance of forest to provide new industrial or urban sites. But this process requires surveys, inventories and evaluation to ensure that each hectare of land is used in the optimum way, to meet both the needs of today and those of future generations. A series of questions which should be answered in deciding whether or not to alter a tropical forest has been proposed by Poore and Sayer (1987):

- *What are the dominant ecological and social constraints in the tropical forest area in question?* Soil analysis is essential; many tropical forest lands (especially on poor, wet soils) can sustain nothing but an intact tropical forest cover.
- *Is there degraded, altered – or at least non-primary – tropical forest available which could also sustain the required enterprise?* Such land usually supports fewer species, provides fewer ecological benefits, is nearer to settlements and infrastructure and can be rehabilitated by development. A notable example is timber production from old secondary tropical forest, which often yields a much higher proportion of desirable species than primary tropical forest.
- *Are there ways in which the enterprise could be sustained within the existing tropical forest?* In other words, can this be done without altering the forest structure.
- *If the tropical forest were to be cleared and a new use established, what would be the net change in costs and benefits (including all tropical forest values, such as watershed protection)?*

These authors suggest that the full range of ecological processes and species diversity of tropical forests can only be maintained if large undisturbed areas are established in perpetuity for conservation. Yet very little of the remaining tropical forest area is currently legally protected as national parks or other reserves, and even if existing proposals are implemented, the proportion will not be greater than 7 per cent of remaining tropical forests. In practice, the amount which is properly protected from intrusion by hunters and settlers, or excluded from oil and mineral prospecting, hydroschemes and highway development is very much less. Conservationists agree that a particular effort

is needed to increase protected areas and to strengthen effective protection. However, totally protected areas can never be sufficiently extensive to provide for the conservation of all ecological processes and all species. In this case, the obvious avenues to explore are ways in which a natural forest cover can be used sustainably, to harvest products without degrading the forest. In this case, sustainable forestry means to harvest forest in a way that provides a regular yield of forest produce without destroying or radically altering the composition and structure of the forest as a whole. In other words, sustainable operations need to mimic natural ecosystem processes.

The obvious product in this respect is timber, but attitudes towards the logging industry have become polarised, with some people suggesting that it is essentially a primary cause of tropical deforestation and that the international trade in tropical timber should therefore be banned. In practice, however, logging in the tropics almost always involves the removal of selected trees rather than the clear-felling which is much more common in temperate forestry operations. Thus, logging is rarely a direct cause of deforestation, although it is often responsible for degradation and does open the way for other agents of deforestation, such as agriculturalists. Hence, others argue that despite the poor track record of loggers in the past, they should be encouraged to move away from a mining approach towards timber harvesting through sustainable management. Although, to date, there are very few examples of a sustainable approach to logging actually being carried out, many foresters believe that most tropical moist forests could be managed as a truly renewable resource if human intervention operated within the inherent limits of the natural cycle of growth and decay found in all forests (Johns, 1992). Unfortunately, logging is most often controlled by entrepreneurs, to whom short-term profits are of prime importance,

FIGURE 3.5 *Processing of tropical rain forest logs at Samreboi in Ghana, where machinery is outdated and inefficient, means wastage rates of 65 per cent for plywood and 70 per cent for planks*

rather than by foresters, whose duty is to the long-term maintenance of the resources. Ironically, a considerable portion of tropical forest trees felled for timber in many areas is wasted due to inefficient processing using outdated equipment (Fig. 3.5).

One key policy change which could encourage such a move would be to lengthen the periods over which concessions were granted. Typically, concessions are granted for periods of between 5 and 20 years, but if they were increased to at least 60 years, it would be in the interest of the logging companies to protect their forest and produce timber sustainably on perhaps 30–70-year rotations. One problem here, however, is that no one can guarantee that there will be a market for tropical timber so far into the future.

Two international initiatives have been established with the aim of managing timber production in a more sustainable way. One is the International Tropical Timber Agreement (ITTA), which is administered by the International Tropical Timber Organization (ITTO), came into being in 1983. Although it is primarily a commodity trade agreement, the ITTA also has a component that is rare for such agreements which states that one of its objectives is to encourage national policies aimed at maintaining the ecological balance in timber-producing regions. The ITTO has established some important projects on multiple-use forestry practices, natural forest management and plantation development which aim to demonstrate appropriate management of tropical forests, but the organisation also needs

TABLE 3.4 *Economic returns from different uses of one hectare of lowland tropical rain forest in eastern Peru*

Use	Annual income (after labour and transport)*	Net present value (20-year discounted)*	Comments
EXTRACTIVE USES			
Latex harvest	$22	$440	Does not include other forest products or tourism
Edible fruits harvest	$400	$6000	
Total (harvesting)	**$422**	**$6440**	
Sustainable selective logging	$15	$490	
Total (harvesting + sustainable logging)	**$437**	**$6930**	
CONVENTIONAL 'DEVELOPMENT' USES			
One-time removal of marketable timber		$1000 (no 20-year discount)	Cutting destroys extractive resources
Reforestation with *Gmelina arborea*	$159	3184	Not sustainable
Total (forestry)		**$4184**	
Intensive cattle ranching on ideal pasture	$148	<$2960	Not sustainable
Total (ranching)		**<$2960**	

*US$
Source: after Peters et al. (1989)

to encourage policy changes at national level, more realistic pricing and the international cooperation needed to encourage these changes.

By far the largest initiative to have ever focused on the conservation and sustainable management of tropical rain forests is the Tropical Forestry Action Plan (TFAP), set up in 1985 by several international bodies under the philosophy that since poverty is the root cause of deforestation, richer countries should fund projects to alleviate poverty in poorer tropical countries and thus reduce the problem. The basic objective of the TFAP is to restore, conserve and manage forests sustainably to benefit rural people, agriculture and the economy, in general, of the countries concerned. It is offered to developing countries rather than forced on them, and, although there is a good deal of scepticism over whether development agencies can effectively discourage deforestation, since their previous track records have not been good, more than seventy countries have joined. To date, however, progress in implementing projects has been slow and a lack of appropriate funding has been a serious problem.

Although logging has received particular attention for sustainable forestry practices, other forest products have also been the subject of some research. Low-intensity harvesting of fruits, nuts, rubber and other produce should result in a minimum of genetic erosion and other forms of disturbance, and a maximum of conservation. One economic study of the Peruvian Amazon (Table 3.4) has shown that this form of sustainable resource use is the most economically profitable in the medium to long term (Peters *et al.*, 1989), although adequate attention needs to be given to appropriate markets and sufficient transport and distribution networks. If the products concerned do not have an international market, such an approach is likely to be less attractive than timber for national governments, because of the hard currency revenues which an internationally tradeable commodity can command. Conversely, some believe that since the root causes of overexploitation of tropical forests are often traced to national and international élites, the highest priority should be given to increasing the control of small-scale forest dwellers over their resources (Dove, 1993). Where existing small-scale harvesting from tropical forests is already carried out for local markets, the diversity of forest products represents a great strength for the producers, allowing them to respond well to the changing needs and tastes of the market and furnishing them with a good annual income (Anderson and Ioris, 1992).

Some observers are confident that countries can develop and still maintain a high forest cover given sufficient investment in agriculture, and government dedication to sustainable forest management and to a strong national network of protected areas (Grainger, 1993a). Inevitably, some further tropical forest areas will still be lost due to economic problems, international disputes, organisational shortcomings, and as policies evolve to make balanced decisions on various alternative land uses. There is a lot of willing, both inside and outside tropical forest countries, to use forests in a sensible manner, but a sustainable future for the world's tropical forests will also depend upon a greater level of international cooperation and equity than we have seen to date.

FURTHER READING

Furley, P.A. (ed.) 1994 *The forest frontier: settlement and change in Brazilian Roraima*. London, Routledge. A collection of papers on both physical and human aspects of forest use in one of Brazil's states.

Grainger, A. 1993 *Controlling tropical deforestation*. London, Earthscan. This book looks at the causes of tropical rain forest destruction and presents an integrated strategy for controlling the losses and using forests sustainably.

Mather, A.S. 1990 *Global forest resources*. London, Belhaven. An assessment of global forest resources, their use through history and in the present day, and policies and management techniques.

Poore, D. 1989 *No timber without trees: sustainability in the tropical forest*. London, Earthscan. This book focuses on the prospects for timber production in the wider context of tropical rain forest conservation. It is based on a study for the International Tropical Timber Organization.

4

DESERTIFICATION

Human societies have occupied deserts and their margins for thousands of years, and the history of human use of these drylands is punctuated with many examples of productive land being lost to the desert. In some cases the cause has been human, through overuse and mismanagement of dryland resources, while in other cases natural changes in the environment have reduced the suitability of such areas for human habitation. Often a combination of human and natural factors are at work. Salinisation and siltation of irrigation schemes in Lower Mesopotamia are thought to have played a significant role in the eventual collapse of the Sumerian civilisation 4000 years ago, and similar mismanagement of irrigation water has been implicated in the abandonment of the Khorezm oasis settlements of Uzbekistan dating from the first century AD. The decline of Nabatean towns in the Negev Desert from about 500 AD has been attributed to Muslim–Arab invasions, but a gradual change in climate towards a more arid regime probably played an important role. The belief that climates in post-glacial times were becoming gradually drier, so-called 'progressive desiccation', generated widespread concern for the future human use of drylands in Central Asia, southern and West Africa during the late nineteenth and early twentieth centuries, but the concept was subsequently disproved. Invasion and settlement of areas by outsiders has often been highlighted as a reason for the

onset of dryland environmental degradation in historical times. The Spanish conquest is cited in South America and a more diverse mix of European settlers were responsible for the dramatic years of the Dust Bowl in the US Great Plains in the 1930s. Desertification became a major global environmental issue in the 1970s after international concern over famine in West Africa focused attention on both drought and desertification as insidious causes. A UN Conference on Desertification (UNCOD) was convened in 1977, at which desertification was seen as a global threat affecting drylands all over the world, in rich and poor countries alike. Since that time, however, desertification has become surrounded in controversy, over its nature, extent, causes and effects, and even its very definition.

DEFINITION OF DESERTIFICATION

The word 'desertification' was first coined by a French forester to describe the change towards more desert-like conditions in humid areas adjacent to the Sahara Desert in West Africa as forests in the region were gradually cleared (Aubreville, 1949). Literally, desertification means the making of a desert, but although the word has been in use for more than 40 years, there is no universally agreed definition of the term. Indeed, a survey of the literature on the subject has revealed more

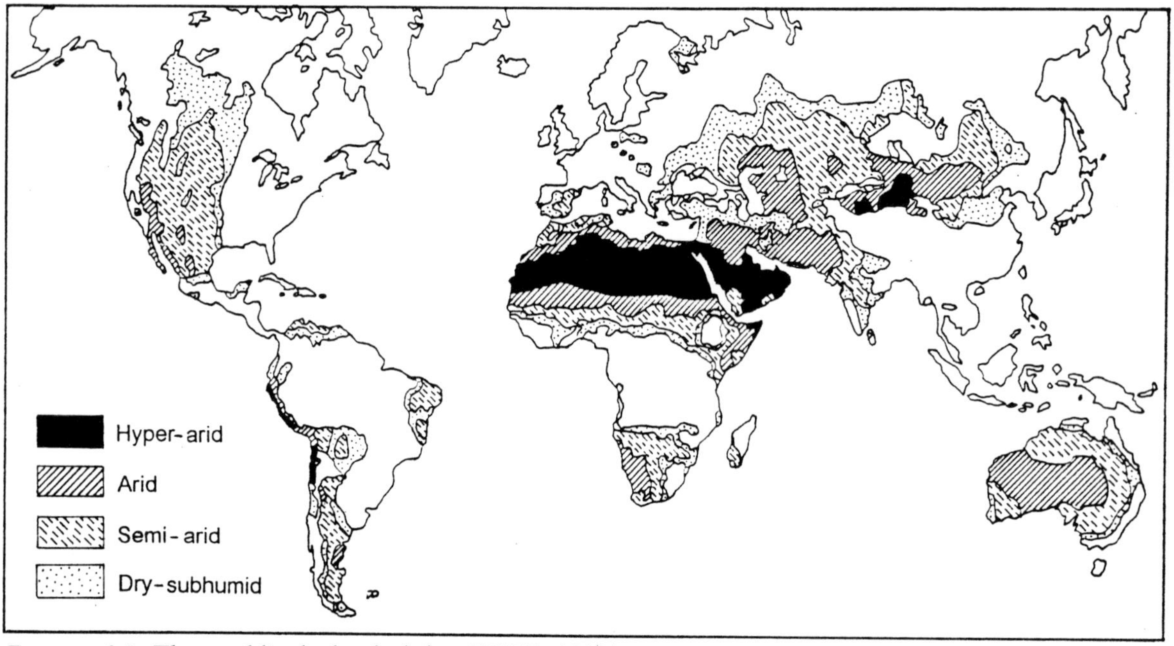

FIGURE 4.1 *The world's drylands (after UNEP, 1992)*

than a hundred different definitions (Glantz and Orlovsky, 1983).

Most of these definitions suggest that the extent of deserts is increasing, usually into desert-marginal lands. The idea of a loss of an area's resource potential, through the depletion of soil cover, vegetation cover or certain useful plant species, for example, is included in many definitions. Some suggest that such losses are irreversible over the timescale of a human lifetime.

Most, though not all, authorities agree that desertification occurs in drylands, which can be defined using the boundaries of climatic classifications (Fig. 4.1) and make up about one-third of the world's land surface. Arid, semi-arid and dry subhumid zones can be taken together as comprising those drylands susceptible to desertification, excluding hyper-arid areas because they offer so few resources to human populations that they are seldom used and can hardly become more desert-like. Recent thinking on desertification refers to it as a process of land degradation in

drylands. The concept of degradation is linked to using land resources – which include soil, vegetation, and local water resources – in a sustainable way. Land which is being used in an unsustainable manner is being degraded, which implies a reduction of resource potential caused by one or a series of processes acting on the land. The most recent official definition adopted by the UN Environment Programme (UNEP) refers to 'desertification/land degradation' which is 'land degradation in arid, semi-arid and dry subhumid areas resulting mainly from adverse human impact'. This definition focuses on human action as the main cause of desertification, but acknowledges that natural climatic variability can affect dryland resources. The relative roles of human and natural factors, such as drought, have long been a subject of debate. Drylands are, by definition, areas which experience a deficiency in water availability on an annual basis, but precipitation in drylands is characterised by high variability in both time and space. Rainfall typically falls in just a few rainfall events and its spatial pattern is often highly localised since much comes from convective cells. In addition to this variability on an annual basis, longer-term variations such as droughts occur over periods of years and decades. Dryland ecology is attuned to this variability in available moisture and is therefore highly dynamic. On the large scale, satellite imagery has been used to observe changes in the vegetational boundary of the Sahara of up to 200 km from one dry year to a following wet one (Tucker *et al.*, 1991).

In practice, it can be difficult to distinguish in the field between the adverse effects of human action and the response of drylands to natural variations in the availability of moisture. A good example is the dramatic rise in the amount of soil lost to wind erosion in parts of the Sahel region of Africa during the 1970s and 1980s as recorded by the annual number of dust storm days. At Nouakchott, the capital of Mauritania, dust storms blew on average less than 10 days a year during the 1960s, but in the mid-1980s the average had increased to around 80 days per year (Middleton, 1985). This increase in soil loss can be explained both by drought, which characterised the area during the 1970s and 1980s, and by human action, but the relative roles of these factors are difficult to quantify.

AREAS AFFECTED BY DESERTIFICATION

In preparation for UNCOD, UNEP produced an estimate of the global area at risk from desertification. The estimate suggested that at least 35 per cent of the Earth's land surface was threatened, an area inhabited by 20 per cent of the world's population. This estimate was based upon what limited data were available at the time and a lot of expert opinion and guesswork. It was a fair first attempt to assess a very difficult problem, but subsequent efforts to assess the global proportions of desertification have not greatly reduced the reliance upon estimates. Two subsequent assessments have been made (Table 4.1) and UNEP has been accused of treating these estimates as if they were based upon scientific data (Thomas and Middleton, 1994). It has even been suggested that since UNCOD, desertification has become an 'institutional fact', one that UNEP wanted to believe for its own purposes, for the continued existence of its desertification control unit, rather than

TABLE 4.1 *UNEP estimates of type of drylands deemed susceptible to desertification, proportion affected and actual extent*

	1977	1984	1992
Climatic zones susceptible to desertification	Arid, semi-arid and subhumid	Arid, semi-arid and subhumid	Arid, semi-arid and dry subhumid
Total dryland area susceptible to desertification (million ha)	5281	4409	5172
Proportion of susceptible drylands affected by desertification (%)	75	79	70
Total of susceptible drylands affected by desertification (million ha)	3970	3475	3592

Source: Thomas and Middleton (1994)

because the evidence was there to support it (Warren and Agnew, 1988). Indeed, the lack of good scientific evidence to support the concept has led some people to question the very existence of desertification. As Hellden (1991: 372) puts it: 'the word "myth" has been mentioned'.

Some people question whether a global assessment of something as poorly understood as desertification is possible, or even useful, given that solutions to the problem should be locally oriented. However, a case can be made for a global assessment if only to put the issue into perspective, to identify specific problem areas and to generate the political and economic will to do something about it.

'Spreading deserts'

Another area in which many authorities have helped to 'propagate the myth' has been in the numerous statements concerning the rate at which desertification is happening. These statements are made in the academic literature and in popular and political circles for publicity and fund-raising purposes. Warren and Agnew (1988) quote several examples, including one made in 1986 in which US Vice-President Bush was being urged to give aid to the Sudan because 'desertification was advancing at 9 km per annum' (Warren and Agnew, 1988: 2). Such pronouncements have become so commonly used that their validity has seldom been questioned, but in reality they are rarely supported by any scientific evidence. They also reinforce an image of sand dunes advancing to engulf agricultural land. Although undoubtedly a powerful vision for publicity purposes, and one that is appropriate in some places, the idea is oversimplified and not a generally accurate image of the way desertification works. It is a much more complex set of processes that is more commonly manifested as patches of degradation in the desert fringe.

One well-known study which did set out to measure the rate of desert advance was made in northern Sudan (Lamprey, 1975). The vegetational edge of the desert was mapped by reconnaissance in 1975 and then compared to the boundary drawn from another survey carried out in 1958. The two boundaries indicated that the desert's southern margin had advanced by 90–100 km between the two dates (Fig. 4.2), at an average rate of advance of 5.5 km per year over 17 years.

An immediate criticism can be made of the methodology used for the study in that given the inherent variability of drylands, two 'snapshot' surveys can only be of very limited use in determining significant changes in vegetation, particularly since the 1970s were a period of drought in the Sudan while the 1950s was a decade characterised by above-average rainfall. Reliable conclusions can only be drawn after long-term monitoring which takes into account seasonal and year-to-year fluctuations in vegetation cover. When Lamprey's conclusions were scrutinised by Swedish geographers they were found to be inaccurate. It transpired that the 1958 desert vegetation line was not based on a ground survey but followed the 75 mm isohyet, and satellite imagery and ground survey data compiled over several years did not confirm Lamprey's findings (Hellden, 1988).

CAUSES OF DESERTIFICATION

The methods of land use which are applied too intensively and, hence, contribute to desertification are widely quoted and apparently well known. They can be classified under the headings of overgrazing, overcultivation and deforestation. Salinisation of irrigated cropland is often viewed as a separate category from these. Although the way in which these inappropriate land uses lead to desertification is also well known in theory, in practice, particular areas deemed to

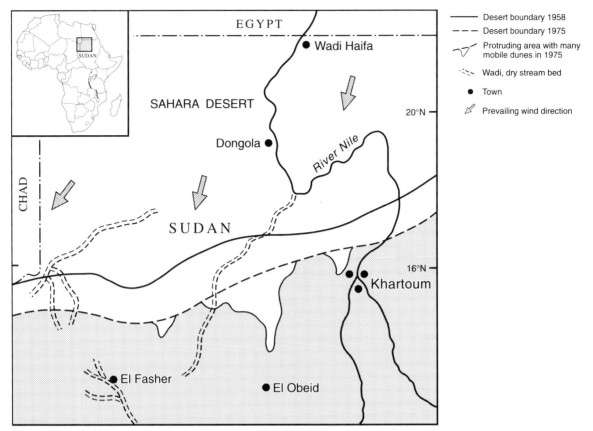

FIGURE 4.2 *Lamprey's (1975) 'advancing Sahara' – a study which purported to show that the desert was advancing at 5.5 km per year. See text for criticism of these findings*

be desertified due to a particular cause have often been so described according to subjective assessments rather than after long-term scientific monitoring. Furthermore, although specific land uses have been the subject of most interest from desertification researchers, it can be argued that permanent solutions to specific problem areas can only be found when the deeper reasons for people misusing resources are identified. Such reasons, many of which are rooted in social, economic and political systems (Table 4.2), enable, encourage or force inappropriate practices to be used. Although an understanding of the physical processes involved in overcultivation, for example, is important, there has generally

been too much emphasis on this physical side of the equation, resulting in an over-reliance upon technical solutions. It is just as important to understand the underlying driving forces, the amelioration of which is of equal significance in the quest to find long-lasting solutions to desertification problems.

Overgrazing

The overuse of pastures, caused by allowing too many animals or inappropriate types of animals to graze, has been the cause of degradation over the largest areas of desertified land on the global scale according to the estimates produced by UNEP. Of the 3592

Table 4.2 *Suggested root causes of land degradation*

Natural disasters	Degradation due to biogeophysical causes or 'acts of God'
Population change	Degradation occurs when population growth exceeds environmental thresholds (Neo-Malthusian) or decline causes collapse of adequate management
Underdevelopment	Resources exploited to benefit world economy or developed countries, leaving little profit to manage or restore degraded environments
Internationalism	Taxation and other forces interfere with the market, triggering overexploitation
Colonial legacies	Trade links, communications, rural–urban linkages, cash crops and other 'hangovers' from the past promote poor management of resource exploitation
Inappropriate technology and advice	Promotion of wrong strategies and techniques which result in land degradation
Ignorance	Linked to inappropriate technology and advice: a lack of knowledge leading to degradation
Attitudes	People's or institutions' attitudes blamed for degradation
War and civil unrest	Overuse of resources in national emergencies and concentrations of refugees leading to high population pressure in safe locations

Source: Thomas and Middleton (1994)

million hectares estimated in 1992 to be suffering from desertification, no less than 2576 ha, just under 72 per cent, were considered to be experiencing degradation of vegetation. Edible plant species are removed by overgrazing, encouraging invasion by inedible species. Further pressure, both from eating, trampling and compaction from hooves, can result in poorer soil structure and erosion, reducing the ability of soil to retain moisture and support plant growth. This degradation of the plant resource, in turn, reduces the number of livestock which are able to graze the area.

Traditional herders, for whom larger numbers of livestock represent greater personal wealth and social standing, have often been seen as the culprits in this process in the developing world. Such arguments have been used as a reason for encouraging settlement of nomadic pastoralists in many dryland countries of Africa, the Middle East and Asia. Overgrazing problems have also been identified on commercial ranches in North America and Australia.

In practice, the reasons for increased grazing pressure on pastures are numerous and complex. They include competition for land as cultivated areas increase, pushing grazers into more marginal pastures, as has happened in such countries as Senegal, Niger and Kenya. Sinking boreholes to provide new and more-reliable water supplies has also led to increased grazing pressures across much of the Sahel and in the Kalahari of Botswana where, between 1965 and 1976, livestock numbers and the accessible grazing resources were increased by about two and a half times (Sandford, 1977). In some parts of the world, degradation has been seen as the result of a change in the approach to pasture use: from a flexible strategy typical of traditional pastoralists in which the natural dynamism of dryland vegetation is used by regular movement of herds and maintaining several different animal species (Fig. 4.3), to a less-flexible Westernised approach in which the pasture may be fenced off and only cattle are grazed.

Although these ideas on the 'overgrazing problem' have been followed by many scientists and policy makers, recent thinking on the nature of dryland vegetation has questioned much of this conventional view. Distinguishing between degradation of vegetation and other forms of vegetation change is very difficult, and the importance of overgrazing as a cause of dryland resource degradation has probably been overexaggerated in many areas (see below).

Overcultivation

Several facets of overcultivation are commonly quoted. Some stem from intensification of farming which can result in shorter fallow periods, leading to nutrient depletion and eventually reduced crop yields. Soil erosion by wind and water is another result of the overintensive use of soil, resulting from a weakened soil structure and reduced vegetation cover. In many cases, erosion has been

FIGURE 4.3 *Mixed herds are traditionally kept by herders in drylands, as here in Mongolia's Gobi Desert. Different animals eat different types of vegetation and have different susceptibilities to moisture availability*

caused by the introduction of mechanised agriculture with its large fields and deep ploughing which further disturbs soil structure and increases its susceptibility to erosive forces. Similar outcomes have also been noted in areas where cultivation has expanded into new zones which are marginal for agricultural use because they are more prone to drought, or are made up of steeper slopes that are more prone to erosion. Some classic examples of overcultivation, their root causes, inappropriate land uses and resulting degradation, are shown in Table 4.3.

TABLE 4.3 *Examples of overcultivation*

Area (date)	Root causes	Land use	Environmental response
US Great Plains (1930s)	Pioneer spirit	Deep ploughing for cereals	Wind erosion
Former USSR Virgin Lands (1950s)	Political ideology, food security	Deep ploughing for cereals	Wind erosion
Niger (1950s and 60s)	Colonial influence, cash crops	Extensification of groundnut cultivation	Nutrient depletion

Deforestation

Clearance of forested land, to make way for the expansion of grazing or cultivation and/or to provide fuelwood, reduces the protection offered to soil by tree cover. Accelerated erosion may result, along with a breakdown in soil structure. Water tables may also be affected. As trees are removed, villagers may supplement wood for fires with dried animal dung which would otherwise be left on the soil to provide valuable nutrients. One study in Ethiopia estimated that the economic value of manure diverted from this fertilising role to the cooking stove was US$123 million annually, which could increase grain harvests by 1–1.5 million tonnes a year (Newcombe, 1984). An expanding human population is the ultimate driving force behind this degradation. In the dryland regions of West Africa, where the so-called 'fuelwood crisis' is particularly acute, the World Bank (1985) has estimated that sustainably harvested fuelwood can support just two-thirds of the human population (Table 4.4). The demand for fuelwood from urban areas, where inhabitants tend to use more fuel per head than their rural counterparts, has left large halos virtually devoid of trees around many cities in the region. Few trees survive within 90 km around the Sudanese capital of Khartoum, and similar treeless zones can be seen around such cities as Dakar in Senegal, Ouagadougou in Burkina Faso, and Niamey in Niger.

Salinisation

Poorly managed dryland irrigation schemes can result in a number of forms of environmental degradation which ultimately feed back on agriculture by reducing yields. Problems are caused by the wasteful use of water: by applying more than can be taken up by plants, through leakage from improperly lined drainage canals, and as a consequence of inadequate drainage. Rising water tables, waterlogging and the build-up of salts and alkalis can result.

The salt tolerance of most cultivated plants is relatively low, so salinisation rapidly leads to declines in productivity. Such problems are found in dryland irrigation schemes throughout the world. Abrol *et al.* (1988) estimate that nearly 50 per cent of the irrigated land in the world's arid and semi-arid regions have some degree of soil salinisation problems, although some of this land is naturally saline. Table 4.5 shows the serious scale of the problem in some individual countries. In Pakistan, where irrigation is used on 80 per cent of all cropland, about 35 per cent of the irrigated

TABLE 4.4 *Population in relation to fuelwood availability in the Sudano-Sahelian region**

Zone	Actual population (millions)	Fuelwood sustainable population (millions)
Saharan and Sahelo-Saharan	1.8	0.1
Sahelian	4.0	0.3
Sahelo-Sudanian	13.1	6.0
Sudanian	8.1	7.4
Sudano-Guinean	4.0	7.1
Total	31.0	20.9

*Data from Senegal, Mali, Niger, Burkina Faso, Mauritania, Chad and Gambia
Source: World Bank (1985)

TABLE 4.5 *Salinisation of irrigated cropland in selected countries*

Country	Proportion of irrigated land affected by salinisation (%)
Australia	15–20
China	15
Egypt	30–40
Iraq	50
Pakistan	35
Spain	10–15
Syria	30–35
USA	20–25

Source: after Gleick (1993)

area suffers from problems of salinisation and more than 40 000 ha of irrigated land are lost each year to waterlogging and salinity (Pakistan, 1990). Overuse of irrigation water can also cause degradation off-site, as the example of the Aral Sea tragedy illustrates (see Chapter 12). Financial constraints, maintenance problems, inexperienced management and farmers are factors common to many such irrigation problems and failures. A lack of appropriate advice and training, and credit and marketing facilities for farmers are again commonly encountered, as the experience of Iraq's Greater Mussayeb Project illustrated in the 1950s (Iraq, 1980).

PROBLEMS WITH DESERTIFICATION THEORY

Although the above causes of desertification and their consequences are often cited, recent research has raised question marks over some of the accepted theories. Two of the most contentious issues surround the idea of overgrazing as a cause of desertification, and the role of desertification in explaining food shortages.

Overgrazing

The idea of overgrazing as a major cause of desertification, by degrading the vegetation on pastures, has been the subject of considerable scrutiny in recent years as our understanding of rangeland ecology has changed. Vegetation in drylands is typically dynamic, responding primarily to the highly variable input of moisture from rainfall, but also to other influences such as fire. Hence, distinguishing between a relatively permanent change brought about by overexcessive grazing pressure and natural variability in response to moisture availability or fire can only be achieved by long-term study of an area before, during and after such disturbances. To date, however, such long-term studies are few and far between.

Another problem with the overgrazing idea stems from its basis in the concept of 'carrying capacity', the theoretical number of livestock a unit area of pasture can support without being degraded. While calculation of a carrying capacity is possible in relatively unchanging environments, it is difficult to apply to dryland pastures because the numbers of animals an area can support varies on several timescales: before and after a rainy day, between a wet season and a dry season, and between drought years and wet years. Grazing the number of livestock appropriate for a wet year on the same pasture during a drought could result in degradation, but several studies cited by Warren and Khogali (1992) show that such disastrous concentrations of livestock are rarely reached because, under conditions of environmental stress, the animals are often the first component of the system to fail, by dying. Traditional pastoralists have also developed social and economic mechanisms for coping with such occurrences, as McCabe (1990) has documented for the Turkana in East Africa.

Another commonly quoted situation considered to be typical of desertified areas is

FIGURE 4.4 *Cattle around a waterhole in Niger*

the loss of vegetation around wells or boreholes, which supposedly grow and coalesce. In fact, several studies have failed to show that these 'piospheres' expand over time (e.g. Hanan *et al.*, 1991). Areas surrounding wells *are* typically bare of vegetation, due to both grazing and trampling (Fig. 4.4), but they are also typified by higher levels of soil nutrients than surrounding areas thanks to regular inputs from dung and urine (Barker *et al.*, 1990) which may balance out any negative effects of possible soil loss by erosion. Such bare areas are perhaps best interpreted as 'sacrifice zones' in which the loss of the vegetation resource is outweighed by the advantages of a predictable water supply (Perkins and Thomas, 1993).

Food production

One of the most important reasons for desertification's position as a major global environmental issue is the link between resource degradation and food production. Desertification reduces the productivity of the land and, hence, less food is available for local populations. Malnutrition, starvation and ultimately famine may result. This link was the central reason behind the calling of the UN Conference on Desertification in 1977 and has been a strong theme in the negotiations leading to the UN Convention to Combat Desertification signed in 1994.

However, the links between soil degradation and soil productivity, and consequently crop yields, are seldom straightforward or clear-cut (Stocking, 1987). Productivity is affected by many different factors, such as the weather, disease, pests, the farming methods used and economic forces. Distinguishing the effects of these various factors is, in practice, very difficult, quite apart from the fact that data on crop productivity are often unreliable and many measurements of soil erosion are also open to question (see Chapter 6). While the theory linking desertification to food production is basically sound, it is often oversimplified and deserves further careful research.

The link between desertification and famine is even more difficult to make. This is not the case, in theory, since food shortages stem from reduced harvests and thinner or even dead animals, although such shortages are also caused by factors of the natural environment, particularly drought. In practice, however, famines typically occur in areas which are characterised by a range of other factors such as poverty, civil unrest and war. Several authors have emphasised the social causes of severe food shortages (e.g. Sen, 1981; Wijkman and Timberlake, 1984). In the African context, Curtis *et al.* (1988: 3) suggest that 'social, not natural or technological, obstacles stand in the way of modern famine prevention'.

Mass starvation is probably more a function of poor distribution of food, and people's inability to buy what is available, than desertification, although desertification, drought and many other natural hazards may act as a trigger. This tentative conclusion is supported by one of the few studies which specifically aimed to assess the relative importance of these factors, in the Sudanese famine of 1984/85 (Olsson, 1993). Millet and sorghum yields in the provinces of Darfur and Kordofan, the worst affected areas, were reduced to about 20 per cent of pre-drought levels, largely due to climatic factors – the amount and timing of the rains. Desertification, by contrast, was calculated to account for just 10–15 per cent of the variation in crop yields. While drought was undoubtedly a trigger for ensuing famine, the profound causes were different. As fears of crop failures mounted, the price of food in Darfur and Kordofan soared to levels that most rural people were unable to afford. While food was available at the national level, the distribution network was inadequate and famine followed.

UNDERSTANDING DESERTIFICATION

Appropriate solutions to desertification problem areas can only be found after more long-term studies of specific areas. Monitoring is necessary to identify where desertification is happening, how it works and why it is occurring. It must also attempt to distinguish between natural environmental fluctuations and impacts which can be attributed to human action, since it is the human aspects we have the ability to alter. Monitoring programmes also need to combine scientific measurements with an understanding of how social systems are related to the environment. Although, to date, there have been few such long-term investigations, work from two areas illustrates the approach.

The US Great Plains

The Dust Bowl of the 1930s on the American Great Plains is one of the best-known dryland environmental disasters in history. Over a period of 50 years, grasslands were turned into wheat fields by a culture set on dominating and exploiting natural resources using ploughs and other machinery developed in western Europe (Worster, 1979). When, in the 1930s, drought hit the area, as it periodically does, wind erosion ensued on a huge scale. By 1937, the US Soil Conservation Service

estimated that 43 per cent of a 6.5 million hectare area at the heart of the Dust Bowl had been seriously damaged by wind erosion. The large-scale environmental degradation, combined with the effects of the Great Depression, ruined the livelihoods of hundreds of thousands of American families.

However, although the lessons learnt from the Dust Bowl years inspired major advances in soil conservation techniques, the environmental health of the area has continued to be degraded. Continued degradation is believed to account for a significant part of declining yields of sorghum and kafir over 30–40 years in the Texas Panhandle (Fryrear, 1981), and soil erosion during the drought in the 1970s was on a comparable scale to that in the 1930s (Lockeretz, 1978). Investigation of the soil losses of the 1970s illustrates some important underlying causes, indicating that political and economic considerations overshadowed the need for sustainable land management. Large tracts of marginal land had been ploughed for wheat cultivation in the early 1970s, driven by high levels of exports, particularly to the then USSR. To encourage farmers to produce, a Federal programme was set up which guaranteed them payment according to the area sown, so that the disincentive to plough marginal areas was removed: the farmers were paid whether these areas yielded a crop or not. At the same time, new centre-pivot irrigation technology was widely adopted to water the new cropland, but the use of the rotating irrigation booms required that wind breaks, planted to protect the soil, be removed (McCauley et al., 1981).

The Sahel

The important role of external factors in the degradation equation can also be illustrated from work on the Sahel catastrophe of the 1970s and 1980s, events which sparked worldwide interest in desertification. The failure of traditional social insurance strategies in

TABLE 4.6 *Traditional indigenous insurance strategies against drought in northern Nigeria*

Food storage	Store food during times of plenty, the oldest insurance system in history
Monetary savings	Work hard at alternative opportunites – off-farm production and labouring at home
Social insurance	Rely on the community network of extended families, supplemented by the moral responsibility under Islam of richer members of society to assist the poor in times of hardship
Ecological insurance	Practise diversified farming and herding – intercropping, spatial fragmentation of holdings, diversification of livestock, grazing mobility – to give some protection against hazards specific to particular crops, animals or places

Source: after Mortimore (1987)

northern Nigeria against times of hardship caused by drought, which can be applied equally to desertification (Table 4.6), have been examined by Mortimore (1987). The changing political economy of the region has undermined these traditional schemes, and illustrates the changing scene faced by inhabitants of many dryland communities in the developing world.

The possibilities for grain storage have been diminished by a number of circumstances, including a growing population and land scarcity. The imposition of taxes payable in currency, the monetisation of the economy and poor terms of trade, changes which began during colonial times, have also encouraged people to produce cash crops in place of subsistence crops.

The generation of savings from activities away from the fields also suffered under changing political policies. Agriculturalists who had moved to urban areas to work in the informal sector were forcibly sent back to the land by the government, while cash savings were further eroded by inflation and sudden devaluations in the national currency.

The prolonged drought, combined with continued population growth, reduced the effectiveness of social insurance, as people who could previously have been relied upon

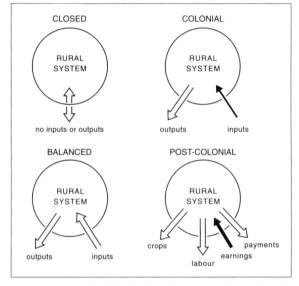

FIGURE 4.5 *Model for the incorporation of rural systems into the market (after Mortimore, 1987)*

to help out their poorer relations did not have enough to feed themselves. The traditional advantages of ecological insurance have also been eroded as colonial and subsequent governments have encouraged moves towards specialisation in crop production and discouraged traditional pastoralism which is inconsistent with modern political policies and boundaries.

A loss of autonomy over their affairs lies at the heart of the consequent problems faced by farming and pastoral communities in developing-country drylands. The absorption of such communities into centralised economies has undermined the traditional ways in which they have coped with environmental stress, increasing their vulnerability (Fig. 4.5). While government welfare or private economic insurance is available to inhabitants of richer countries, such safety nets are seldom available in the Third World and it is in these regions where the need to more fully comprehend the nature and extent of desertification is the most pressing.

FURTHER READING

Beaumont, P. 1989 *Drylands: environmental management and development*. London, Routledge. This book gives an overview of the nature of drylands and the history of their use by human society. It uses numerous detailed case studies to illustrate different management approaches.

Grainger, A. 1990 *The threatening desert*. London, Earthscan. A good review of the theories and evidence put forward to describe and explain desertification, and actions to combat it.

Mortimore, M. 1989 *Adapting to drought: farmers, famines and desertification in West Africa*. Cambridge, Cambridge University Press. A detailed case study of northern Nigeria focusing on the physical effects of drought and desertification, and the social responses.

Thomas, D.S.G. and Middleton, N.J. 1994 *Desertification: exploding the myth*. Chichester, Wiley. This book focuses on the political aspects of the desertification issue and highlights many of the difficulties of identifying desertification in the field.

5

FOOD PRODUCTION

The origins of agriculture, which includes both crop and livestock production, are traced back 10 000 years to the start of the Holocene. Before that, the genus *Homo* had lived for more than 2 million years by gathering plants and hunting wild animals for food and *Homo sapiens* continued this practice for 100 000 years. For the last 5000 years, virtually the entire world's population has been reliant on farmers and herders who consciously manipulate natural plants and animals to provide food. As global population has grown, the area used for food production has expanded and production techniques have become increasingly sophisticated. Both the extensification and the intensification of food production have thrown up numerous environmental issues. Some of these are covered in this chapter, but other environmental impacts of farming are dealt with elsewhere. These include the problems associated with the clearance of forests for farming (Chapter 3), the environmental impacts of heavy grazing in drylands (Chapter 4), soil erosion on agricultural land (Chapter 6) and some of the impacts associated with irrigation (Chapters 4, 12 and 13).

AGRICULTURAL CHANGE

Agricultural management enables more calories of food to be obtained from a given area of land over a more predictable period of time than the simple collecting of wild foods. Some people believe that agriculture developed in response to rising human populations (e.g. Cohen, 1977), while others have argued that new methods of food production were a cause rather than a consequence of population growth (e.g. Hassan, 1980). It is likely that social, economic, technological and environmental factors all contributed in varying combinations to the early transition from hunters and gatherers to herders and farmers, and this 'Neolithic Revolution' was associated with the emergence of urban civilisations on some of the world's great rivers (the Nile, Tigris–Euphrates, Indus and Huang Ho) between 3500 and 5500 years ago and in MesoAmerica and the Central Andes around 3000 years ago.

Part and parcel of the start of agriculture was the process of domestication, the deliberate selection and breeding of wild plants and animals which leads to genetic change. By cultivating plants, for example, farmers chose particular seeds which offered the best results and by planting more of them increased their chances of survival. The dominance of particular characteristics was thus deliberately increased. In crops valued for their edible seed, such as wheat, a shift towards non-shattering occurred quickly, and hence the

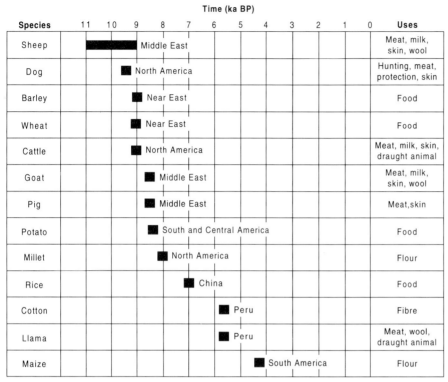

Time (ka BP)

Species	11	10	9	8	7	6	5	4	3	2	1	0	Uses
Sheep	■■■■■ Middle East												Meat, milk, skin, wool
Dog			■ North America										Hunting, meat, protection, skin
Barley			■ Near East										Food
Wheat			■ Near East										Food
Cattle			■ North America										Meat, milk, skin, draught animal
Goat				■ Middle East									Meat, milk, skin, wool
Pig				■ Middle East									Meat, skin
Potato				■ South and Central America									Food
Millet				■ North America									Flour
Rice					■ China								Food
Cotton						■ Peru							Fibre
Llama						■ Peru							Meat, wool, draught animal
Maize								■ South America					Flour

FIGURE 5.1 *Origins of domesticated plants and animals (after Williams* et al., *1993)*

natural mechanism for seed dispersal was lost and the plant became dependent on humans for survival. The origins of domestication, which can be seen as one of the most important human impacts on the natural environment, have been traced to diverse parts of the world using archaeological evidence (Fig. 5.1). The process of selection and breeding continues today in a more sophisticated manner using such techniques as tissue culture and genetic engineering (see below).

Irrespective of the original causes of the emergence of agriculture, there is no doubt that the ability to provide more food from a unit area of land per unit time has been associated with an exponential growth of the world's human population. A global population thought to have been around 5 million 10 000 years ago increased by a factor of twenty to 100 million in 5000 years (May, 1978). The 500 million mark was passed about 300 years ago and by 1850, 1000 million (1 billion) people inhabited the planet. By this time, the exponential rate of population growth was clear and world population had surpassed 4 billion by 1978 and 5 billion by 1990.

Until the early 1950s, world agricultural production had kept pace with population growth chiefly through expansion of the cultivated area. Expansion subsequently slowed and had all but stopped by the end of the 1980s. Today, after 10 000 years of agriculture, cropland occupies about one-sixth of the world's biologically productive land area (Gilland, 1993), while the area used as pasture for livestock is much greater. By clearing natural vegetation and replacing it usually with fewer domesticated species, or by

FIGURE 5.2 *A landscape transformed by agriculture on the Portuguese North Atlantic island of Faial in the Azores*

grazing livestock, agriculturalists have been responsible for one of the most widespread changes wrought by humans on the natural environment. Forests and grasslands of the temperate latitudes have been the most severely altered, but all the world's major biomes, with the exception of the polar ice caps, have been affected (Fig. 5.2).

Continued increases in food production since the 1950s, to keep up with population growth which has continued at unprecedented rates, has been achieved chiefly by increasing crop yields. In fact, the rising trend in yields has a long history in many parts of the world and has been the result of numerous improvements in agricultural efficiency. The tonnage of wheat harvested from an average hectare of English cropland in the 1980s, for example, was three times that in the 1930s, five times that in 1800 and ten times that of 1300. The increases were achieved first by the replacement of open-field farming by mixed farming, and then by the replacement of mixed farming by increasingly mechanised chemical farming (Grigg, 1992). One of the outcomes of using more machinery on arable land has been the removal of hedgerows to enlarge field sizes, which in England is akin to a reversion to the large open fields of the Medieval period as illustrated in Table 5.1. But modern cultural attachment to hedgerows with their flora and fauna, which only became established in the eighteenth century, has caused strong reactions.

Worldwide, it is the adoption of modern chemical farming techniques, both in the developed countries and in parts of the Third World during the so-called Green Revolution, which has enabled the continued increase in yields. Large fields and mechanisation have come hand-in-hand with several other key components of these techniques:

- application of increasing amounts of chemical fertilisers to supply plant nutrients (nitrogen, phosphorus and potassium);

TABLE 5.1 *Changes in the length of hedgerows in three parishes in Huntingdonshire, England since the fourteenth century*

Date	Length of hedges (km)	Date	Length of hedges (km)
Pre-1364	32	1780	93
1364	40	1850	122
1500	45	1946	114
1550	52	1963	74
1680	74	1965	32

Source: Moore *et al.* (1967)

- expansion and improvement of irrigation systems;
- protection of crops from diseases, pests and weeds with synthetic chemicals; and
- development of new varieties of crop plants which are more responsive to fertilisers, more resistant to pests and with a higher proportion of edible material.

The environmental impacts which have come with these developments will form the focus of this chapter, but parallel intensification has occurred over the last 50 years in the livestock industry, particularly in the industrialised world where livestock products provide 30 per cent of the calorie intake, in contrast to the less-developed countries where they provide less than 10 per cent (Grigg, 1993). Factory-style farms are commonplace as pigs and poultry have been moved from the farmyard to indoor feeding facilities and cows are fattened with special feeds. Abundant feed grain, drugs to prevent disease, and improved breeds have enabled animals to be raised faster and leaner as meat production for consumers rich enough to eat it regularly has become the focus of livestock production, taking over from the multiple benefits of manure, draught power, milk and eggs.

The intensification of modern livestock farming has raised a number of environmental issues, not least of which are the concerns over animal welfare that has engendered a minor shift towards 'free-range' animal products in some Western countries (Fig. 5.3). An increasingly serious issue is that of pollution from farm wastes. While the traditional mixed farm operated as a closed system, with manure being spread on cropland to provide valuable nutrients, specialisation in the livestock industry has meant a trend towards concentration in areas distant from centres of arable production. Leakage of manure, with a high biochemical oxygen demand, from storage tanks has become an increasingly

FIGURE 5.3 *Pigs on a free-range farm in the English Midlands*

serious point-source pollutant, depleting the dissolved oxygen in groundwater, wells and rivers (see Chapter 12). Nitrogen in manure is also lost to the atmosphere as ammonia, contributing to acid deposition: the livestock industry is the single largest source of acid deposited on soils in The Netherlands, for example (Durning and Brough, 1991).

FERTILISER USE

Although manure and other materials have long been used as natural fertiliser, and are still applied to cropland in many parts of the world, the rise of modern farming techniques has been underpinned by a rapid increase in the use of chemical fertilisers to supply plant nutrients. Specially manufactured chemical fertilisers date back to the mid-nineteenth century when a factory producing superphosphates by dissolving bones in sulphuric acid was opened in the English county of Kent, but their use has become globalised only in the last 50 years. World production of nitrogenous fertilisers in 1913 was 0.77 million tonnes nitrogen content and had risen to 3.6 million tonnes in 1950. By 1975, production had increased more than ten times to 42 million

TABLE 5.2 *Increase in use of chemical fertilisers, selected countries*

Country	Fertiliser used on cropland (kg/ha/year)		Change (%)
	1964–66	1989–91	
Chile	26	69	265
China	24	284	1183
Bulgaria	82	153	187
Egypt	117	361	309
India	5	73	1460
The Netherlands	582	614	105
Nigeria	0	12	—
Mexico	15	69	460
Philippines	11	65	591
Syria	3	51	1700
UK	288	350	122
USA	63	99	157
Venezuela	10	116	1160

Source: from data in WRI (1986 and 1994)

tonnes and continued to rise to 85 million tonnes in 1990 when its manufacture absorbed about 1.3 per cent of world energy consumption. Similarly rapid increases have occurred in the quantities manufactured of the other two main fertilisers: phosphates and potash. The increase in use of all fertiliser types over a 25-year period for a selection of countries is shown in Table 5.2. This rapid increase in supply was a key component of the increase in average world cereal grain yields from 1.13 t/ha in 1948–52 to 2.76 t/ha in 1990, and the maintenance of cereal production per head at the present level to the year 2030 will require a further doubling of chemical nitrogen consumption (Gilland, 1993).

This projected increase in use has fuelled concerns over the environmental effects of fertilisers which are not taken up by crops. In practice, fertilisers are lost from a field because of overexcessive or poorly timed applications, and they can become pollutants in three ways:

- excess nitrate and phosphate is lost by runoff and leaching and enters rivers, lakes, groundwater and coastal waters where it can cause eutrophication, and, in the case of nitrate, may cause 'blue baby syndrome' and stomach cancer;
- excess nitrate can be converted by denitrification into nitrous oxide which is released into the atmosphere (nitrous oxide is a greenhouse gas and also contributes to stratospheric ozone depletion);
- excess nitrogen can be converted by volatilisation into ammonia which is released into the atmosphere where it can contribute to the effects of acid rain.

Rising levels of nitrates in surface waters have been reported from many areas where modern intensive farming is practised (see Chapters 10 and 12). Groundwater levels are also rising. Sampling in Denmark indicated that mean groundwater concentrations of around 1 mg/l NO_3-N in the 1940s and 1950s had risen to 3 mg/l in 1980, with just under 10 per cent of waterworks tested exceeding the maximum limit for drinking water of 11 mg/l (Forslund, 1986). In the USA, the US Geological Service found that in a third of all counties surveyed, 25 per cent of wells had nitrate concentrations that exceeded background levels and in 5 per cent of counties 25 per cent of wells exceeded the Federal drinking water standard of 10 mg/l (Goodrich *et al.*, 1991). Even if field losses are reduced by changes in land use and fertiliser application, groundwater concentrations are likely to continue to rise in the short to medium term in response to past losses which reach the water table with a time delay dependent upon the depth of the unsaturated zone. Analysis of pore water in the unsaturated zone of chalk in southern England has revealed a front of nitrate-rich water moving towards the water table at 1–2 m per year (Burt *et al.*, 1993).

IRRIGATION

Irrigation probably first began in the sixth millenium BC when two river-basin civilizations, on the Tigris–Euphrates of Lower Mesopotamia and on the Nile in Egypt, first began to sustain settled agriculture. In modern times, irrigation is often one of the key tools used to intensify agriculture and the global area of irrigated cropland has more than doubled in the second half of this century (Fig. 5.4). In many dryland countries, the national area under cultivation would be considerably smaller without irrigation, and in the extreme case of Egypt it would be almost zero (Fig. 5.5).

Unfortunately, however, poorly managed irrigation systems can lead to a wide range of environmental problems. Rising water tables, which cause salinisation and waterlogging, reduce crop yields on irrigation schemes all over the world (see Chapter 4) and can

accelerate the weathering of buildings as Akiner *et al.* (1992) have documented for numerous examples of ancient Islamic architecture in Uzbekistan. Conversely, overpumping of groundwater resources can result in rapid depletion of supplies which can threaten the long-term viability of the irrigation schemes themselves. Many examples involving centre-pivot irrigation technology can be quoted, such as the depletion of the Ogallala aquifer beneath the High Plains in the south-west US (Riebsame, 1990) and the rapid drop in groundwater level of the Minjur aquifer, central Saudi Arabia where rates of 1.8 m per year were recorded between 1984 and 1990 as irrigated wheat cultivation expanded in the Tebrak area (Al-Saleh, 1992).

An increased incidence of water-associated diseases is another problem often associated with the introduction or extension of irrigation schemes. Amin (1977) reports that the prevalence of schistosomiasis in Sudan's Gezira Project rose from 5 per cent in 1945 to 60 per cent in 1973 (see also Chapter 13). Other detrimental health effects can stem from

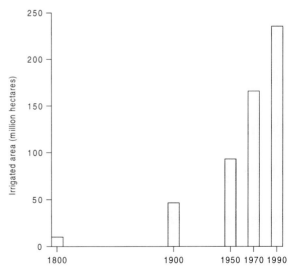

FIGURE 5.4 *Rise in the global area of irrigated cropland, 1800–1990 (from data in FAO, 1993; Gleick, 1993)*

FIGURE 5.5 *Cropland irrigated with water from the River Nile in Egypt. Irrigation first began on the Nile 6000 years ago and agriculture in modern Egypt is no less dependent upon the river (courtesy of Stephen Stokes)*

the overexcessive use of pesticides on irrigation schemes, as exemplified by Central Asian cotton plantations (see below), an example which also illustrates the scale of potential off-site environmental effects on the Aral Sea (see Chapter 12).

AGRICULTURAL PESTS

Losing a portion of a field's crop to pests has been a problem for farmers since people first started cultivating soils. Worldwide, an estimated 67 000 different pest species attack agricultural crops and about 35 per cent of crop production is lost to them each year. Insects cause an estimated 13 per cent of the

loss, plant pathogens 12 per cent, and weeds 10 per cent (Pimental, 1991). Hence, pests represent a serious problem for farmers and for human society in general.

Pesticides

The use of chemical sprays to control agricultural pests and diseases dates back to the 1860s when they were used to control Colorado beetle which damaged potato crops in the USA, and vine mildew in France. Pesticides (which include insecticides, herbicides and fungicides) entered general use in many of the world's agricultural areas along with chemical fertilisers in the 1950s, following the discovery of DDT in Switzerland in

1939 and 2,4-D in the USA during the Second World War. This development has been a response to an age-old problem but also an attempt to limit the escalation of pest problems which came about as a result of changing cultivation practices, such as shortened fallow periods, narrow rotation and the replacement of mixed cropping by large-scale monocultures of genetically uniform varieties.

Worldwide, annual sales of pesticides have grown from around US$7 billion in the early 1970s to US$25 billion in 1990 (Tolba and El-Kholy, 1992). Most are used in the industrialised countries, but while consumption in these countries has levelled off in recent years, sales to developing countries continue to grow. Most pesticides are used in agriculture, with about 10 per cent used in public health campaigns such as the fight against mosquitoes which transmit malaria.

One unfortunate aspect of the increasing use of pesticides is the development of resistance to the chemicals. While most individuals of a pest population die on exposure to a pesticide, a few individuals survive by virtue of their genetic make-up and pass on this resistance to future generations. Increasing resistance is often countered by stronger doses of chemicals or more frequent applications, at increasing cost to the farmer, which also speeds up the trend towards resistance, so encouraging the development of new chemicals. Elimination of one pest can also result in the rapid growth of secondary pest populations, however. Widespread applications of DDT in Egypt, for example, to control the American bollworm in cotton plantations, caused a species of whitefly, which had previously been a secondary pest, to explode in numbers in the late 1970s. DDT treatment actually stimulated whiteflies to produce a higher number of eggs than unsprayed flies and the whitefly supplanted bollworm as the country's worst cotton pest (Dittrich et al., 1985).

Pesticides are designed to kill living things, whether they be insects, weeds or plant diseases, but any amount of chemical which misses its intended target can affect numerous other aspects of the environment. In practice, the proportion may be large: some estimates suggest that for many insecticides as little as 0.1 per cent of the amount applied to a particular field may reach the target organism (Pimental and Levitan, 1988); the remainder becomes a contaminant. The resultant effects on non-target plants, animals, soil and water organisms and the pollution of land, air and water which can also affect human populations has caused considerable concern.

Natural predators of the problem pest may be destroyed, for example, thereby increasing the need for more pesticide applications. Other creatures useful to the farmer, such as bees and other pollinators may be affected. The concentration of some compounds has been seen to accumulate up the food chain, reaching damaging levels in higher species. DDT, a particularly persistent compound, is the classic example (Table 5.3) which at high

TABLE 5.3 *Bioaccumulation of DDT residues up a salt marsh food web in Long Island, USA*

Organism	DDT residues (parts per million)
Water	0.000 05
Plankton	0.04
Silverside minnow	0.23
Sheephead minnow	0.94
Pickerel (predatory fish)	1.33
Needlefish (predatory fish)	2.07
Heron (feeds on small animals)	3.57
Herring gull (scavenger)	6.00
Fish hawk (osprey) egg	13.8
Merganser (fish-eating duck)	22.8
Cormorant (feeds on larger fish)	26.4

Source: after Woodwell *et al.* (1967)

concentrations has caused considerable increases in mortality among seabirds and birds of prey (see Chapter 7) and has also been reported at dangerous concentrations in human mother's milk. Similar bioaccumulation has been documented for methyl mercury.

On the national scale, the excessive use of pesticides has left large agricultural areas poisoned. In Romania, more than 900 000 ha are chemically polluted, of which 200 000 ha have been rendered totally unproductive through the excessive use of pesticides, consumption of which rose from 5900 t in 1950 to 1.2 million tonnes in 1985 (Turnock, 1993). Contamination of groundwater was first recognised in the USA in 1979 when pesticide residues were found in ninety-six wells on Long Island and in more than 2000 wells in California (Goodrich et al., 1991). The dangers posed to human health by high concentrations of residues are illustrated in Central Asia, where Glantz et al. (1993) suggest that pesticide applications have been on average up to ten times the levels used elsewhere in the former USSR or USA. These high levels have been related to deteriorating human health via contaminated surface water and groundwater. A sharp increase in oesophageal cancers, high rates of congenital deformities, outbreaks of viral hepatitis and a life expectancy in some areas of about 20 years less than that for the former USSR in general, have been cited for Central Asian regions of cotton monoculture (Feshbach and Friendy, 1992), and these impacts upon human health have been exacerbated by the dearth of medical and health facilities in the Aral Basin.

More direct exposure to dangerous levels of pesticides occurs in numerous ways. Agricultural workers are particularly at risk if not adequately protected during pesticide applications, as is often the case on Third World plantations. In one tragic incident in the 1970s, 1500 male banana-plantation workers in Costa Rica became sterile after repeated contact with dibromochloropropane (Thrupp, 1991). A number of cases have also been documented in which seeds treated with pesticides have been eaten rather than planted. Wheat seeds treated with mercury-based fungicides and distributed as food in Iraq caused the hospitalisation of 370 people in 1960 and 6530 in 1972, with 459 deaths in the latter case (Bakir et al., 1973). These authors note similar cases in Guatemala in the early 1960s and Pakistan in 1969. One estimate suggests that as many as 25 million people in developing countries may be affected by pesticide poisonings each year, although this number also includes suicide attempts (Jeyaratnam, 1990).

Realisation of the dangerous effects of many pesticides has encouraged some governments to introduce programmes to reduce their use in agriculture. Some of the countries most heavily dependent upon chemicals, such as Denmark, The Netherlands and Sweden, have been the most active in this regard (Hurst, 1992). These moves follow the introduction of complete bans on the use of some of the more toxic and persistent pesticides in many developed countries. Long-lasting organochlorine insecticides such as DDT, aldrin and dieldrin, and herbicides such as 2,4-D and 2,4,5-T, are among the pesticides now rarely used in more-developed countries, although these compounds are still common inputs to agriculture in Latin America (UNECLA, 1990) and many other developing countries.

However, in some cases, such bans have not been without their critics. The use of dieldrin in locust control programmes is a good example. It was successfully used in Africa throughout the 1960s and 1970s by spraying strips of land near locust hatching areas. Although dieldrin breaks down quickly in tropical sunlight, the strips remained lethal to hoppers for 6 weeks. Concern over dieldrin's lethal effects on birds, its persistence in the food chain, and its toxicity to people, led to bans in developed countries who also refused to donate the pesticide to African countries. A much shorter-lived replacement, fenitrothion, was advocated, but fenitrothion's 3-day active period means that spraying has to be delayed until mature locusts are swarming, and then spray them directly. This approach requires precise monitoring and timing. Swarms also need to be sprayed much more frequently, making the environmental advantages over dieldrin marginal. Using fenitrothion also takes more work, more spray and more money to stop a locust swarm (Skaf, 1988).

Alternatives to pesticides

Despite the widespread use of pesticides, about one-third of total crop production is still lost to insects, weeds and plant diseases, a similar proportion to that lost in pre-chemical times (Pimental, 1991). Although this percentage has been maintained during a period of rapidly increasing crop production, the success of chemicals in controlling pests has not been as great as was once hoped. Alternatives to synthetic chemicals have long been used and, given the problems of resistance and contamination associated with chemical use, these alternatives are being given more attention in both developing and developed countries.

There are five main alternative approaches to chemical pest control (El-Hinnawi and Hashmi, 1987):

- environmental control
- genetic and sterile male techniques
- biological control
- behavioural control
- resistance breeding.

A combination of these alternatives is often employed as a form of 'integrated pest management' which acknowledges that both crops and pests are part of a dynamic agricultural ecosystem and aims to limit pest outbreaks by enhancing the effects of natural biological checks on pest populations, and only resorting to chemicals in extreme cases.

Environmental control measures incorporate all human alterations to the environment which range from simple measures, such as digging up egg pods of pests and the planting of trap crops to lure pests away from the main crop, to larger-scale efforts such as the draining of wetlands which harbour pests. The special breeding for release of large numbers of genetically altered, often sterile individuals has been used in a number of cases, the aim being to decimate pest populations by preventing reproduction. The technique was pioneered in the control of the screw-worm fly, a pest affecting cattle. Behavioural control techniques which aim to control pests by use of sex pheromones and by attractant or repellant chemicals are in the early stages of development, while the selected breeding of organisms resistant to pests has long been practised and is now a more precise science thanks to modern biotechnology (see below).

The most widely practised and successful method of biological control, which uses living organisms as pest control agents, has been the introduction and establishment of appropriate species to provide permanent control of a pest. A well-known historical example is the introduction of a South American moth to Australia in the 1920s to combat the explosive spread of prickly pear across the east of the country. The prickly pear had itself been introduced from South America in the late eighteenth century by the first Governor of Australia in the mistaken belief that it might form the basis for a dye industry. Having become established on a new continent, the prickly pear spread very rapidly through eastern Australia in the early decades of the twentieth century, and at the peak of infestation occupied an area larger than Great Britain, about 24 million hectares (Fig. 5.6), ruining many farmers and ranchers and threatening the rural economy of the entire continent.

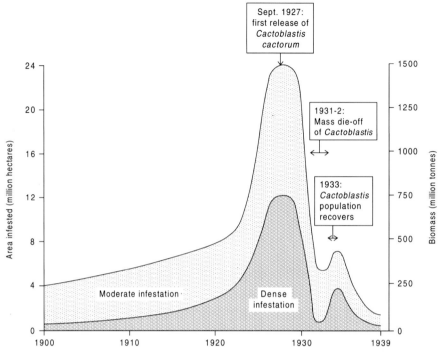

FIGURE 5.6 *Growth and collapse of prickly pear infestations in eastern Australia, 1900–39 (after Freeman, 1992. Reproduced with modifications by permission of the American Geographical Society)*

Having identified the *Cactoblastus* moth larvae as a natural predator of the prickly pear, eggs were first placed on plants in September 1927 with dramatic effect. After a temporary decline in moth numbers in 1931–32, when the prickly pear regained some of its former territory, the *Cactoblastus* population resurged to contain the pest. Since 1940, a balance between pear biomass and insect numbers has been established, at a level slightly less than 1 per cent of the peak prickly pear biomass (Freeman, 1992).

Introductions for biological control can have unwanted effects, however. A particularly damaging example occurred on the French Polynesian island of Moorea, where the giant African snail was introduced in the nineteenth century by colonial governors with a yearning for snail soup. However, the snail spread across the island and caused widespread destruction to crops, so a predaceous snail, the Ferussac, was brought in to control the giant African snail in 1977. Unfortunately, the Ferussac found native snail species to be more palatable targets and by 1987, seven species endemic to Moorea had become extinct in the wild, although six survive in captivity (Murray *et al.*, 1988).

BIOTECHNOLOGY

Although the term biotechnology is a recent one, people have manipulated the genetic make-up of crops and livestock for thousands of years by selecting and breeding according to their needs. Similarly, the production of bread, cheese, alcohol and yoghurt are age-old examples of the deliberate manipulation of organisms or their components to the benefit

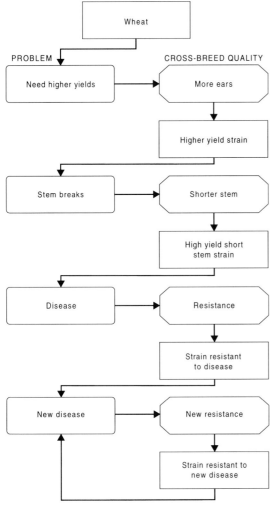

FIGURE 5.7 *Steps in the breeding of new wheat strains*

further cross-breeding. Such procedures can now be conducted much more precisely using genetic engineering, which transfers useful DNA material from one organism into the genetic make-up of another organism. These recombinant DNA techniques allow researchers to incorporate selected genes into a particular crop or animal. The implanted genes may come from other species and can cross between the animal and plant kingdoms. The scope for modifying the properties of commercial food sources is very wide. Among the many applications, DNA can be transferred to increase resistance to pests and diseases, to modify the timing of fruit ripening, to enhance nutritional value, and to increase tolerance of altitude or temperature regimes. These techniques are now beginning to yield new commercial food products.

Once a new variety of plant has been created using recombinant DNA techniques, it can be mass-produced using cell or tissue culture, another important biotechnology technique. The approach enables a complete plant to be regenerated in the laboratory from a single cell exposed to nutrients and hormones, and is commonly used for clonal propagation, the mass production of clones or genetic duplicates which can then be planted like conventional seedlings.

Although biotechnology offers great potential for world food production, there are fears that many of the more expensive techniques will continue to be used predominantly on high-value crops grown in industrialised countries, with limited application in more needy areas of the Third World. Developing countries may be disadvantaged in other ways: some see the virtual monopoly over biotechnology held by transnational companies as inherently unfair to poorer countries, and those whose economies are heavily reliant on such crops as sugar and cocoa may suffer as substitutes are found (Hobbelink, 1991). Even in developed countries, there may be resistance from consumers against genetically

of society and are, therefore, examples of biotechnology. Some of the recent steps in the cross-breeding of wheat strains are shown in Fig. 5.7. Higher yields have been attained by cross-breeding with varieties which had more ears per stem, but the extra weight often caused the stem to break, making it impossible to harvest. Cross-breeding with short-stemmed varieties solved this problem. Diseases and pests are a constant problem which can be combated with pesticides and

engineered foods in the same vein as the backlash against the inhumane treatment of animals in intensive factory farms. Already the use of some growth hormones, such as BST in cattle, has been banned in the USA, for example, due to safety fears from consumers but also due to milk producers' worries that increased production might depress prices.

Developments in biotechnology have also raised environmental concerns. Genetically altered plants or micro-organisms might have a competitive advantage which could disturb natural ecology if they were released into the wild. The potential effects could echo those caused by unintended releases of foreign insects and other pests into environments where they have proliferated to the detriment of many other species, as the above examples of the prickly pear in Australia and the Ferussac snail in Moorea illustrate (see also Chapter 15).

The new techniques also highlight the need to maintain the reservoir of genetic diversity in the wild, to provide new genetic material which could be of use to humankind for food production and in many other ways. The dangers of our increasing reliance on a small number of food crops are enhanced by their limited genetic diversity, a situation exacerbated by crop cloning, as examples of major crop failures through history illustrate (Table 5.4). The constant need for new genetic material, much of it found in wild relatives of domesticated crops, can be shown by just one example: Plucknett and Smith (1986) demonstrated that a new variety of sugar-cane is required in Hawaii about every 10 years to adapt to pests and to maintain yields. While realising the need to protect and maintain genetic diversity in the wild, regulating the use of such material, thereby enabling compensation to be paid to the countries of

TABLE 5.4 *Crop failures attributed to genetic uniformity*

Date	Location	Crop	Cause and result
900	Central America	maize	Anthropologists speculate that collapse of Mayan civilisation may have been due to a maize virus
1846	Ireland	potato	Potato blight led to famine in which 1 million people died and 1.5 million emigrated
Late 1800s	Sri Lanka	coffee	Fungus wiped out homogeneous coffee plantations on the island
1943	India	rice	Brown spot disease, aggravated by typhoon, destroyed crop starting 'Great Bengal Famine'
1960s	USA	wheat	Stripe rust reached epidemic proportions in Pacific North-West
1974–77	Indonesia	rice	Grassy stunt virus destroyed over 3 million tonnes of rice (the virus plagued South and South-East Asian rice production from 1960s to late 1970s)
1984	USA	citrus	Bacterial disease caused 135 nurseries in Florida to destroy 18 million trees

Source: modified after WCMC (1992: 428, Table 27.22)

origin, is another pertinent issue, particularly for developing countries. It is an aspect of the biodiversity debate which the Convention on Biodiversity signed at the Earth Summit in 1992 specifically recognises and aims to tackle.

AQUACULTURE

A notable recent development in global food production is the rise in importance of aquaculture, which is dominated by fish farming but also includes the production of other aquatic organisms such as seaweeds, frogs and turtles. Aquaculture currently accounts for about 13 per cent of the global production of fish, and while marine capture fisheries have been declining since 1989 and are thought to be close to their maximum sustainable yield (see Chapter 10), aquacultural production has continued to grow over this period (FAO, 1993). China is by far the world's largest producer with over 5 million tonnes annually, about half the global total, followed by Japan which produces about 1.5 million tonnes (Coull, 1993). Most of the Chinese production comprises freshwater species such as carp and tilapia, but globally aquacultural output is split about 50/50 between inland and marine production.

Fish farming has a long history in parts of East and South Asia, dating back to 3000 BC in China, and has been practised in Europe since before Medieval times. At its simplest level, fish are raised in ponds, where they find their own food, from eggs collected in the wild, and are harvested when grown to a certain size. More sophisticated approaches involve a series of ponds or cages for different levels of development, including special nursery ponds and/or hatcheries. Provision of feed and control of natural predators, competitors and diseases is also common practice. Particular advances in fish farming techniques have occurred in

the last 50 years: selective breeding of stock is now widely practised in more-advanced farms, reproduction is controlled by hormonal injection and disease by antibiotics. Such intensive management has been used mainly in more-developed countries for species which command a high market price. The spectacular development of the Atlantic salmon farming industry in Norway, initiated in the early 1970s and reaching a 1990 production of 140 000 t, is an important example (Coull, 1993). Such intensive fish farming is, however, associated with a number of adverse impacts on the environment. These include the displacement of native fish species, control of wildlife which might prey on the farmed species, and local contamination of waters. In a review of the issues, Iwama (1991) highlights the output of organic particulate wastes from feed and faeces, which can cause eutrophication and can smother sea or lake bed organisms, as the most pressing area of concern. On a wider scale, the conversion of natural habitats for fish and prawn farming has been a primary cause of mangrove destruction in many coastal areas of the tropics (see Chapter 11).

Some predictions suggest that global production from aquaculture could reach 50 million tonnes in the next 25 years, which has engendered talk of a 'Blue Revolution' to rival the agricultural Green Revolution (Doumenge, 1986). In essence, this would amount to a major shift from capture to farming or husbandry in water, which would parallel that which occurred when hunting and gathering gave way to agriculture on land. Such a significant shift is unlikely to be based on the intensive approaches noted above for simple reasons of cost. Cheaper and more ecologically sound methods are long established in China, however, where carp are reared alongside pigs with reciprocal recycling of nutrients between land and water. Other species, such as tilapia, a prolific breeder that can also be fed mainly on wastes, are widely farmed in South Asia. Here, and in many other parts of the world, aquaculture has the additional benefit of helping to deal with problems of waste disposal. The largest of many such systems which are based on sewage is in Calcutta, where 7000 t of fish are produced annually for the local market. Water and sewage are channelled into two lakes covering about 2500 ha, and carp and tilapia are introduced after an initial bloom of algae. Thereafter, the lakes are fed with additional sewage on a monthly basis (Edwards, 1985). Health concerns over fish raised in this manner can be eliminated by retaining sewage in settling ponds for 20 days before introducing it into fish ponds, or by moving fish to clean-water ponds before harvesting. To date, however, the wide employment of these methods in rural Asia has been thanks to the traditional dietary importance of fish, and adoption elsewhere in the Third World may need considerable active encouragement.

FURTHER READING

Conway, G.R. and Pretty, J.N. 1991 *Unwelcome harvest: agriculture and pollution*. London, Earthscan. A comprehensive look at the main forms of agricultural pollution (pesticides, fertilisers and farm wastes) and the impacts of all types of pollution on agriculture.

Goodman, D. and Redclift, M. 1991 *Refashioning nature: food, ecology and culture*. London, Routledge. This book takes an interesting look at food as one of the chief links between humankind and nature and how modern farming has tended to further widen the gap between the two.

National Research Council 1993 *Sustainable agriculture and the environment in the humid tropics*. Washington, DC, National Academy Press. A practical discussion of twelve major land use options for boosting food production while protecting the natural resource base. The book also covers technology and policy and contains seven country profiles.

van Emden, H.F. 1989 *Pest control*. London, Edward Arnold. A concise account of the pros and cons of chemical pest control and of the alternatives to pesticides.

6

SOIL EROSION

Soil erosion is a natural geomorphological process which occurs on most of the world's land surface, with the principal exception of those areas on which soil eroded from elsewhere is deposited. Since soil erosion rarely occurs in dramatic events, perception of its seriousness, measurement of its progress and persuasion of the need to take ameliorative action has been a long process. A fundamental distinction is made between natural or geological soil erosion and rates which are accelerated by human activity, and it is accelerated erosion that has received most scientific attention. Such attention is not new, however. Ancient Greek and Roman observers of the natural world commented on the effects of such human activities as agriculture and deforestation on soil loss. Since soils are an integral part of the support system for ecosystems and human communities, primarily through their importance as a medium in which crops and other plants grow, their erosion remains an important environmental issue. This importance is all the more pressing when we consider the fact that the world's human population is increasing, along with its use of resources, and we know that natural rates of soil formation are slow, essentially making it a finite resource.

by mass movement. Water erosion occurs both by the action of raindrops, which detach soil particles on impact and move them by splashing, and by runoff which transports material either in sheet flow or in concentrated flows which form rills or gullies. Wind moves soil particles by one of three processes depending on the size and mass of the particles: the largest particles move along the surface by the process of surface creep; sand-sized particles usually bounce along within a few metres of the surface by saltation, and finer dust particles are transported high above the surface in suspension. Mass movement occurs principally by landsliding and various forms of flow, depending upon the amount of moisture in the soil profile.

When and where soil erosion occurs is determined by the mutual interaction of the erosivity of the eroding agent and the erodibility of the soil surface. These variables of erosivity and erodibility change through time and space, at varying rates and differing scales, so that the relationship between the variables is in a constant state of flux. The various factors affecting erosivity and erodibility in relation to water erosion are shown in Fig. 6.1. Most of the numerous human activities that affect the erosion system do so by altering the erodibility of the soil surface (see below).

FACTORS AFFECTING SOIL EROSION

Soil is eroded by the forces of water or wind acting on the soil surface, and on steep slopes

MEASURING SOIL EROSION

A number of techniques are commonly used to measure soil erosion and they do not all

EROSIVITY FACTORS

ERODIBILITY FACTORS

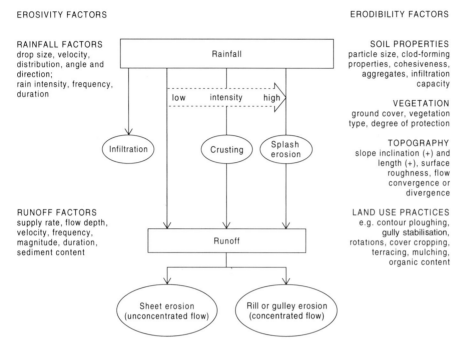

RAINFALL FACTORS
drop size, velocity,
distribution, angle and
direction;
rain intensity, frequency,
duration

SOIL PROPERTIES
particle size, clod-forming
properties, cohesiveness,
aggregates, infiltration
capacity

VEGETATION
ground cover, vegetation
type, degree of protection

TOPOGRAPHY
slope inclination (+) and
length (+), surface
roughness, flow
convergence or
divergence

RUNOFF FACTORS
supply rate, flow depth,
velocity, frequency,
magnitude, duration,
sediment content

LAND USE PRACTICES
e.g. contour ploughing,
gully stabilisation,
rotations, cover cropping,
terracing, mulching,
organic content

FIGURE 6.1 *Main factors affecting types of soil erosion by water. Generally, erosion will be reduced if the value of an erosivity factor is reduced and/or the value of an erodibility factor is increased, with the exception of factors shown with a (+) where the reverse is the case (after Cooke, R.U. and Doornkamp, J.C. 1990,* Geomorphology in environmental management, *2nd edition. Reproduced by permission of Oxford University Press)*

measure the same thing. Some are designed to monitor the redistribution of soil particles over distances of a few centimetres or metres on a field surface. Others measure soil loss from a field or the quantity of soil which leaves a catchment in a river. Measurements of the rate of soil loss from an area over long periods of time – up to thousands of years – are calculated from the depth of sediment accumulated in lakes or on the sea bed. As a general rule, the larger the area under the consideration, the lower will be the rate obtained.

Many older estimates of erosion are based on studies of experimental plots, usually of a standard size (22 × 1.8 m), on which controlled experiments are undertaken by varying such factors as slope angle, crop type

and management practice. This approach suffers from the serious drawback that actual landscapes have much more complex topography than standard plots, and certain types of erosion forms, such as rills and gullies, may not develop on such a small scale. Some measurements on real fields have concentrated on these important forms, with field workers repeatedly visiting an area to measure rills and gullies (e.g. Boardman, 1990a), an exercise which can be supplemented by aerial photography.

Measurement of suspended sediment loads in rivers is another commonly used approach which gives a rough estimate of current erosion on slopes and fields in the catchment area. Not all eroded soil reaches the river, however, and there have been many attempts to estimate 'sediment delivery ratios'. The technique is further hampered by the fact that a proportion of a river's sediment load is eroded from the river's own banks. A recent technique used to estimate both erosion and deposition measures the radionuclide caesium-137 present in a soil profile, which can be compared to amounts in nearby undisturbed profiles. The caesium-137, most of which has been released by atmospheric bomb testing, has been deposited on soils in the last 30 years or so and the technique can provide average rates over that period and patterns of erosion and deposition at particular sites.

Field measurements and experiments have also provided data from which general relationships between the factors affecting soil erosion have been derived, and these relationships have been incorporated into models used to predict erosion. The most widely used is the Universal Soil Loss Equation (USLE), developed in the USA to predict soil loss by runoff from US fields east of the Rocky Mountains under particular crops and management systems. The USLE is calculated as:

$$E = R \times K \times L \times S \times C \times P$$

where E is mean annual soil loss, R is a measure of rainfall erosivity, K is a measure of soil erodibility, L is the slope length, S is the slope steepness, C is an index of crop type, and P is a measure of any conservation practices adopted on the field.

The USLE can be adapted for use outside the temperate plains of North America, although it has often been used inappropriately, without such adaptations (Wischmeier, 1976). An equivalent wind erosion equation has also been developed (Woodruff and Siddoway, 1965), and both of these relatively simple models have been further refined to produce more complex computer-run models such as EPIC, one of the few erosion models

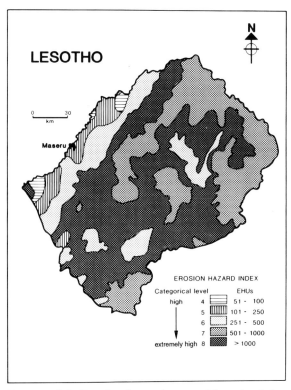

FIGURE 6.2 *Erosion hazard map of Lesotho (after Chakela and Stocking, 1988)*

which can be used to predict erosion by both water and wind. Models have also been used to produce soil erosion hazard maps such as that shown in Fig. 6.2 for Lesotho, the country with the highest erosion hazard in southern and central Africa thanks to its steep slopes, high rainfall totals, poor soils and average vegetation covers.

Despite the availability of numerous measurement and prediction techniques, the fact remains that there is virtually no soil erosion data for most of the world's land surface. An attempt to overcome this deficiency has been made with the Global Assessment of Human-Induced Soil Degradation (GLASOD) which employed more than 250 local soil experts around the world to give their opinion on soil degradation problems to supplement what data are

TABLE 6.1 *GLASOD estimates of land area affected by human-induced soil erosion*

Continent	Area affected in million ha (% total continental area)	
	Water erosion	**Wind erosion**
Africa	227.4 (7.7)	186.5 (6.3)
Asia	439.6 (10.3)	222.1 (5.2)
Australasia	82.9 (9.4)	16.4 (1.9)
Europe	114.5 (12.1)	42.2 (4.4)
North America	106.1 (4.8)	39.2 (1.8)
South America	123.2 (7.0)	41.9 (2.4)

Source: Deichmann and Eklundh (1991)

available. The project, carried out by the International Soil Reference Center in conjunction with the UN Environment Programme, followed a strict methodology by dividing the world's land surface into mapping units corresponding to physiographic zones, and for each unit an assessment was made of water and wind erosion as well as chemical and physical degradation processes. The degree of degradation, its extent and causes were assessed to produce an overall estimate of degradation severity in each unit (Oldeman *et al.*, 1990). Table 6.1 shows the continental areas affected by water and wind erosion.

EFFECTS OF EROSION

The movement of soil and other sediments by erosive forces has a large number of environmental impacts which can affect farmers and many other sectors of society. Many of these effects are consequent upon natural erosion, but are exacerbated in areas where rates are accelerated by human activity. The environmental effects associated with erosion occur due to the three fundamental processes of entrainment, transport and deposition. This three-fold division is used in Table 6.2 to illustrate the

TABLE 6.2 *Some environmental consequences and hazards to human populations caused by wind erosion and dust storms*

ENTRAINMENT
Soil loss
Nutrient loss
Crop root exposure

TRANSPORT
Sand-blasting of crops
Air pollution
Radio communication problems
Local climatic effects
Transport disruption
Disease transmission (human and plants)

DEPOSITION
Nutrient gain (soils, plants and oceans)
Salt deposition and groundwater salinisation
Burial of structures
Rainfall acid neutralisation
Machinery problems
Reduction of solar power potential
Electrical-insulator failure

Source: after Goudie and Middleton (1992)

hazards posed to human populations by wind erosion, while the following sections are divided into on-site and off-site effects.

On-site effects

The loss of topsoil changes the physical and chemical nature of an area. Deformation of the terrain due to the uneven displacement of soil can result in rills, gullies, mass movements, hollows, hummocks or dunes. For the farmer, such physical changes can present problems for the use of machinery, and in extreme cases such as gullying, the absolute loss of cultivable land (Fig. 6.3). Uneven displacement of soil may also result in burial of plants and seedlings; loss of soil may expose roots; and sand-blasting by wind-eroded material can both damage plants and break down soil clods, impoverishing soil structure and rendering soil more erodible. Splash erosion can cause compaction and crusting of the soil surface, both of which may hinder germination and the establishment of seedlings, while exposure of hardpans and duricrusts presents a barrier to root penetration.

Erosion has implications for long-term soil productivity through a number of processes. Reduced soil thickness diminishes the depth through which roots can penetrate and the soil's capacity for holding water. The effects on soil structure, by preferential removal of organic material and fine particles, can also reduce the water-holding capacity of the soil. These problems are most serious on thin soils where plants root to depths greater than that of available topsoil. Erosion also carries away soil nutrients, including fertilisers, and even seeds.

Some researchers sound a note of caution in directly relating erosion rates to losses in productivity (e.g. Larson et al., 1983), not least because the relationship is highly variable depending upon soil type and crop type and because numerous other factors can affect crop yields. However, many experiments have shown that as erosion proceeds, crop yields do decline. Figure 6.4 illustrates this for two staple crops in south-west Nigeria for a range of slope angles on the same soil type. Yield declines can, of course, be compensated for by adding fertilisers, assuming the farmer can afford to do so. The additional fertiliser needed to maintain corn yields for an area of the USA are shown in Table 6.3.

Off-site effects

The off-site effects of eroded soil are caused by its transport and deposition. Material carried in strong winds can cause substantial damage to structures such as telegraph poles, fences and larger structures by abrasion. Material transported in dust storms, which can affect very large areas (Fig. 6.5), severely reduce visibility, causing a hazard to transport, and adversely affect radio and satellite communications. Inhalation of fine particles

FIGURE 6.3 *A 15-m deep gully in central Tunisia*

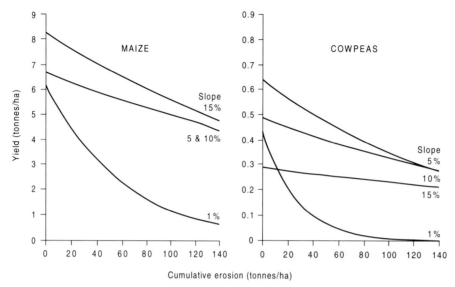

FIGURE 6.4 *Decline in yields of maize and cowpea with cumulative loss of soil in south-west Nigeria (reproduced from Lal, R. 1993 in Pimental D (ed.),* World soil erosion and conservation, *by permission of Cambridge University Press)*

TABLE 6.3 *Increase in fertiliser needs for corn as soil erodes in southern Iowa, USA*

Change in erosion phase	Additional fertiliser needed (kg/ha)		
	Nitrogen	**Phosphate**	**Potash**
Slight to moderate	11	2	7
Moderate to severe	34	1	8

Source: after Rosenberry et al. (1980)

can aggravate human diseases such as bronchitis and emphysema, and the transport of soil pathogens spreads human and plant diseases. The nutrients attached to soil particles transported by water erosion can cause the eutrophication of water bodies (see Chapters 11 and 12).

The deposition of eroded material can also cause considerable problems for human society. The flood hazard can be increased due

to river bed infilling, and the siltation of reservoirs, harbours and lakes presents hazards to transport and loss of storage capacity in reservoirs. The economic costs thus imposed are often considerable. In Java, the off-site costs due to siltation of irrigation systems and reservoirs, and harbour dredging were estimated to be US$58 million in 1987 (Magrath and Arens, 1989). About 25 per cent of the sediment deposited in lakes and reservoirs in the USA

FIGURE 6.5 *Severe dust storm at Melbourne, Australia in February 1983 (courtesy of Australian Bureau of Meteorology)*

is thought to originate from cropland. The resulting damage, which contributes to a 0.22 per cent annual loss in national water storage capacity, has been valued at from $144 million to $197 million per year (Crowder, 1987). Deposition of sediment reaching the coastline can adversely affect marine environments used by local populations, including coral reefs and shellfish beds.

Despite the numerous negative aspects of sediment deposition for human society, the positive side of the equation should also be highlighted. Relatively flat, often low-lying areas of deposition provide good, fertile land for agriculture and many other human activities. Floodplains are an obvious example, as are valley bottoms which receive material transported from valley slopes. Wind-deposited dust, or loess, also provides fertile agricultural land – the loess plateau of northern China is the country's most productive wheat-growing area, for example – although

it requires careful management because it remains highly erodible. Similarly, inputs of soil nutrients to the oceans enhance productivity in marine ecosystems.

ACCELERATED EROSION

Although many of the adverse effects of soil erosion for human society outlined above may occur due to natural rates, the most pressing problems are found in areas where accelerated rates occur. Natural rates vary enormously depending on such factors as climate, vegetation, soils, bedrock and landforms. Information on natural erosion rates, combined with our limited knowledge of the rates of soil formation, are used as yardsticks against which to measure the degree to which human action exacerbates natural processes and the sort of rates which soil conservation measures might aim to achieve (see below).

Natural rates of soil formation are generally taken to be less than 1 t/ha/year, and for many practical purposes acceptable target rates for soil conservation are commonly set between 2.5 and 12.5 t/ha/year, depending on local conditions (Cooke and Doornkamp, 1990). In some countries, however, much lower limits have been established. The Swedish Environmental Protection Board, for example, considers a soil loss of 0.1–0.2 t/ha/year as a recommended limit for preventative measures to be applied on arable land (Alström and Åkerman, 1992).

The two most common effects of human activity which lead to accelerated erosion are modifications to, or removal of, vegetation, and destabilisation of natural surfaces. Such actions have a variety of motives: vegetation may be cleared for agriculture, fuel, fodder, or construction; vegetation may be modified by cropping practices, or deforestation for timber; land may be disturbed by ploughing, off-road vehicle use, military manoeuvres, construction, mining or trampling by animals. The effects of several of these disturbances on soil erosion by water in central Oahu, Hawaii are shown in Table 6.4. Other processes of erosion also follow such disturbances: an example of the effects on slope stability and resulting mass movement problems caused by highway construction is given in Chapter 15 (see Table 15.3). New transport routes can cause accelerated landsliding by increasing disturbing forces acting on a slope, both during construction when cuts and excavations remove lateral or underlying support, and through earth stresses caused by passing vehicles. Other human activities which can increase the chance of slope failure do so by decreasing the resistance of materials which make up slopes. This can occur if the water content is increased, as happens when local water tables are artificially increased by reservoir impoundment, for example.

The initial impact of certain activities may be reduced when a new land use is established, however. Many construction activities initially cause marked increases in soil loss, but erosion rates can be reduced to below those recorded under natural conditions when a soil surface is covered by concrete or tarmac. Conversely, soil erosion problems may become displaced by the effects of construction, with water flows from drainage systems causing accelerated soil loss where they enter the natural environment.

TABLE 6.4 *Effects of human activities on erosion rates in Oahu, Hawaii, USA*

Initial land cover	Disturbance	Increase in erosion rate
Forest	Planting of row crops	×100–1000
Grass	Planting of row crops	×20–100
Forest	Building logging road	×220
Forest	Woodcutting and skidding	×1.6
Forest	Fire	×7–1500
Forest	Mining	×1000
Forest	Construction	×2000
Pasture	Construction	×200
Row crops	Construction	×10

Source: El-Swaify *et al.* (1982)

There is widespread agreement that the prime causes of accelerated soil erosion are deforestation and agriculture. Deforestation removes the protection from raindrop impact offered to soil by the tree canopy, and reduces the high permeability humus cover of forest floors, a permeability which is enhanced by the many macropores produced by tree roots. Cultivation also removes the natural vegetation cover from the soil, which is particularly susceptible to erosion when bare after harvests and during the planting stage. Some crops, such as maize and vines, usually leave large portions of the ground unprotected by vegetation even when the plants mature. Furthermore, mechanical disturbance and compaction of the soil by ploughing and tilling can enhance its erodibility.

An illustration of the dramatic effects on soil loss instigated by deforestation of a tropical forest area is given by experimental work on plots in French Guiana (Fritsch and Sarrailh, 1986). Suspended sediment yields from undisturbed rain forest plots of less than 0.7 t/ha/year increased by up to fifty times after mechanised clear-cutting. When the plots were planted with grasses, erosion rates fell, but were still two to three times higher than under forest cover. A similar pattern of events over a longer period, derived from analysis of sediment and pollen in a lake bed, is shown for an area cleared of woodland in the mid-nineteenth century in Michigan, USA (Fig. 6.6). The initial response to clearance was a sharp rise in erosion by thirty to eighty times the rates derived for the pre-settlement period. Fluctuating rates characterised a period of about 30 years as a new steady state was reached under farmland, which is nonetheless some ten times greater than under undisturbed woodland.

Some of the most significant cases of accelerated erosion by wind have occurred in dry grassland areas used for grain cultivation. In the Maghreb countries of North Africa in the nineteenth century, French settlers brought

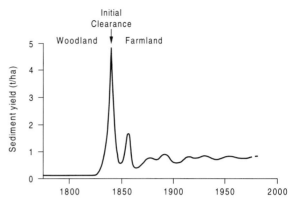

FIGURE 6.6 *Historical reconstruction of sediment yield at Frains Lake, Michigan, USA (reprinted with permission from Davis, M.B. 1976, Erosion rates and land used history in southern Michigan.* Environmental Conservation *3: 139–48. Copyright © 1976 Elsevier Science)*

European agricultural machinery which turned the soil to twice the depth of the traditional hoes and removed shrubs and weeds, baring the soil to erosive forces. By the early twentieth century, the cropland area of the 'telle' zone had quadrupled and traditional cultivators had been pushed onto steeper slopes and to the climatic limits of dry cereal crops. The situation has changed little since the independence of Algeria and Tunisia, with the desire to enlarge cropland pushing the tractor and multidisc plough further into the steppe. Increasing frequency of wind erosion and a general degradation of the former grasslands has been the result (Dresch, 1986).

Probably the most infamous case of wind erosion in the Western world came after widespread ploughing of the grasslands of the US Great Plains, which created the Dust Bowl of the 1930s (see Chapter 4), but similar environmental mishaps occurred in the former USSR in the 1950s and in a copycat exercise in the Mongolian steppes in the 1960s. Forty million hectares of Virgin Lands were put to the plough in northern

TABLE 6.5 *Effect of the Virgin Lands scheme on the frequency of dust storms in the Omsk region of western Siberia*

Station	Mean annual number of dust storm days		Increase
	1936–50	1951–62	
Omsk, steppe	7	16	×2.3
Isil '-Kul'	8	15	×1.9
Pokrov-Irtyshsk	4	22	×5.5
Poltavka	9	12	×1.3
Cherlak	6	19	×3.2
Mean	6.8	16.8	×2.84

Source: Sapozhnikova (1973)

Kazakhstan, western Siberia and eastern Russia between 1954 and 1960. Deep ploughing was employed, removing the stubble from the previous year's crop to allow planting earlier in the year, so reducing the chance of losses to early snows at harvest time. Land was also used more intensively, doing away with alternate years when land was traditionally left fallow under grass. Wind erosion soon began to take its toll (Table 6.5), coming to a head in the early 1960s when drought hit the region.

The introduction, from outside, of agricultural techniques which are inappropriate to local conditions is a widely cited cause of erosion. In many parts of Latin America, management techniques brought by Spanish and Portuguese colonists have been blamed for widespread soil degradation. In Ecuador, for example, farmers have neglected the main principles of pre-colonial agriculture which were more suited to the mountainous areas, and today, accelerated soil erosion is estimated to affect 50 per cent of the national territory at rates which reach 200–500 t/ha/year in the great basin of Quito (De Noni *et al.*, 1986). In addition to the Spanish conquest, poor

agricultural reform and the population explosion at the beginning of the twentieth century are highlighted as significant factors behind the soil erosion situation.

However, evidence from lake sediment cores in central Mexico suggests caution over the ease with which invading Iberians are blamed for unbalancing traditional, supposedly harmonious systems of soil use. Analysis of the cores indicates that pre-Hispanic agriculture in the Lake Patzcuaro Basin was not as conservationist in practice as was previously thought. Several periods of accelerated erosion have been identified which occurred before the arrival of the Spanish and were of comparative magnitude to those during colonial times. This suggests that the introduction of the plough did not have a greater impact on soil erosion than traditional methods (O'Hara *et al.*, 1993).

Despite questions over where exactly to lay the blame, a long history of accelerated erosion in some regions has meant that a state of system collapse has been reached, as in parts of the Caribbean where erosion has been significant since plantation monoculture was introduced in the early eighteenth century. Large-scale deforestation, the use of fire for

TABLE 6.6 *Estimated rates of soil loss on slopes in Ethiopia by type of land cover*

Land cover/use	Proportion of national land area (%)	Soil loss (t/ha/year)
Cropland	13.1	42
Perennial crops	1.7	8
Grazing and browsing	51.0	5
Totally degraded	3.8	70
Currently uncultivable	18.7	5
Forest	3.6	1
Wood and bushland	8.1	5

Source: after Hurni (1993)

land clearance and clean weeding of cropland, which continued after the emancipation of plantation slaves, enhanced rates of degradation. Population increase became a significant factor from the latter years of the nineteenth century, the pressure on resources being exacerbated by fragmentation of land holdings. Some of the worst affected areas are on the island of Hispaniola (Haiti and Dominican Republic). In Haiti, large areas of marginal land are thought to be irreversibly degraded and an estimated 6000 ha of land are abandoned to erosion each year (Paskett and Philoctete, 1990).

Another country where accelerated erosion is widely accepted to have reached crisis proportions is Ethiopia. Erosion rates which average 42 t/ha/year on Ethiopian cropland (Table 6.6) reach as high as 300 t/ha/year on some fields in western parts of the highlands where rainfall erosivity is highest. Hurni (1993) suggests that this level of soil loss reduces soil productivity in Ethiopia by 1–2 per cent per year and that at current erosion rates most of the country's cropland soils will be completely lost within 150 years. The crucial factor behind the degradation of Ethiopia's soils and other land resources is the country's large population, and Hurni believes that whatever scenario for soil conservation is proposed, sustainable use of land resources can only be achieved if population growth rates are reduced to zero within 50 years.

Although serious soil erosion is often associated with the humid tropics and semi-arid areas, it does give cause for concern in more temperate regions. The intensification of agriculture in the UK, for example, has brought erosion problems to the fore, with about 36 per cent of arable land in England and Wales classified as being at moderate to high risk of erosion (Evans, 1990). Sandy and peaty soils in parts of the Vale of York, the Fens, Breckland and the Midlands suffer from deflation during dry periods, but water is the most widespread agent of soil loss. The central reason for the increase in erosion in the last 20 years or so has been a shift in the sowing season, from spring to autumn, for the main cereal crops of wheat and barley. The change has come about because autumn-sown 'winter cereals' produce higher yields, but the shift in sowing season leaves winter cereal fields exposed in the wettest months – October and November – in many parts of the country, since arable fields are at risk from erosion until about 30 per cent of the ground is covered by the growing crop.

Several other aspects of more intensive arable farming have also contributed to the enhanced erosion rates (Boardman, 1990b).

- The expansion of arable crops onto steeper slopes made possible by more powerful tractors.
- The creation of larger fields by removal of walls, hedges, grass banks and strips has produced longer slopes and larger catchment areas which generate greater volumes of water on slopes and in valley bottoms.
- More powerful machinery is able to work the land even when damp, thus compacting soils along wheel tracks which tend to channel runoff.
- The tendency to break up soil into a fine tilth using powered harrows to aid seed germination – a technique which also increases erodibility.
- Decreasing aggregate stability of the soil, due to reduced organic matter content, renders soils more likely to form crusts under raindrop impact.

SOIL CONSERVATION

The numerous techniques available for protecting cultivated soils from erosion can be classified into three groups:

- **agronomic measures** which manipulate vegetation to minimise erosion by protection of the soil surface;
- **soil management techniques** which focus on ways of preparing the soil to promote good vegetative growth and improve soil structure in order to increase resistance to erosion; and
- **mechanical methods** which manipulate the surface topography to control the flow of water or wind.

Maintaining a sufficient vegetative cover on a soil is sometimes referred to as the 'cardinal rule' for erosion control. Mulching, the practice of leaving some residual crop material – such as leaves, stalks and roots – on or near the surface, is a widely used agronomic technique. It is successful in reducing erosion, but crop residues also provide a good habitat for insects and weeds so that the method often requires higher chemical inputs of pesticides and herbicides which is expensive and increases hazards of off-field pollution and the killing of non-target species. Soil management techniques are concerned with different methods of soil tillage, including strip or zone tillage which leaves protective strips of untilled land between seed rows. Minimum tillage, which incorporates the idea of stubble mulching mentioned above, is another successful anti-erosion practice. Mechanical methods include such techniques as the building of terraces and the creation of protective barriers against wind, such as fences, windbreaks and shelter belts. The beneficial effects of some of these techniques for erosion control and crop yields are shown in Table 6.7, in which the range of percentage figures are derived from a review of more than 200 case studies. The effects on soil and water loss of various cultivation practices for a range of crop types in the humid zone in Brazil are shown in Table 6.8.

Implementation of soil conservation measures

Since we have a good understanding of the factors affecting soil erosion, the problems caused by the process, and the methods for its control, the question 'Why is soil erosion still a problem?' is an obvious one to ask. Innumerable soil conservation projects have been implemented, particularly in the poorer areas of the world, but often with disappointing results. An analysis of the reasons for the failure of such projects has been made by Hudson (1991), who found that the poor

TABLE 6.7 *Effect of low-cost soil conservation techniques on erosion and crop yields*

Method	Decrease in erosion (%)	Increase in yield (%)
Mulching	73–98	7–188
Contour cultivation	50–86	6–66
Grass contour hedges	40–70	38–73

Source: Doolette and Smyle (1990)

TABLE 6.8 *Soil and water losses for different crops and cultivation practices in the humid zone of north-east Brazil*

Crop and cultivation practice	Soil loss (t/ha/year)	Water loss (% of rainfall)
Cotton planted down slope	32	15
Cotton planted across slope	13	11
Mixed cotton, maize and beans planted across slope	24	13
Mixed cotton, maize and beans planted across slope with herbaceous anti-erosion strips	15	12
Mixed maize and beans planted across slope	13	12
Beans planted across slope	12	13
Maize planted down slope	10	9
Maize planted across slope	11	10
Permanent planted pasture	1	3
Bare soil	94	26

Source: Leprun *et al.* (1986)

design of projects was a fundamental flaw. He concluded that errors are mainly a function of incorrect assumptions made at the design stage, by both donor agencies and host governments. The main donor errors are:

■ overoptimism, including overestimating the effect of new practices,

■ overestimating the rate of adoption of new practices,

■ overestimation of the ability of the host country to provide backup facilities,

■ underestimation of the time required to mobilise staff and materials for the project, and

■ frequently a quite unreal estimate of the economic benefits.

Some of the main problems arising from the assumptions of the host governments are:

- overestimating their capacity to provide counterpart staff and the funds for the recurrent costs arising from the project,
- a tendency to underestimate the problems of coordination among different ministries or departments, and
- a tendency to overestimate the strength of the national research base and its ability to contribute to the project.

Such an analysis provides useful guidelines for future soil conservation projects, but other observers of the soil erosion problem believe that attention must also be focused on the central reasons for people misusing soils, which lie at a deeper level. The Haitian and Ethiopian examples cited earlier illustrate some of the underlying driving forces behind accelerated erosion rates which are emphasised by social scientists (e.g. Blaikie, 1985). These driving forces are social, economic and political in nature and include such factors as population growth, unequal distribution of resources, land tenure, terms of trade and subsidies, colonial attitudes and legacies, and class struggles. These factors limit the options available to poorer sectors of society who may be forced into degrading their soil resource simply in order to survive.

In less-critical situations, erosion may result because it is not perceived as a serious problem since it often proceeds at relatively imperceptible rates, or because its effects are masked by fertilisers. Allied to this problem, the benefits of soil conservation may be long-term, whereas farmers may have a short-term view of the relative importance of maintaining their income or repaying their debts. The costs involved may also be high and there is an ethical and practical question of who should pay: the individual farmer or society as a whole?

FURTHER READING

Blaikie, P. 1985 *The political economy of soil erosion.* London, Longman. This book looks at soil erosion from a social perspective, emphasising the political and economic reasons behind accelerated rates of soil loss.

Hallsworth, E.G. 1987 *Anatomy, physiology and psychology of soil erosion.* Chichester, Wiley. The author highlights the similarities between traditional and modern methods of soil conservation, proposes reasons why modern techniques are not always adopted and suggests how these problems can be overcome.

Morgan, R.P.C. 1986 *Soil erosion and conservation.* London, Longman. A good basic text detailing the processes of soil erosion, methods of measurement and modelling, and the techniques developed for its control.

Pimental, D. (ed.) 1993 *World soil erosion and conservation.* Cambridge, Cambridge University Press. A collection of papers on numerous aspects of soil erosion, its measurement, the problems it causes, and its control.

7

THREATENED SPECIES

Biodiversity, a term that refers to the number, variety and variability of living organisms, has become a much-debated environmental issue in recent times. It is commonly defined in terms of genes, species and ecosystems, corresponding to three fundamental levels of biological organisation. Genetic and ecosystem diversity are dealt with in more detail elsewhere (see Chapters 3 and 5), while here the focus is at the species level.

The key issue at stake in the biodiversity debate is its loss, which can take many forms, the ultimate of which is the extinction of a species. Extinction is a natural process, with the fossil record indicating that over geological time all species have a finite span of existence. Indeed, fossil evidence indicates that in the last 570 million years of the Earth's history five 'mass extinction events' have occurred. The most severe was during the late Permian period some 245 million years ago, and the most recent was at the end of the Cretaceous period 65 million years ago when the dinosaurs and several other families of species were extirpated. Currently, however, there is widespread fear that another mass extinction event is occurring, one in which the Earth's human population is playing the key role.

EXTINCTION RATES

Calculation of the exact rates of species extinction is fraught with difficulty. Part of this difficulty is due to the fact that there is still considerable ignorance about the total number of species present on the Earth today. While the numbers of species in some groups of organisms are relatively well known, our knowledge of others is extremely imprecise. The number of bird species, for example, is close to 10 000 (Sibley and Monroe, 1990), but estimates of the number of insect species vary tremendously (from 8 million to 30 million) with about 1 million having been described (WCMC, 1992).

Hence, it is likely that many species that we have never known about have become extinct. Even among those organisms that have been described, problems emerge in documenting their disappearance. Just because no member of a species has been documented for some time does not necessarily mean that it has become extinct: absence of evidence is not evidence of absence.

Despite these and other difficulties, estimates of the rate at which species are becoming extinct, mostly due to human action, indicate that the rate has been growing exponentially since about the seventeenth century. Current and projected estimates of species loss are based upon the rate at which habitats are being destroyed, modified and fragmented – the most serious threats to species diversity – coupled with biogeographical assumptions relating numbers of species and area of habitat. We should be aware, however, that estimates for habitat loss in tropical forest areas, the most diverse ecosystems, are themselves subject to wide variations (see Chapter 3). While some earlier projections suggested that 20–50 per cent of species would be lost by the end of the century (Myers, 1979; Ehrlich and Ehrlich, 1981), these now seem exaggerated. Reid (1992) estimates a 1–5 per cent loss per century, and figures of 100 000 species lost per year are frequently quoted (WCMC, 1992).

THREATENED SPECIES

Some species are particularly at risk from the threat of extinction simply because they are only found in a narrow geographical range, or they occupy only one or a few specialised habitats, or they are only found in small populations. Other factors may also affect the degree of risk faced by certain species. These include:

- low rates of population increase
- large body size (hence requiring a large range, more food and making the species more easily hunted by humans)
- poor dispersal ability
- need for a stable environment
- need to migrate between different habitats
- perceived to be dangerous by humans.

Combinations of some of these characteristics are found in species known as *K*-strategists, and it is these species which are generally more likely to become extinct because they tend to live in stable habitats,

delay reproduction to an advanced age and produce only a few, large offspring. By contrast, species that produce many offspring at an earlier age and have the ability to react quickly to changes in their environment, are known as *r*-strategists, and it is their speedier turnover and flexibility that make *r*-strategists less likely to experience extinction.

The wider ecological implications of the loss of a certain species varies between species. Another important aspect of the extinction issue is the fact that certain keystone species may be important in determining the ability of a large number of other species to persist. Hence, the loss of a certain keystone species could potentially result in a cascade of extinctions. Such fears have been expressed over tropical insects, many of which have highly specialised feeding requirements.

Species known to be at risk are documented according to the severity of threat they face and their imminency of extinction (Table 7.1) by the International Union for Conservation of Nature and Natural Resources (IUCN). The general term 'threatened' is used to refer to a species of fauna or flora considered to belong to any of the categories shown in Table 7.1.

THREATS TO FLORA AND FAUNA

Most of the factors currently threatening species of both fauna and flora are induced or influenced by human action. Such actions may be deliberate, as in the case of destruction by hunting, or inadvertent, as in the case of destruction or modification of habitats in order to use the land for other purposes. In practice, many species are at risk from more than one threat and some threats tend to combine: the clearance of forests, for example, makes the hunting of large mammals easier. The threats that a particular species faces may also vary through time. The decline of the New Zealand mistletoe (*Trilepidea adamsii*) began as its habitat was reduced by deforestation, first by

TABLE 7.1 *Categories of threatened species*

Extinct
A species not definitely located in the wild during the past 50 years

Endangered
A taxon in danger of extinction and whose survival is unlikely if the causal factors continue to operate

Vulnerable
A taxon believed likely to move into the Endangered category in the near future if the causal factors continue operating

Rare
A taxon with a small world population that is not at present Endangered or Vulnerable but is at risk

Indeterminate
A taxon *known* to be Endangered, Vulnerable or Rare but where there is not enough information to say which of the three categories is appropriate

Insufficiently known
A taxon that is *suspected* but not definitely known to belong to any of the above categories because of lack of information

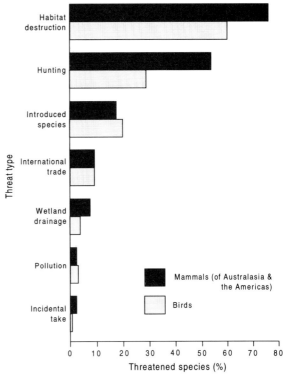

FIGURE 7.1 *Threats to mammals and birds (after Figure 17.2, p. 236 in WCMC 1992,* Global biodiversity: status of the Earth's living resources. *Chapman & Hall)*

the Maoris and at an accelerating rate by British settlers in the late nineteenth century. The population was further reduced by collectors, and the decline of bird populations that were responsible for seed dispersal, due to forest clearance. The final specimens, which disappeared from North Island in 1954, may have been eaten by brush-tailed possums deliberately introduced from Australia during the 1860s to establish a fur trade (Norton, 1991). Although the detailed nature of threats may be complex and variable, some indication of the relative importance of the different threats is given in Fig. 7.1, which was compiled for the birds of the world and mammals in Australasia and the Americas.

All of these human-induced reasons for the demise of species have operated in the past. In Britain, for example, the combination of hunting and habitat destruction by deforestation has put paid to many original animal species over the centuries (Table 7.2).

Habitat loss and modification

The destruction of habitats is widely regarded to be the most severe threat to biological diversity, while fragmentation and degradation, which are often precursors of outright destruction, also present significant cause for concern. In many countries, particularly on islands (see below) and where human population densities are high, most of the natural habitat has been

TABLE 7.2 *Dates when the last wild member of selected animal species was killed in Britain*

Animal	Year
Beaver (*Castor fiber*)	Late 1100s
Wild swine (*Sus scrofa*)	1260
Wolf (*Canis lupus*)	1743
Goshawk (*Accipiter gentilis*)	1850
Wildcat (*Felis catus*)	1870
White-tailed sea eagle (*Haliaetus albicilla*)	1918

Sources: from information in Rackham (1986); Peters and Lovejoy (1990)

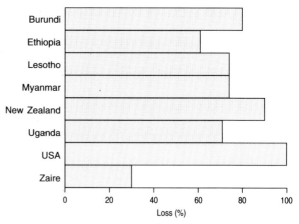

FIGURE 7.2 *Proportion of grasslands and savanna lost since pre-agricultural times in selected countries (from data in WRI, 1990)*

destroyed to provide farmland, rangeland, and land for settlement and industry. No fewer than forty-nine out of sixty-one countries surveyed in the African and Asian tropics are thought to have lost more than 50 per cent of their wildlife habitats (IUCN/UNEP, 1986a,b). In the tropical African countries, 65 per cent of the original wildlife habitat has been lost, with particularly high rates of destruction reported from Gambia (89 per cent), Liberia (87 per cent), Rwanda (87 per cent), Burundi (86 per cent) and Sierra Leone (85 per cent). In the Indomalayan countries the overall loss is 68 per cent, with particularly severe losses reported from Hong Kong (97 per cent), Bangladesh (94 per cent), Sri Lanka (83 per cent), Viet Nam (80 per cent) and India (80 per cent).

The destruction of tropical rain forests, coral reefs, wetlands and mangroves, documented elsewhere in this book, is particularly serious given the high biodiversity of these habitats. Hence, the large majority of current human-induced extinctions are occurring in the world's tropical rain forest areas, and insects are the order of species most at risk. Nevertheless, other habitats have suffered equally severe destruction; Fig. 7.2 shows the percentage of grasslands and savannas lost since pre-agricultural times in selected countries.

There are numerous examples of species that have been driven to the edge of extinction because of the loss of their habitat. In the UK, for example, the distribution of the threatened green winged orchid (*Orchis morio*) has been severely reduced by the loss of 40 per cent of unintensified lowland grassland between 1932 and 1992 (DoE, 1992). Internationally, one of the best-known threatened species, the giant panda (*Ailuropoda melanoleuca*), has also suffered from progressive human encroachment into its habitat. Once found throughout much of China's high-altitude regions and beyond, the species is now confined to a few sites near Chengdu (Fig. 7.3). The panda's heavy reliance upon its specialised bamboo diet, which occasionally requires forays into lowland regions during natural bamboo die-offs, puts the normally shy creature in direct conflict with encroaching human populations. Despite huge efforts to conserve the panda, these factors, combined with the extreme difficulties encountered by attempts at captive breeding, look set to add the panda to a long list of species whose ultimate fate can be traced to destruction of their habitat by humans.

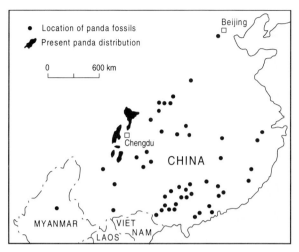

FIGURE 7.3 *Past and present distribution of pandas (reprinted with permission from Roberts, L. 1988, Conservationists in Panda-monium. Science 241: 529–31. Copyright 1988 American Association for the Advancement of Science)*

Fragmentation of habitats not only reduces the area of habitat but also affects normal dispersion and colonisation processes, and reduces areas for foraging. In addition, fragmentation increases the ratio of edge to total area, hence increasing dangers from outside disturbances such as the effects of hazards (e.g. fire and diseases) and invasion by competitors and/or predators. The recent decline in numbers of songbirds throughout North America, for example, is at least partly explained by increased predation and parasitism of nests in fragmented nesting tracts (Terborgh, 1992). Species–area relationships derived from the study of oceanic islands suggest that the number of species a fragment is able to support is directly related to the fragment area (MacArthur and Wilson, 1967).

Research in the Brazilian Amazon on dung beetles shows that they are significantly affected by fragmentation, with small forest fragments having fewer species of beetles, lower population densities for each species and smaller-sized beetles than undisturbed forest areas. Since dung beetles are a keystone species in the forest ecosystem, through their role in burying dung and carrion as a food source for their larvae – thus facilitating the rapid recycling of nutrients and the germination of seeds defecated by fruit-eating animals, and reducing vertebrate disease levels by killing parasites that live in the dung – the effects on community interactions and ecosystem processes are probably very widespread (Klein, 1989).

Human-induced modifications to habitats can also lead to species loss either directly or in a more pervasive manner through stress or subtle effects on ecological processes. Pollution in its various forms is probably the most significant problem in this respect. Examples of such effects are referred to in many of the chapters in this book; hence, just a few illustrations will be made here.

The detrimental effects of agrochemicals have been widely documented (see Chapter 5), with some of the most deleterious being observed in creatures at the top of the food chain where some materials bioaccumulate. Raptors are typical in this respect, suffering steep decline in population numbers in the postwar years with the rise in use of organochlorine pesticides, particularly DDT which had the effect of thinning egg shells, leading to breakage in the nest, and alterations in breeding behaviour. Use of the highly toxic dieldrin and aldrin in cereal seed dressings and sheep dip, which occurred from 1956 in the UK, also led to widespread declines through increased adult mortality.

However, the phasing out of organochlorine pesticides in many Western countries has resulted in bird of prey populations rising once more. In the UK, the population of peregrine falcons (*Falco peregrinus*) numbering around 850 pairs in the 1930s, had fallen to 360 pairs by 1962, but had risen to 1050 pairs in 1991 (DoE, 1992).

Numerous forms of water pollution can have detrimental effects on aquatic flora and fauna. Acidification of lakes and streams, for example, has affected waterbirds and invertebrates (see Chapter 9). An example of a species extinction due to pollution is the splitmouth (*Lagochila lacera*), discovered in the Chickamauga River in Tennessee just over 100 years ago and named as a new genus and species. The splitmouth was very common in the region and found in the Tennessee, Cumberland, White and Ohio river drainages to Lake Erie. The species was last seen in the Auglaize River, Ohio in 1893. Its extinction – the first North American freshwater fish species known to die out entirely due to human action – appears to have been caused by continuous silting and pollution of its habitat (Maitland, 1991).

Air pollution can also have insidious effects on plants and animals. Atmospheric acidification has been linked to numerous impacts on trees and other floral species (see Chapter 9); the widespread impoverishment of lichens, which are very sensitive to air pollution, across large parts of Europe and North America has been linked to sulphur dioxide, and, to a lesser extent to photochemical smogs (Hawksworth, 1990). Atmospheric pollutants from heavy industries such as smelting can be devastating to surrounding ecology (see Chapter 18).

Overexploitation

Humans have long been implicated in the extinction of species through overexploitation, particularly by hunting for food. Some believe that such effects can be traced back to the Stone Age and to the Late Pleistocene, with the loss of large mammals such as the mammoth and sabre-toothed tiger being intimately linked to overhunting, although the role of abrupt and substantial climatic change has also been suggested as being instrumental in the extinction of such creatures (Martin and Klein, 1984).

Larger creatures have been suffering from overexploitation ever since. In Roman times, hunters and trappers extirpated the elephant, rhino and zebra in Africa north of the Sahara, and lions from Thessaly, Asia Minor and parts of Syria. A similar fate befell tigers in Hercynia and northern Persia. In many instances, the loss of a species has knock-on effects for other organisms. The death of the last dodo (*Raphus cucullatus*), killed by seafarers on Mauritius in 1681, has meant that another of the island's endemic species, the tambalocque tree (*Sideroxylon sessilisiorum*), has been unable to reproduce for the past 300 years because the fruit-eating bird prepared the fruit in its gizzard for germination.

The advent of the firearm greatly enhanced humans' ability to exterminate creatures in large numbers, resulting in the decimation of the buffalo and the extinction of the passenger pigeon in North America, for example. In some parts of the world, migrating birds continue to be killed in huge numbers by hunters. Some of the worst culprits are the European Mediterranean countries such as Italy and Malta (Fig. 7.4).

Overzealous hunting is still a very serious threat for numerous large mammals, particularly for those whose products fetch high prices in local and international markets. The African elephant population, for example, was cut by half during the 1980s, falling from 1.3 million to under 600 000 (Cohn, 1990) before a ban on international trade in ivory was introduced in 1989 (although even this ban has not been 100 per cent effective – see below). Similarly, the market for whale products fuelled the hunting of oceanic cetacean species, many of which currently have indeterminate chances of survival (see Chapter 10).

The market for certain floral species is also an underlying factor behind their overexploitation. Cacti and orchids are particularly at risk from collectors and many species of tree have been dramatically reduced in their

FIGURE 7.4 *Hide for shooting migratory birds in rural Umbria, Italy. Italian hunters are thought to kill as many as 3 million birds per year*

range by selective logging. Prance (1990) notes the complete loss of the scented sandalo tree (*Santalum fernandezianum*) from Juan Fernández islands, all of which had been felled and shipped to Peru and Chile by 1908, and the bois de prune blanc (*Drypetes caustica*) of Mauritius and Réunion, renowned for its excellent hard timber, of which just twelve trees remain.

Island species

Although islands make up only a minor proportion of the world's land surface area, they feature prominently in any account of threatened flora and fauna. Island species are considered to be especially vulnerable for a number of reasons. Many are endemic species which have evolved in isolation and are thus also particularly susceptible to introduced competitors, diseases and predators. The physical size of islands also means that human activities can rapidly degrade large portions of their ecosystems and can have a great impact on relatively small island populations.

The susceptibility of island species to human-induced change is indicated by the fact that at least 75 per cent of vertebrate extinctions to date have occurred among island fauna (Diamond, 1984). For obvious reasons, it is often the largest species on islands which first suffer from direct human action. On Madagascar, six genera and at least fourteen species of lemurs have become extinct, along with several species of flightless bird and land tortoises, since humans arrived there less than 2000 years ago. There is good evidence to suggest that people were responsible, and hunting was probably the primary cause (Mittermeir *et al.*, 1992).

Severe ecological damage on most islands dates from the arrival of explorers and colonisers from the European maritime powers, for whom oceanic islands had strategic importance. Goats were introduced to the South Atlantic island of St Helena in 1513, and within 75 years, vast herds roamed the island. St Helena's endemic plants had evolved without large grazing animals to contend with and so had few defences against them. Botanists first reached St Helena in the early 1800s, so we can only guess at the damage done by then. Today, forty-six endemic species are known, of which seven are extinct, but some estimates put the original number of endemics at more than a hundred. St Helena remains an island with severely threatened endemic flora (Table 7.3). The most serious threat on the other islands shown in this table is introduced plants which have outcompeted endemic species.

TABLE 7.3 *Selected examples of islands known to have severely threatened endemic flora*

Island (ocean)	Status of endemic flora
St Helena (Atlantic)	7 species already extinct; all of 39 remaining are also threatened
Mauritius (Indian)	21 species already extinct; 89% of 215 remaining are also threatened
Juan Fernández (Pacific)	1 species already extinct; 81% of 122 remaining are also threatened
Hawaii (Pacific)	108 taxa already extinct; 39% of 877 remaining are also threatened

Source: modified after WCMC (1992: 246, Table 17.6)

European colonisers are by no means the only culprits, however. Damage to the wildlife of the Hawaiian islands, where almost 100 per cent of the native insects are endemic, as well as 98 per cent of the birds, 93 per cent of the flowering plants, and 65 per cent of the ferns, began long before the arrival of Captain Cook, the first European, in 1778. Polynesians, who had settled the island by 750 AD, cleared extensive lowland areas for cultivation and introduced rats, dogs, pigs and jungle fowl. Palaeontological evidence suggests that at least half of the known total of eighty-three species of Hawaiian birds became extinct in the pre-European period (Olson and James, 1984). However, further introductions and disturbance of the natural habitat since 1778 has resulted in the loss of at least twenty-three bird species and 177 species of native plants.

In some cases, the introduction of one exotic species to an island may result in the local extinction of numerous native species. An example here is the brown tree snake (*Boiga irregularis*) which has been introduced onto a number of Pacific islands with devastating effects on endemic birds. Savidge (1987) documents how the snake has pushed ten endemic bird species to the point of extinction on the island of Guam. Similarly, the introduction of the predaceous Ferussac snail (*Euglandina rosea*) to several Pacific islands in an attempt at biological control of a previously introduced snail which had attained pest status, the giant African (*Achatina fulica*), has had severe repercussions for endemic snail species on Moorea (see Chapter 5).

CONSERVATION EFFORTS

The two ends of the spectrum of arguments in favour of conserving the biodiversity of species are rooted in morals and pragmatism. Some believe that the destruction of any living organism is morally unacceptable, and that people who share the planet with plants and animals have no right to exterminate other species. More pragmatic arguments point out that the extermination of other species is not in the interests of humankind. Biodiversity is useful to us in the wide perspective, in maintaining the biosphere as a functioning system of which we are a part, and at a more functional level in providing resources for agriculture, industry, medicine and other utilitarian needs.

Conversely, however, certain species are considered detrimental to human activities and are therefore the subject of deliberate attempts to control or exterminate their populations. Species regarded as pests are obvious examples here. While deliberate efforts to make an organism extinct are not common, the smallpox virus (although not strictly a species) is a case in point. Subject to a successful international effort to eradicate the disease by vaccination, the virus now remains only in laboratory stocks. Several recommendations to destroy these stocks have been made, although to date they remain to be used for research purposes.

Habitat protection

Since the destruction, modification and fragmentation of habitats is the primary threat to species, the protection of habitats from the causes of their destruction is the most obvious conservation method. Currently, protected areas cover just over 5 per cent of the Earth's land area, although the proportion varies widely at the national level (Table 7.4), and not all of these areas are designated primarily for species conservation. In practice, some types of habitat are better protected than others, largely due to the fact that the actual designation of sites is often based more upon socio-economic and political factors than on conservation ideals. Biomes such as mixed mountain systems and island systems, which are typically not intensively used by humankind, are better represented than systems such as temperate grasslands and lakes. It comes as no surprise to learn, therefore, that the world's largest protected area is the Greenland National Park, which covers 972 000 km², and that the area proposed for the first World Park is Antarctica.

The types and quality of protection and management in these areas also vary greatly. A lack of political and financial support limits the success of many protected areas. The need to maintain levels of protection is well illustrated by the Operation Tiger reserves set up in several Asian countries in the early 1970s to protect the Asian tiger from local people faced with the loss of their domesticated livestock, and in some cases fellow villagers. Threats from the tiger have been enhanced in many areas due to the loss of its forest habitat to agricultural lands.

While Operation Tiger has long been hailed as a conservation success story, the story's most recent chapter has not been a happy one. Poor management and continued human encroachment into protected areas threatens Thailand's remaining tiger population, estimated to be around 250 (Rabinowitz, 1993). Russian experts think that between fifty and a hundred Siberian tigers were poached in 1992 in the Russian Far East, where poaching of all wildlife is rampant due to the breakdown in law and order following the demise of the USSR. The Siberian tiger population had fallen to less than a hundred in the early twentieth century, and a few years of poaching are rapidly negating a half-century of conservation efforts which had allowed the Russian population to approach an estimated 350. Poaching for skins and bone smuggling has also rapidly increased in the 1990s in the Indian subcontinent where over half of the remaining world tiger population of between 5000 and 7000 is located (Fig. 7.5). Many known tigers have disappeared in India and Nepal in the present decade and guards in Indian Operation Tiger reserves have been killed in clashes with poachers (IUCN, 1993).

Conflict between the aims of conservation and those of local people is a key issue in the threatened species debate since without local involvement in both the design and management of protected areas, adequate protection can only be achieved if the park agency has the authority and ability to enforce regulations. This is both undesirable and often unattainable, and this realisation represents a

TABLE 7.4 *Proportion of land areas protected by continent and in selected countries*

Country	Proportion of national area protected (%)		
	Total protection	Partial protection	All categories
WORLD	**3.04**	**2.12**	**5.17**
ASIA	**1.31**	**2.95**	**4.26**
Bhutan	1.41	18.03	19.44
Iran	1.81	2.76	4.57
Japan	3.54	9.07	12.61
EUROPE	**0.99**	**7.01**	**8.00**
Austria	0.00	24.93	24.93
Greece	0.59	0.20	0.79
UK	0.00	18.96	18.96
NORTH AND CENTRAL AMERICA	**7.03**	**3.79**	**10.82**
Cuba	3.79	2.20	5.99
Costa Rica	9.50	2.74	12.24
USA	4.10	6.38	10.49
SOUTH AMERICA	**3.35**	**2.61**	**5.96**
Chile	11.14	7.12	18.26
Surinam	0.53	3.96	4.49
Uruguay	0.08	0.09	0.17
OCEANIA	**7.98**	**1.94**	**9.91**
Australia	3.99	2.09	6.09
New Zealand	9.76	1.22	10.97
Tuvalu	0.00	0.00	0.00
AFRICA	**2.99**	**1.50**	**4.49**
Algeria	5.28	0.05	5.33
Botswana	15.28	2.15	17.43
Equatorial Guinea	0.00	0.00	0.00

Source: after WCMC (1992: 460–463, Table 29.6)

major recent shift in practical conservation philosophy.

Even in those areas which receive maximum protection, however, active conservation management is usually needed because often it is not sufficient simply to cordon off an area and leave it be. Such management is not always an easy task, however. A ban on hunting, for example, is theoretically easy to introduce, and grazing goats, sheep, pigs and even rabbits are relatively easy to control and even eliminate, but a threat from an introduced plant is much more intractable.

Effective management is also dependent upon a sufficient understanding of the ecology of the organisms in question, which is

FIGURE 7.5 *Tiger* (Panthera tigris) *in Rhanthambore National Park, an Operation Tiger reserve in northern India (courtesy of Mark Carwardine)*

not always the case. Although the gradual decline in British colonies of the large blue butterfly (*Maculinea arion*), a dry grassland species, was largely due to the fact that fifty of its former ninety-one sites were ploughed or otherwise fundamentally changed in the period 1800–1970 (Fig. 7.6a), its eventual extinction from Britain in 1979 was due to ignorance. Conservationists were aware of the large blue's highly specialised lifecycle: eggs are laid on *Thymus praecox* on which the larvae feed briefly before being adopted and raised by *Myrmica* spp. of ant. But the disappearance of the large blue from another forty-one sites, including four nature reserves where *Thymus* and *Myrmica* remained abundant, proved enigmatic. The realisation that only one species of *Myrmica* ant could act as host, and that regular heavy grazing was necessary to maintain appropriate soil surface temperatures came too late to save the large blue.

A similar pattern of events occurred with the heath fritillary (*Mellicta athalia*), a species which has declined as its woodland clearing habitat has disappeared and specially designated nature reserves were not adequately managed (Fig. 7.6b). The heath fritillary is still the most endangered British butterfly, but a clearer understanding of the make-up of its narrow niche may yet save it from extinction (Thomas, 1991). Likewise, the lessons learned over the loss of the large blue are being put to good use in recent attempts to reintroduce a similar subspecies from Sweden.

Bans on hunting and trade

Legislative bans on threatening activities have been widely implemented to safeguard certain species. In Russia, for example, hunting of the Saiga antelope during the nineteenth century (Fig. 7.7) pushed the

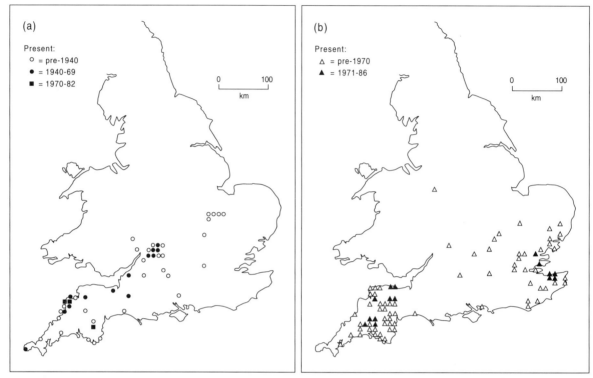

FIGURE 7.6 *Decline in distribution of butterflies in southern Britain: (a) the large blue (*Maculinea arion*); (b) the heath fritillary (*Mellicta athalia*) (after Heath* et al.*, 1984. Reproduced with modifications by permission of Penguin Books Ltd)*

SIBERIAN ANTELOPES.

FIGURE 7.7 *Saiga antelopes (*Saiga tatarica*) fleeing a hunter in Siberia. This engraving was made in the late 1800s when the Saiga was widely hunted for its meat, hide and horn. Driven to the brink of extinction, the species became protected in the 1920s, and numbers have since risen to more than 1 million*

species to the brink of extinction before being banned in the 1920s. Subsequently, controlled cropping has allowed the Saiga to recolonise most of its original range. Similarly, a ban on catching and selling of coelacanths has been introduced in the Indian Ocean island state of the Comoros as part of a conservation plan to protect the small population of this fish, first found in 1938, which had been previously thought to have become extinct 65 million years ago (Fricke and Hissman, 1990).

However, a ban on hunting is, in practice, like any legislation, only as effective as the ability of states to uphold it. If enforcement is weak, and incentives are large enough, poachers will continue to operate in spite of the law. The seemingly intractable problem of African elephant poaching during the 1980s inspired drastic conservation measures, with some countries implementing shoot-on-sight policies to deal with poachers in protected areas. But the placement of the African elephant on Appendix I of the Convention on International Trade in Endangered Species (CITES) in 1989 has dramatically reduced the world price of ivory, resulting in a downturn in poaching of the African elephant. Nonetheless, the listing on Appendix I was not welcomed by all conservationists. In the southern African countries of Botswana, Malawi, Namibia, South Africa and Zimbabwe, well-managed elephant herds are large and expanding, to the extent where they have to be culled to offset the risk of degradation in their ranges. The ban has meant that ivory from the culled animals cannot be sold to maximum profit.

Problems still occur with trade bans, and the case of the Old World fruit bats (or flying foxes) is an interesting one in this respect. The flying foxes constitute a single family (Pteropodidae) with forty-one genera and 161 species, which play an important ecological role among Pacific islands as pollinators and seed dispensers for hundreds of plant species, many of which have important economic and subsistence value to humans (Fujita and Tuttle, 1991). One of the central threats to the fruit bats is from hunters. Indigenous people on Guam have eaten them for at least 2500 years, but Guam's bat population was decimated in the years after the Second World War with the proliferation of firearms used in hunting. Many other south Pacific islands' fruit bat populations have subsequently suffered declines as exporters have supplied the Guam market. Listing on CITES Appendix I has had some success in curbing the threat to some of the species, but problems have arisen because of the lack of wildlife inspectors to uphold the law and the fact that several other islands in the region are, like Guam, effectively part of the USA, and therefore the trade is considered domestic rather than international (Sheeline, 1993).

Some conservationists argue that bans can only be a short-term measure to protect certain species. Long-lasting protection can only be envisaged if the people who demand certain products can be persuaded against them. Concerted publicity campaigns by animal rights groups have had considerable effect in this respect in certain European countries where ornamental furs are concerned.

Off-site conservation practices

While the conservation of species is best achieved by their maintenance in the wild, through protected area programmes and legislative measures, other practices may be necessary for species whose populations are too small to be viable in the wild or are not located in protected areas. Maintenance of species in artificial conditions under human supervision is a strategy known as off-site or *ex situ* conservation. The off-site approach includes game farms, zoos, aquaria and captive breeding programmes for animals – although in reality few of these are actively involved in the conservation of endangered species – while plants are maintained in botanical gardens, arboreta and seed banks.

In practice, off-site techniques complement on-site approaches in a number of ways, such as by providing individuals for research, the results of which can be fed back into management techniques in the wild. Perhaps the ideal way in which the maintenance of captive individuals of species can help in the biodiversity issue is by providing individuals which can be reintroduced into the wild.

Such programmes are expensive, logistically demanding and require long-term management and monitoring to assure and assess their success. Plant reintroductions are generally viewed as a high-risk strategy with uncertain indications of their long-term success (WCMC, 1992). Reintroduction of animals is also a difficult task, but one notable success has been the reintroduction of the Arabian oryx (*Oryx leucoryx*) to Oman, a species which is thought to have become extinct in the wild in 1972. Individuals kept in captivity in the Middle East and elsewhere have been released to the wild in Oman in batches through the 1980s and now seem to be established. A similar programme was initiated in Mongolia in 1992 when some Przewalski's horses (*Equus caballus przewalski*)

were taken to Mongolia from zoos and game parks in Canada, The Netherlands and Russia. The horses, which are extinct in the wild, will be kept under close observation in corrals for a period before being released. Given the long-term commitment and substantial funding that such programmes require, their contribution to the maintenance of species diversity can only be limited. Perhaps the greatest value of such programmes is in their symbolic and educational importance.

FURTHER READING

McNeely, J. 1988 *Economics and biological diversity*. Gland, IUCN. This book highlights the pragmatic reasons for conserving biological resources and focuses on the economic incentives for so doing.

Primack, R.B. 1993 *Essentials of conservation biology*. Sunderland, MA, Sinauer Associates. A comprehensive review of biodiversity, the threats it faces, its value and strategies for its conservation.

Wilson, E.O. 1992 *The diversity of life*. Harvard, Harvard University Press. This book covers the evolution and creation of biodiversity and human impact.

8

CLIMATIC CHANGE

Global climatic change due to increasing atmospheric concentrations of greenhouse gases has dominated the environmental agenda since the mid-1980s and has engendered considerable international political debate. There is no doubt that over the last 100 years or so, human action has significantly increased the atmospheric concentrations of several gases which are closely related to global temperature. It seems likely that these increased concentrations, which are set to continue to build up in the near future, are already affecting global climate, but our poor knowledge and understanding of the workings of the global heat balance make the current and future situation uncertain. Predictions of the climatic nature of the Earth into the next century, and the effects of potential climatic changes on other aspects of the natural and human environment, are thus tentative. However, many of the responses that could ameliorate the impacts of potential climatic change can be argued for on the grounds of their other benefits.

HUMAN IMPACTS ON THE ATMOSPHERE

Human activities inadvertently affect the workings of the atmosphere in numerous ways, in many cases with possible effects on climatic regimes (Table 8.1). Direct inputs of gases, small particles (aerosols) and heat energy can all affect the operation of climate on various scales. Emissions of aerosols and heat are responsible for local 'heat islands' around urban areas and the creation of photochemical smogs (see Chapter 14), and enhanced inputs of soil aerosols, particularly from agricultural areas in drylands, affect the radiation properties of the atmosphere, possibly resulting in

TABLE 8.1 *Possible mechanisms for inadvertent human-induced climate change*

DIRECT ATMOSPHERIC INPUTS
■ Gas emissions (carbon dioxide, methane, chlorofluorocarbons, nitrous oxide, krypton 85, water vapour, miscellaneous trace gases)
■ Aerosol generation
■ Thermal pollution

CHANGES TO LAND SURFACES
■ Albedo change (dust addition to ice caps, deforestation, overgrazing)
■ Extension of irrigation

ALTERATIONS TO THE OCEANS
■ Current alterations by constricting straits
■ Diversion of fresh waters into oceans

Source: after Goudie (1993b)

decreased rainfall locally (Bryson and Barreis, 1967). On larger scales, gas emissions are believed to be responsible for an enhanced global greenhouse effect (see below) and depletion of the concentrated layer of ozone present in the stratosphere.

Stratospheric ozone plays a key role in climatic processes through its capacity to absorb incoming solar ultraviolet radiation. This warms the stratosphere and maintains a steep inversion of temperature between about 15 and 50 km above the Earth's surface, affecting convective processes and circulation in the troposphere below. Just how the human-induced depletion of stratospheric ozone, which was first identified as a 'hole' over the Antarctic in 1984 (Farman *et al.*, 1985), may affect global climate is unclear. To date, most concern has focused on the possible ecological effects of increased ultraviolet radiation reaching the Earth's surface. Modifications to the rate of photosynthesis are likely to have significant impacts on many living organisms, reducing productivity in aquatic life such as plankton, and terrestrial plants. A direct impact on human health is also likely, through increases in the incidence of skin cancers.

A number of human modifications to the land surface are suspected of inducing more localised climatic effects by changing the reflectivity or 'albedo' of ground surfaces. Relatively long-lived albedo changes can be brought about by actions which severely modify natural vegetation covers, such as deforestation and heavy grazing, and similar effects have been noted when large amounts of dust particles blown from agricultural soils are deposited on glacier ice. Such albedo changes can also come about through natural processes, however. Drought can alter vegetation cover, and volcanic eruptions can eject huge quantities of dust into the atmosphere. Changes in surface albedo affect the amount of solar energy absorbed by a surface, and hence the amounts of heat energy the surface subsequently releases. This, in turn, affects

such atmospheric processes as convection and rainfall. In simple terms, a greater surface albedo results in a cooler land surface which, in turn, reduces convective activity and hence rainfall. Such effects may cause a positive feedback, as some researchers have suggested as an explanation for the prolonged drought period in the Sahel since the late 1960s (Charney *et al.*, 1975). In this case, widespread loss of vegetation cover caused by a combination of natural drought and overgrazing could be acting to prolong the drought through the positive feedback induced by albedo change. Warnings that similar feedbacks could operate in areas of widespread tropical deforestation have also been made (Potter *et al.*, 1975), although proving such hypotheses is by no means simple.

Direct human modifications to the hydrological cycle may also have climatic impacts. The rise in the area of irrigated agriculture in many parts of the world (see Chapter 5) and the creation of numerous large water bodies behind dams (see Chapter 13) both modify local albedos and enhance local evaporation and transpiration rates. Impacts on the nature and workings of the oceans might also have local climatic effects. Changes in coastal salinity due to modifications of river regimes, for example, alters the heat capacity of marine waters.

PAST CLIMATIC CHANGE

The Earth's climate has never been static and the human impact on climate has been minor relative to the large-scale perturbations brought about by natural processes. Instrumental or documentary records of climatic parameters are only available for a few thousand years at most, but even on such short timescales some marked variations in climate have been noted (see below). Our knowledge of Earth's climate over longer time periods is derived from 'proxy' indicators.

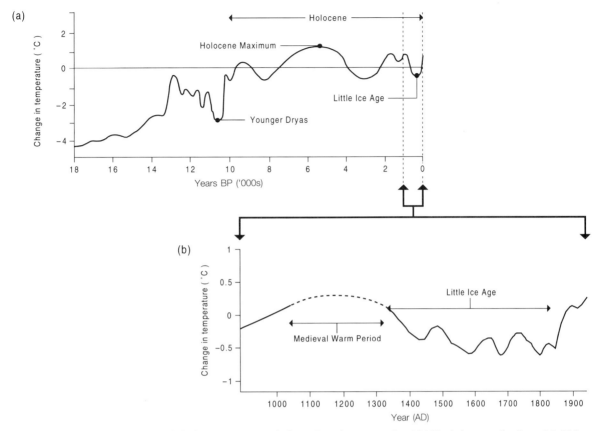

FIGURE 8.1 *Variations in global temperature (after Houghton* et al.*, 1990): (a) over the last 18 000 years; (b) over the last 1000 years*

The first evidence of the existence of former 'ice ages' came from glacial landforms and buried soils and fossils which indicated that climate in the Pleistocene (1 800 000–10 000 years BP) had alternated between cold glacials and warmer interglacials, with interglacial conditions in most mid- and high-latitude regions being similar to those of today. These lines of evidence have subsequently been supplemented with information gleaned from other sources, most notably deep-sea cores.

Theories developed to explain these changes in the Earth's climate have focused on variations in the amount of solar radiation received by the atmosphere. A composite sequence of these variations due to perturbations in the Earth's orbit around the sun was calculated in the early years of the twentieth century by a Serbian mathematician named Milankovitch. Three types of variation, operating with different periodicities were identified:

- orbital eccentricity with a periodicity of 100 000 years;
- axial tilt or obliquity with a periodicity of 41 000 years;
- precession of the equinoxes with a periodicity of about 21 000 years.

These periodic variations, which have been confirmed from deep-sea core evidence (Hays *et al.*, 1976), help to explain the occurrence of glacial and interglacial periods on Earth. Over

TABLE 8.2 *Atmospheric trace gases that are significant to global climatic change*

	Carbon dioxide (CO$_2$)	Methane (CH$_4$)	Nitrous oxide (N$_2$O)	Chlorofluoro-carbons (CFCs)	Tropospheric ozone (O$_3$)	Water vapour (H$_2$O)
Greenhouse role	Heating	Heating	Heating	Heating	Heating	Heats in air; cools in clouds
Effect on stratospheric ozone layer	Can increase or decrease	Can increase or decrease	Can increase or decrease	Decrease	None	Decrease
Principal natural sources	Balanced in nature	Wetlands	Soils; tropical forests	None	Hydrocarbons	Evapo-transpiration
Principal anthropogenic sources	Fossil fuels; deforestation	Rice culture; cattle; fossil fuels; biomass burning	Fertiliser; land use conversion	Refrigerants; aerosols; industrial processes	Hydrocarbons (with NO$_x$); biomass burning	Land conversion; irrigation
Atmospheric lifetime	50–200 years	10 years	150 years	60–100 years	Weeks to months	Days
Pre-industrial concentration (1750–1800) at surface (ppb)	280 000	790	288	0	10	Unknown
Present atmospheric concentration in parts per billion by volume at surface	360 000 360 000	1720	310	CFC-11: 0.28 CFC-12: 0.48	20–40*	3000–6000 in stratosphere
Present annual rate of increase	0.5%	0.9%	0.3%	4%	0.5–2.0%	Unknown
Global warming potential	1	11	270	3400–7100	—	—
Relative contribution to the anthropogenic greenhouse effect	60%	15%	5%	12%	8%	Unknown

*Northern hemisphere
Source: after Earthquest (1991)

the last 0.8–0.9 million years, for example, eight major glacial–interglacial cycles have occurred, each lasting about 100 000 years. The last glacial is generally thought to have been at a maximum about 18 000 years ago, and since then the Earth's climate has warmed (Fig. 8.1a). The most recent phase of the Quaternary is known as the Holocene, which began 10 000 years ago, and even during this period the Earth's climate has been by no means constant. Indeed, the last 2000 years have seen a relatively warm period during Medieval times and a so-called Little Ice Age during the fifteenth to eighteenth centuries (Fig. 8.1b). The end of the Little Ice Age coincided with the era when human activity began to have a significant impact on global temperatures through enhancement of the greenhouse effect.

GREENHOUSE TRACE GASES

The atmospheric warming caused by so-called greenhouse gases present in the atmosphere in trace amounts is a natural phenomenon caused by the effect of these gases being mainly transparent to incoming short-wave solar radiation but absorbant of re-radiated long-wave radiation from the Earth's surface. The most important greenhouse gases – carbon dioxide, methane, water vapour and nitrous oxide – all occur naturally in the atmosphere, and without their greenhouse properties the Earth's mean temperature would be about 33°C lower than at present (Houghton *et al.*, 1990). However, the atmospheric concentrations of some of these gases have increased dramatically over the last 100 years or so due to modifications of natural cycles by human populations, while new gases with greenhouse properties, notably chlorofluorocarbons or CFCs, have also been added. Table 8.2 summarises the main atmospheric gases that influence the operation of the greenhouse effect, many of

which also impact another atmospheric issue of potentially global significance: the destruction of the stratospheric ozone layer.

Most interest has focused on carbon dioxide as the most important greenhouse gas to have been increased by human action, atmospheric concentrations of which have risen by 25 per cent over the last 100 years, with about half of this increase occurring in the past 25 years. The burning of fossil fuels is the most significant source of human additions to atmospheric carbon dioxide (Fig. 8.2), while cement manufacture and land use changes – primarily deforestation – are also important. Global carbon dioxide emissions from fossil fuel combustion and cement manufacture have more than trebled since 1950 and these sources now release about 6 Gt of carbon a year. As Fig. 8.3 indicates, North America and Europe are by far the largest sources of these industrial emissions, although the trend in these areas has levelled off or fallen in the last decade, while emissions continue to increase rapidly in other areas, particularly in parts of Asia. On a national per capita basis, emission rates are closely tied to economic prosperity and closely related to energy use. The highest emission rates are from the oil-rich Gulf states

FIGURE 8.2 *Coal-fired power stations like this one at Didcot in southern England are major contributors to the human-induced increase in global atmospheric carbon dioxide concentrations*

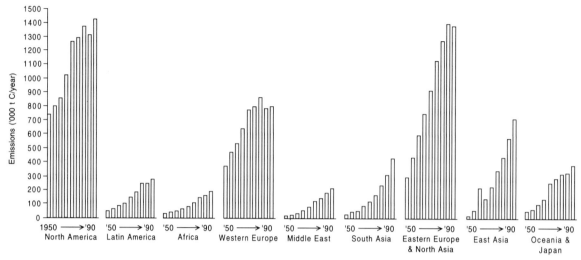

FIGURE 8.3 *Trends in carbon dioxide emissions from industrial sources in major world regions, at five year intervals, 1950–90 (after UNEP, 1993)*

(e.g. Qatar >10 t of carbon per year), while equivalent figures for the OECD member states average from 2–5 t/year and for many African countries they are < 0.5 t/year.

Several estimates of carbon dioxide emissions due to land use changes have been made, ranging from 0.6–2.6 Gt/year (Intergovernmental Panel on Climate Change, 1992). The considerable uncertainty reflects the problems of assessing deforestation rates in the tropics (see Chapter 3), which is thought to be the most significant source of increased carbon to the atmosphere from this activity.

The rapid rise in the atmospheric concentration of methane, which is today more than double its pre-industrial concentration, is linked to both industrialisation and increases in world food supply. Estimates of the global methane budget shown in Table 8.3 indicate that methane produced by anaerobic bacteria in the standing waters of paddy fields and the guts of grazing livestock (by so-called 'enteric fermentation'), particularly cattle, is almost equal to the total natural production, an output dominated by rotting vegetation in

wetlands. Like carbon dioxide emissions, industrial sources also contain an important fossil fuel element. Overall, anthropogenic sources are currently thought to contribute more than twice as much methane to the atmosphere as natural sources.

CFCs and other halocarbons are compounds which do not occur naturally. Their development, which dates from the 1930s, was for use as aerosol propellants, foam-blowing agents and refrigerants, and their release into the atmosphere has been inadvertent. Although national regulations on the use of CFCs in aerosol sprays during a time of economic recession in the developed world led to a fall in CFC emissions in the late 1970s and early 1980s, emissions climbed again subsequently as economies improved. While aerosol propellants accounted for almost 70 per cent of the market for CFCs in the mid-1970s, by the late 1980s refrigerants and foam-blowing agents accounted for 60 per cent of the market (McFarland, 1989). Concentrations of these compounds are far lower than those of other greenhouse gases, but the greenhouse warming properties of

TABLE 8.3 *Estimates of methane source strengths and sinks*

Sources/sinks	Best estimate (million tonnes per year)	Range (million tonnes per year)
SOURCES	**515**	
Natural	155	
Wetlands	115	100–200
Termites	20	10–50
Oceans	10	5–20
Fresh water	5	1–25
Methane hydrate	5	0–5
Anthropogenic	360	
Coal mining, natural gas and petrochemical industries	100	70–120
Rice paddies	60	20–150
Enteric fermentation	80	65–100
Animal wastes	25	20–30
Domestic sewage treatment	25	?
Landfills	30	20–70
Biomass burning	40	20–80
SINKS	**500**	
Atmospheric removal	470	420–520
Removal by soils	30	15–45
ATMOSPHERIC INCREASE	**32**	28–37

Source: Intergovernmental Panel on Climate Change (1992)

CFCs are several thousand times more effective than carbon dioxide. Although production of eight halocarbons, including the most abundant CFC-11 and CFC-12, has been severely curtailed by the Montreal Protocol adopted in 1987 (see below), their concentrations continue to rise as those already in use are released to the atmosphere.

GLOBAL WARMING

The theory relating increased atmospheric concentrations of greenhouse gases and global warming is strongly supported by evidence from ice-core data, which shows that natural fluctuations in the atmospheric concentrations of greenhouse gases through geological time have oscillated in close harmony with global temperature changes over the past 150 000 years, indicating that the two are almost certainly related (Lorius *et al.*, 1990). The change in global mean temperature since the mid-nineteenth century shown in Fig. 8.4, indicates that, overall, the planet has warmed at the surface by about 0.5°C over that period. Although it is possible that, in part, this recent warming trend reflects the end of the Little Ice Age (see Fig. 8.1b), most researchers suggest that the trend shown in Fig. 8.4 reflects the operation of an enhanced greenhouse effect due to human pollution of the atmosphere.

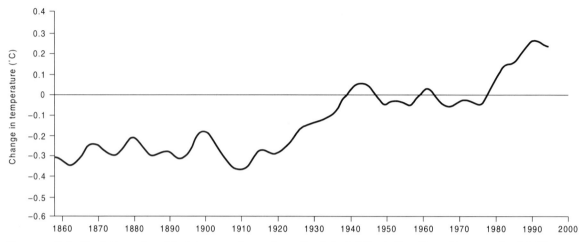

FIGURE 8.4 *Variations in global temperature since the late 1850s (after Houghton* et al.*, 1990)*

Nevertheless, the warming trend over the past thirteen or fourteen decades has not been continuous through either time or space. Two periods of relatively rapid warming (from the 1910s to the 1940s and again from the mid-1970s to the present) contrast preceding periods which were respectively characterised by fairly unchanging (1860s to 1900s) and slightly declining (1940s to 1970s) temperature. Spatially, too, global warming has been discontinuous: the two hemispheres have not warmed and cooled in unison and our understanding of the warming effects of enhanced greenhouse gases is further complicated by the observation that highly industrialised areas appear to be warming at a slower rate than less-industrial regions. The large emissions of sulphate particles to the atmosphere from industry appear to be retarding warming by reflecting solar radiation back into space (Charlson and Wigley, 1994).

Predicting impacts

The widespread acceptance of the fact that human action has increased the atmospheric concentrations of several greenhouse gases,

and the strong probability that these gases are acting to warm the Earth, has engendered a great deal of research into predicting the conditions of a warmer planet. Most of this research uses theoretical numerical models known as general circulation models, or GCMs, run on powerful computers. Higher mean temperatures will, of course, affect many other aspects of the climate, such as winds, evapotranspiration, precipitation and clouds, and by simulating the processes in the atmosphere, the GCMs can be used to predict changes in the distribution of these phenomena through both time and space.

Like all models, GCMs are simplifications of the real world and have numerous deficiencies (Henderson-Sellers, 1994). The models are slow to run, costly to use and their results are only approximate. They can predict changes in climatic conditions on continental scales, but predictions at finer resolutions, for a country the size of the UK, for example, are impossible at present. Part of the problem is the fact that we still do not understand fully all the processes of the climatic system, although we do realise its complexity. Different processes operate on different spatial scales and over varying timescales, and there are numerous feedbacks between different elements of the system which could enhance or dampen particular environmental responses to change. Present GCMs are poor at incorporating the processes in oceans, which are intimately linked to those of the atmosphere, and can only give a very simple representation of the influences of clouds, for example. Both of these elements may have key feedback effects which are poorly understood and inadequately modelled at present. Cloud cover, for example, is expected to be increased due to higher levels of atmospheric moisture consequent upon higher surface temperatures. However, whether greater cloud cover will amplify or dampen warming overall is unclear. Clouds reflect incoming solar radiation, so reducing warming on the ground, but they also transfer heat from their upper to lower surfaces, effectively acting like heat pumps (Maddox, 1990).

The performance of GCMs is tested against our knowledge of present and past climates. The models are constantly being improved, but there is still a long way to go. Our relatively poor ability to simulate the operation of the climatic system is illustrated by the fact that GCMs predict that global temperatures should have risen by about 1°C since about 1880, yet as Fig. 8.4 shows, they have only risen by half that amount.

Despite the many problems of GCMs, however, most people agree on perhaps the most important aspect of global climatic change from the viewpoint of human societies: the rate of change will be faster than anything previously experienced. This being the case, the approximate predictions produced by the GCMs are being used to gain some insight into the nature and conditions of the world that we will be inhabiting in the next few generations, since pre-industrial CO_2 levels are expected to have doubled sometime in the twenty-first century, with consequent significant changes in global climate.

The impacts

Since the atmosphere is intimately linked to the workings of the biosphere, hydrosphere and lithosphere, the projected changes in climate will have significant effects on all aspects of the natural world in which we live. The potential for disruption to socio-economic systems has been illustrated through history. The collapse of the Subir civilisation in northern Mesopotamia around 2200 BC, for example, has been linked to the stresses imposed by natural climatic change, largely in the form of increased aridity in this case (Weiss *et al.*, 1993). Shifts in the Earth's climatic zones over several hundred kilometres during the next 50–100 years would affect terrestrial ecosystems, with different species able to adapt or migrate with the changes to varying degrees of success. Such shifts would also affect patterns of agricultural land use, making the production of certain crops and livestock more, or less, suitable in the changed climate, which would not only affect climatic growing conditions but also the hazards from pests and diseases. One study of the impact on world food supply (Fischer *et al.*, 1994) suggests that a doubling of pre-industrial CO_2 levels would decrease world cereal production by up to 5 per cent, depending upon the climatic scenario, even assuming some adaptation to the changing conditions by farmers and allowing for the fact that greater atmospheric carbon dioxide will increase global plant productivity. The largest falls in output would occur in developing countries and this loss of production, combined with rising agricultural prices, is predicted to increase the number of people at risk from hunger in the order of 5–50 per cent depending on the climatic scenario.

Relatively small changes in climate can also influence the availability of water, particularly in semi-arid regions and more humid areas where demand or pollution has created water scarcity. In high latitudes, where global

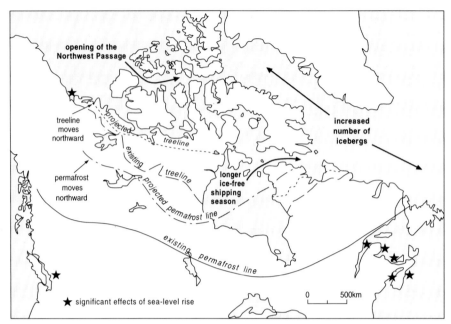

FIGURE 8.5 *Projected changes in northern Canada following climate warming (Slaymaker and French, 1993. Reproduced by permission of McGill-Queen's University Press)*

greenhouse warming is expected to be greatest, significant changes are predicted in glacial and periglacial processes, affecting glacier ice, ground ice and sea ice, which, in turn, would affect vegetation, wildlife habitats and human structures and facilities. The predicted pattern of these and other changes is shown for northern Canada in Fig. 8.5. There is a strong possibility that the Arctic Ocean's ice cover will disappear, facilitating marine transport and oil and gas exploration on the one hand, but also increasing the dangers from icebergs. The northward movement of the permafrost line would have many implications for roads, buildings and pipelines now constructed on permafrost, necessitating reinforcement and new engineering design and construction techniques. A serious feedback aspect of high-latitude permafrost melting is the consequent release of methane, a greenhouse gas.

Numerous other effects of global climatic change on geomorphological processes can be expected. These would occur through the direct effects of warming, and as a consequence of related changes in precipitation and temperature regimes and their impacts on such geomorphologically significant variables as vegetation. In many semi-arid areas, for example, soil moisture is predicted to be reduced by larger losses to evapotranspiration and decreased summer runoff, and increased rates of soil erosion by wind can be expected as a consequence. In many parts of the world, dry periods are normally associated with high frequencies of dust storms and this natural effect has been exacerbated in some cases by human disturbances to vegetation and soil surfaces (Goudie and Middleton, 1992). Many cases of such synergy between climatic change effects and pre-existing human impacts can be predicted and Goudie (1993c) gives a number of examples (Table 8.4).

One aspect of a warmer world which has received considerable attention is that of

TABLE 8.4 *Examples of synergy between climate change and other human impacts on geomorphology*

Phenomenon	Current human abuse	Potential global warming impact
Groundwater reduction in High Plains, USA	Overpumping by centre-pivot irrigation	Increased moisture deficit
Desiccation of Aral Sea and associated dust storms	Excessive irrigation offtake and interbasin water transfers	Increased moisture deficit
Permafrost subsidence	Vegetation and soil removal, urban heating, etc.	Warming
Coastal retreat	Sediment starvation by dam construction and coastal engineering structures	Sea-level rise
Coral reef stress	Pollution, siltation, mining, overexploitation	Overheating, more hurricanes, fast sea-level rise
Coastal flooding	Groundwater and hydrocarbon mining	Sea-level rise and more frequent storms

Source: after Goudie (1993c)

higher sea levels caused both by thermal expansion of the oceans and the added input from melting ice. As Schneider (1989a) suggests, sea-level rise will undoubtedly be the most dramatic and visible effect of global warming into the greenhouse century. Given the very large concentration of human population in the coastal zone (see Chapter 11), the consequences for low-lying and subsiding regions are potentially very severe. Although sea levels change under the influence of natural processes, the recent rise, which averages about 1.5 mm/year over the past 80 years in the Pacific (Fig. 8.6), may herald the early stages of a greenhouse rise which is predicted to be up to 1 m over the next century (Houghton *et al.*, 1990).

Increased flooding and inundation are the most obvious threats and many low-lying settlements on continental coasts, such as Bangkok, Lagos, London and Tokyo to name but a few, would be affected. A great number of islands which existed in the Pacific 18 000 years ago have since been drowned (Gibbons and Clunie, 1986) and a continued rise of just a few metres would greatly alter the present map of the central Pacific and Indian oceans. Several small island countries in the Pacific, such as Tokelau, the Marshall Islands, Tuvalu, the Line Islands and Kiribati, could cease to exist if worst-case scenarios for sea-level rise are realised (Pernetta, 1989). A wide range of

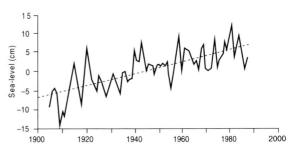

FIGURE 8.6 *Variations in sea level at Oahu, Hawaii, USA, 1905–90 (Nunn, 1994)*

TABLE 8.5 *Threats posed to the Maldives by projected global climatic change*

1.	Increased rates of coastal erosion and alteration of beaches with increased impacts from high waves
2.	Changes in aquifer volume associated with increased saline intrusion
3.	Increased energy consumption (e.g. air-conditioning)
4.	Coral deaths as a result of increased seawater temperature
5.	Accelerated interisland migration due to declining stability and habitability of islands
6.	Loss of capital infrastructure on smaller tourist resort islands
7.	Changes in reef growth and current patterns
8.	Increased vulnerability of human settlements due to aggregation and increasing size

Source: after Pernetta and Sestini (1989)

other physical and consequential socio-economic impacts can also be expected for other oceanic islands, as Table 8.5 illustrates for the Maldives.

Sea-level rise poses a particular problem to much larger, more populous low-lying countries such as Bangladesh. Inundation and salinity intrusion were identified as two of the most serious threats to the national economy (the other was drought) in a recent climatic change projection for the country which analysed the impacts under two scenarios:

- a lower limit of 2°C change associated with 30 cm sea-level rise;
- an upper limit of 4°C change associated with 100 cm sea-level rise.

Other significant primary physical effects were found to be increased incidence of cyclonic storm surges, flash floods and low water flow. The impact on agricultural production, for example, was predicted to be a fall in foodgrain production due to a combination of drought-induced stress on crops, caused by increased evapotranspiration, and lower water availability. Most development efforts planned for the coming three decades would be negated by sea-level rise and further threatened by the lack of a water sharing agreement with India. The study also highlighted the danger to the Sundarbans mangrove forest: the ecosystem's ability to adapt to the pace of projected sea-level rise is limited, not least by a lack of space inland (BCAS, 1993).

The threats from increased frequencies of tropical storms highlighted in the Bangladesh study is another serious facet of recent warming trends in tropical latitudes where tropical cyclones develop only in areas with sea surface temperatures exceeding 27–8°C. A rising trend in tropical cyclone frequencies for a number of areas of the Pacific Basin has been noted (Nunn, 1994). Tropical cyclones can cause severe disturbances to coastal and island ecosystems, quite apart from the damage to social structures, and recurrence intervals for particularly damaging storms may already have been substantially reduced in some areas of the Pacific, although evidence from other oceanic regions is less conclusive, as Walsh and Reading (1991) show for the Caribbean. Nonetheless, the extreme intensities of superhurricanes Gilbert (1988), Hugo (1989) and Andrew (1992), as well as the series of very severe gales in Europe in early 1990, are regarded by some as indications of global warming beginning to have a noticeable effect on extreme climatic events. The implications for society and the insurance industry are clear (Berz, 1991).

Responses to global climatic change

The formidable economic, social and political challenges posed to the world's governments and other policy-makers by impending global climatic change are unprecedented. Policy responses can be categorised broadly into those which aim to prevent change and those which accept the changes and focus on adapting to them. While the issue is a truly global one, since all greenhouse gas emissions affect climate regardless of their origin, the costs and benefits of measures to mitigate the effects of global climatic change are likely to be spread unevenly across countries. The issue raises important questions of international equity since, at present, the major proportion of greenhouse gas emissions comes from the industrialised countries which contain only about 25 per cent of the world's population. Developing-world leaders have called for reductions in emissions from the industrialised countries to make more of the planet's capacity for assimilation of greenhouse gases available to those countries that are industrialising now, a plan which should be facilitated by transfers of finance and technology from the North to the South (Tolba and El-Kholy, 1992). It has also been pointed out that some of the countries most at risk, the small island states, are effectively subsidising the economies of industrial countries which are net producers or exporters of carbon dioxide, since more natural carbon dioxide is fixed by the tropical rain forests, oceans, coral reefs and mangroves of small islands than is emitted locally from these islands (Pernetta, 1992).

Most countries have accepted the need to make some effort to prevent change, or at least to slow its pace, by reducing greenhouse gas emissions. A contribution has been made in this respect by the Montreal Protocol which was signed in 1987 and amended in 1990. Governments have committed themselves to reduce consumption and production of substances which deplete the stratospheric ozone layer, many of which also contribute to global warming. CFCs and other halocarbons are due to be phased out by the year 2000. Most attention has been focused on carbon

dioxide, however, an initiative that has a global forum in the UN Framework Convention on Climate Change. The Convention was signed by more than 150 governments at the Earth Summit in 1992 and came into force in March 1994 following the fiftieth country ratification. In effect, signatory states have agreed to reduce emissions to earlier levels, in many cases the voluntary goal being a reduction of carbon dioxide emissions to 1990 levels. The practical implementation of this pledge, however, will entail large financial investments to insure against future events which are far from certain. This being the case, the additional beneficial aspects of these and other mitigating policies are being widely promoted (Table 8.6). Realistically, however, no government is likely to sacrifice economic growth for reductions in carbon dioxide emissions, so that long-term strategies to reduce emissions must uncouple economic growth from growing fossil fuel consumption. Reducing the amount of energy used per unit of GDP will be one element in such a strategy, but there is also a need to shift away from fossil fuels to using more renewable energy sources (see Chapter 17).

Two scenarios based on projections for world energy demand are shown in Fig. 8.7. If fossil fuels continue to be used for more than 80 per cent of world energy, as they were in 1990 (the 'business as usual' scenario), carbon dioxide emissions are projected to more than triple by the year 2050, but the phasing in of more renewable energy sources is projected to bring carbon dioxide emissions back to 1990 levels soon after 2050. Even this scenario, which envisages the contribution of renewable sources to world energy demand to rise from 10 per cent to 60 per cent in 60 years, would represent an unprecedentedly rapid shift (World Bank, 1992a). It would also still entail a significant increase in atmospheric carbon dioxide concentrations.

The projections in Fig. 8.7 suggest, therefore, that warming is still likely to take place

TABLE 8.6 *Examples of 'no-regrets' initiatives to combat global warming*

Initiative	Effect on greenhouse gases	Other benefits
Energy conservation and energy efficiency	Reduced CO_2, methane and N_2O emissions	Conservation of non-renewable resources for current and future generations, reduction of other problems such as acid rain
CFC emission control	Reduced CFC emissions	Reduced stratospheric ozone layer depletion, reduced surface skin cancer and blindness
Tree planting	Increased biosphere carbon sink strength	Improved microclimate, reduced soil erosion, reduced seasonal peak river flows

Source: modified after Tolba and El-Kholy (1992: 77, Box 7)

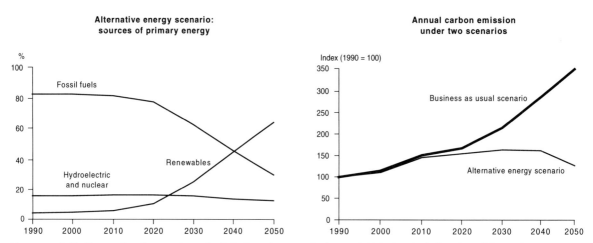

FIGURE 8.7 *Scenarios for projected global carbon emissions, 1990–2050 (after World Bank, 1992a. Reproduced by permission of The World Bank)*

in the next few decades, and thus a need to prepare for adaptive responses remains. An integral part of this strategy is to continue the funding of scientific research and data collection which can help to reduce the uncertainty surrounding possible climate change impacts. Nevertheless, the many question marks which remain over the nature, magnitude, patterns and timing of change make it necessary to emphasise again the 'no-regrets' aspects of preparing for adaptive responses: 'Developing alternative energy sources, revising water laws, searching for drought-resistant crop strains, negotiating international agreements on trade in food and other climate-sensitive goods – all these steps could also offer widespread benefits even in the absence of any climatic change' (Schneider, 1989b: 46).

FURTHER READING

Department of the Environment 1991 *The potential effects of climate change in the United Kingdom*. London, HMSO. A survey of possible impacts on many aspects from flora and fauna to industry and water supply.

Roberts, N. (ed.) 1994 *The changing global environment*. Blackwell, Oxford. This collection of papers contains three on climate change and several others on the effects of global warming on various aspects of the global environment.

Wyman, R.L. (ed.) 1991 *Global climate change and life on earth*. London, Chapman & Hall. A collection of papers which focuses on the greenhouse effect and its relationship to other environmental issues such as deforestation, overpopulation and hunger, pollution, sea-level change and biodiversity loss.

9

ACID RAIN

The effects of atmospheric acidity on biology and the urban fabric were first noted in England nearly 150 years ago when a Scottish chemist found that local rainfall in Manchester was unusually acidic. Smith (1852) suspected a connection with sulphur dioxide from local coal-burning factories and commented on the effects the deposits had on buildings and vegetation. It is Smith who is credited with the first use of the term 'acid rain'.

Interest in the phenomenon was awakened a century later when Gorham (1958) observed that air masses passing over industrial regions influenced the acidity of rain downwind over the English Lake District, and that the consequent acid deposition could affect the ecology of upland lakes. The effects of acidification on lake and river ecology in Sweden pushed the issue onto the international agenda in the 1960s when Swedish researchers suggested that much of the enhanced acidity of precipitation in Sweden was due to the long-range transportation of pollutants from other countries, including Britain (Odén, 1968). Although these claims were rejected by the British Government for many years, the concept of transboundary transportation of pollution is now widely accepted and the widespread and extensive acidification that has occurred in areas such as southern Scandinavia, northern Britain and parts of northeastern North America is generally agreed to have been chiefly caused by

atmospheric acidic deposition. Nevertheless, the effects of this deposition on various aspects of aquatic, terrestrial and built environments are still the subject of some controversies.

NATURE OF ACID RAIN

Acid rain is a misleading term. In practice, it refers to acidic deposits from the atmosphere both by wet deposition (rain, snow, sleet, hail, mist and dew) and dry deposition (by gravitational settling and through contact with surfaces). Hence, some researchers prefer the term 'acid deposition'.

Acidity is commonly expressed using the pH scale which reflects the concentration of hydrogen ions (H^+) in a solution. The scale ranges from 0 to 14, with a value of 7 indicating a neutral solution. Values less than 7 indicate acid solutions (e.g. lemon juice is pH 2.2), while values greater than 7 indicate basic solutions (e.g. baking soda is pH 8.2). The pH scale is logarithmic, so that a solution of pH 4 is ten times more acidic than one with a pH of 5.

The chemistry of the atmosphere is affected by numerous sources of gases and particulate matter which can alter pH. Volcanic emissions, fires, deflation of soil particles and various biological sources all affect atmospheric chemistry as part of the natural cycles of the Earth's matter, but precipitation

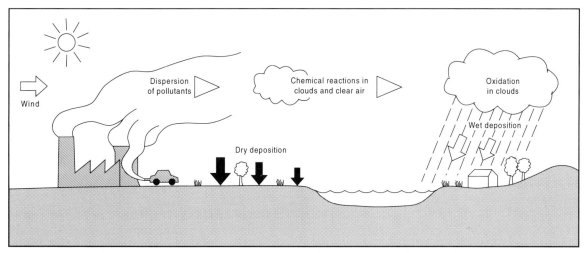

FIGURE 9.1 *Emission, transport, transformation and deposition of pollutants known as acid rain*

in an 'unpolluted' atmosphere commonly has a pH of 5.6. This slight acidity is due to the ubiquitous presence of carbon dioxide in the atmosphere which forms a carbonic acid. As with numerous forms of human-induced pollution, the acid rain debate centres around the role of human action in accelerating some of the pathways in the cycles of certain elements: primarily sulphur and nitrogen. Large quantities of oxides of sulphur and nitrogen are emitted into the atmosphere by the combustion of fossil fuels (coal, oil and natural gas) and industrial processes (princi-pally primary metal production), and it is these gases which are either deposited close to the source, as dry deposition, or are converted into sulphuric and nitric acids and deposited as wet deposition at distances up to thousands of kilometres from the source (Fig. 9.1). While these industrial sources are the largest emitters of acid rain pollutants, a range of other sources also contribute. One of the most widespread is agriculture, which makes important contributions in some areas when ammonia from nitrogenous fertilisers and manure escapes to the atmosphere by volatil-isation.

GEOGRAPHY OF ACID DEPOSITION

The absolute quantities of atmospheric sulphur and nitrogen produced by human action have risen dramatically over the last 100 years or so, to the point at which human-induced emissions are now of a similar order of magnitude to natural sources. However, the uneven global distribution of industrialisation means that the concentration of human-induced acid deposition is highly uneven. Data for the mid-1980s, for example, indicate human-induced sulphur emissions to the atmosphere to be about 60 per cent of the total, but over the industrialised regions of Europe and North America the human-induced flow is thought to be twelve times the natural flow (Bates *et al.*, 1992). At present, sulphur dioxide (SO_2) emissions are consid-ered to present the most serious problems in Europe and the northeastern region of North America, although in Japan and on the North American west coast, nitrogen oxides (NO and NO_2, which are collectively referred to as NO_x) are of greater importance.

Maps such as the one of North America shown in Fig. 9.2 are frequently cited to

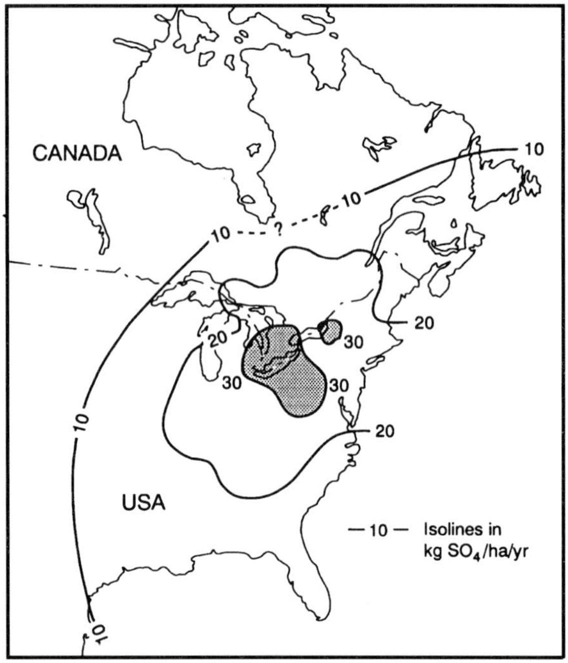

FIGURE 9.2 *Mean annual deposition of SO_4 over eastern North America for the period 1982–87 (after Tolba and El-Kholy, 1992)*

highlight the seriousness of the acid rain problem, and although such maps have been criticised for the way in which they generalise patterns and give no indication of changes through time (e.g. Kallend *et al.*, 1983) they do nonetheless give an impression of the scale of the problem. Changes in the patterns of emissions and deposition of sulphur in Europe have been described by Mylona (1993) who notes that areas of maximum emissions have shifted over the last century away from coalfields to oil-refining areas and regions with high densities of motor vehicles, and the traditional core areas of western and central Europe have been augmented by new sources in the north, east and south. In 1880, the maximum concentrations of SO_2 (4–7 g S/m³) were confined to the English Midlands, moving over to the western and central parts of the continent in the early twentieth century. Since the 1950s, maximum emissions are

found in a core region which comprises South Saxony and North Bohemia, where emission rates range from 25 to 40 g S/m³. Areas of maximum deposition have also shifted, from northwestern Europe to the German/Czech/Slovak/Polish border region where deposition of about 2 g S/m² in 1880 had risen to six times that figure by 1991. Deposition of sulphur over Europe for the period 1880–1991 is shown in Fig. 9.3, indicating that a broad swath from Wales to Poland has received more than 200 g S/m² over that period (Mylona, 1993).

While Europe and North America remain the world's most severely affected large-scale regions, an increasing body of evidence indicates the presence of smaller acid rain 'hotspots' in many other areas with concentrations of particularly polluting industry. Pape (1993), for example, cites several huge metal-smelting centres in Siberia responsible for large emissions of sulphur dioxide, and emissions of both SO_2 and NO_x are high in many of the world's largest urban areas (see Chapter 14). Indeed, while European emissions of sulphur decreased by 30 per cent between 1980 and 1992 (EMEP, 1993), emissions elsewhere have continued to rise, particularly in Asia where those from fossil fuel combustion doubled between 1970 and 1986 (Hameed and Dignon, 1992). Parts of Japan, North and South Korea, southern China and the mountainous portions of South East Asia and southwestern India are all currently vulnerable to acid rain (Bhatti *et al.*, 1992). Recent steady rises in NO_x emissions have also been recorded in all continents except North America and Europe where they have remained relatively constant.

Another important aspect of the acid rain issue is its movement across national political boundaries; indeed it has been noted previously that the problem first became an international issue when links between acid rain in Sweden were made with polluting industries in other parts of Europe. The long-range

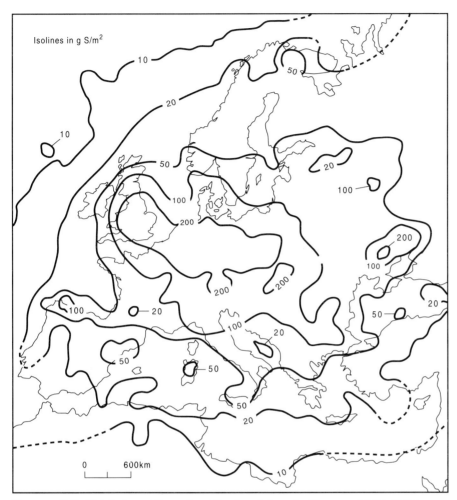

FIGURE 9.3 *Total deposition of oxidised sulphur in Europe, 1880–1991 (after Ågren, 1993, Acid News 5: 1–4)*

transportation of acid rain pollutants is a relatively recent phenomenon. In Britain, for example, reaction to increasing urban air pollution, including the infamous London 'pea souper' smogs of the early 1950s, resulted in the building of higher industrial and power generation smoke stacks in an attempt to combat local pollution problems. Higher chimneys were indeed successful in reducing pollutants locally, but in doing so they transformed a local pollution problem into a regional and international one.

The acid rain pollutant budget of the Scandinavian countries is illustrated for Norway and Sweden in Table 9.1 which has been compiled from data recorded as part of the European Monitoring and Evaluation Programme (EMEP). As the table indicates, only about 11 per cent of the sulphur deposited in Norway and Sweden in the early 1990s came from domestic sources, while significantly larger contributions were transported from Germany to the south and Great Britain to the south-west. The great distances

TABLE 9.1 *Sources of sulphur deposited (wet and dry) in Norway and Sweden, thousand tonnes per year in 1991/92*

Source	Annual deposition (thousand tonnes)
Germany	50
Great Britain	48
Sweden and Norway	33
Poland	21
Russia	15
Denmark	14
Czech and Slovak Republics	9
Ocean trade	7
Biogenic sea emissions	6
France	5
Finland	5
16 other countries	21
Indeterminate	76
Total	310

Source: data from EMEP (1993)

travelled by secondary pollutants which fall in acid precipitation is indicated by the fact that up to 10 per cent of the 76 000 t which cannot be attributed to any known source is thought to have been carried from North America.

EFFECTS OF ACID RAIN

The actual effects of acid rain on the Earth's surface have been, and continue to be, a subject for some debate. It is important to note, for example, that some environments have naturally acidic pH levels due to podsolic soils and peats, and that the ecological effects of acid rain depend very largely upon the ability of an ecosystem to neutralise incoming acids. This ability, the so-called 'buffering capacity', is largely determined by the nature of the bedrock:

environments on hard, impervious igneous or metamorphic rock, where calcium and magnesium content is low, are most at risk of acidification due to inputs of acid rain. Hence, attention has necessarily focused on areas with low buffering capacity where recent changes in acidification have occurred. Even in these areas, however, different degrees of acidification will have different effects and so the concept of 'critical loads' is now commonly used to determine the susceptibility of a particular environment. A critical load is a measure of the amount of acid deposition that an ecosystem can take without suffering harmful effects, and, to date, critical loads have been defined for soil and fresh water.

The fact that numerous ecosystems have experienced large decreases in their pH is now well documented. Some of the most convincing long-term evidence comes from work on lake sediments which analyses the species composition of fossil diatoms (single-cell algae which are sensitive to pH) found in sediment cores. Dramatic increases in acidity over the last 100 years or so have been documented from numerous lakes in North America, northern Britain and Scandinavia.

Linking cause and effect is still by no means straightforward, however, since other human impacts may also affect the pH of soils and water, for example. Changes in land use are particularly important here. Human-induced acidification of soils occurs through agricultural practices such as the liberal use of ammonia fertilisers and the production of forage legumes, and through afforestation practices, particularly where conifers are concerned. In South Africa, for example, soil acidification caused by afforestation may proceed faster than that due to the worst recorded acid rain on the eastern Transvaal highveld (Fey *et al.*, 1990). Long-term diatom data for Swedish lakes indicate the occurrence of a pH increase from about 5.5 to 6.5

dating from the Iron Age (*c.* 2500 BP). This alkalinisation was due to extensive burning, agricultural forestry, grazing and other agricultural practices which increased the base saturation and pH of soils, leading to increased transportation of base cations and nutrients from soils to surface waters. Following this period of alkalinisation, the pH of many lakes has fallen to about 4.5 this century and the cessation of many of these practices could account for some of this change (Renberg *et al.*, 1993). Nevertheless, these authors consider acid rain to be more important in producing the unprecedentedly low pHs this century, and in a similar vein, Patrick *et al.* (1990) have looked at the land use and management history around six Scottish lakes over the past 200 years and concluded that acidification can only be explained by acid rain.

Another facet of the acid rain debate stems from the fact that both sulphur and nitrogen are essential minerals for life, so that additional inputs can, in certain circumstances, be beneficial for organic growth. Moderate inputs of acid rain can stimulate forest growth, for example, in soils where cation nutrients are abundant and sulphur or nitrogen is deficient. Similarly, experiments on peanuts indicate that acid rain can increase the dry matter weight per plant and enhance economic yields (Chen *et al.*, 1990). A further arguably positive aspect of the large quantities of sulphate particles emitted to the atmosphere by human populations is the suggestion that they may reflect solar radiation back into space and, therefore, offset the atmospheric warming associated with the greenhouse effect over some parts of the Earth (Charlson and Wigley, 1994).

Nevertheless, despite the difficulties and provisos outlined above, there is a large body of evidence which implicates acid rain in numerous deleterious effects upon ecosystems, human health and the built environment.

Aquatic ecosystems

Much of the work on acid rain's impact on aquatic ecosystems has focused on fish, and numerous studies have documented the loss of fish from acidified waters. One survey of 1679 lakes in southern Norway, for example, showed that brown trout were absent or had only sparse populations in more than half the lakes, and the proportion of lakes with no fish increased with declining pH (Sevaldrud *et al.*, 1980). Different species of fish are affected to varying degrees by reduced pH levels: many cold water salmonids are more sensitive than other species, and among the salmonids, the rainbow trout is the most sensitive.

Fish have been affected both by gradual long-term changes in pH and the sudden 'acid shock' which follows the first heavy autumnal rains or the early spring snowmelt release of acids accumulated over winter. Another factor affecting fish mortality is the increased mobilisation of toxic metals by acid deposition. Hydrogen cations introduced into soils by acid rain may exchange with heavy metal cations formerly bound to colloidal particles, thereby releasing the metals into the soil and watercourses. Aluminium is a particular problem which affects fish by obstructing the gills. A gradual loss of fish stocks in many freshwater systems is often more a result of recruitment failure as opposed to outright mortality, as evidenced by higher proportions through time of more mature individuals. The loss of fish from the top of the food chain consequently has knock-on effects at lower trophic levels.

Other aquatic organisms are also affected by acidification. Battarbee (1994) notes that populations of algae, zooplankton and benthic invertebrates are all affected, and that at higher trophic levels, studies have documented the negative effects on amphibians (e.g. the UK's rare natterjack toad) and birds (e.g. the dipper). The sequence of biological changes caused by acidification monitored

TABLE 9.2 *Sequence of biological changes during experimental acidification of the Ontario Lake 223*

Year	pH	Changes observed
1976	6.8	Normal communities for an oligotrophic lake
1977	6.13	Increased chironomid emergence
1978	5.93	Fathead minnow reduced abundance; *Mysis relicta* near to extinction; further increased chironomid emergence
1979	5.64	*Orconectes virilis* recruitment failure; further increased chironomid emergence
1980	5.59	Lake trout recruitment failed; *Orconectes* population decline, parasitism increase, egg attachment failure
1982	5.09	White sucker recruitment failed; pearl dace declined; *Orconectes* near to extinction; chironomid emergence falls back to 1976 levels
1983	5.13	*Orconectes* extinct; mayfly (*Hexagenia* sp.) appears to be extinct

Source: modified after Howells, (1990: 163, Table 10.1)

in a Canadian lake deliberately dosed with sulphuric acid over an eight-year period is shown in Table 9.2. The crustacean *Mysis relicta* was an early casualty, while several species of fish were seen to decline and eventually disappear, and an increasing growth of filamentous algae was also noted at about pH 5.6.

Although there are few studies, to date, on acid rain effects on marine ecology, there is some evidence to suggest that increased leaching of trace metals from acidified areas can influence phytoplankton growth in coastal waters (Granéli and Haraldson, 1993).

Terrestrial ecosystems

Acid rain may damage vegetation communities in a number of ways:

- increasing soil acidity
- decreasing nutrient availability
- mobilising toxic metals
- leaching important soil chemicals
- changing species composition and decomposer micro-organisms in soils.

The widespread loss of lichens across many of Europe's industrial areas has been attributed to atmospheric pollution, chiefly by sulphur dioxide, to which nearly all lichens are very sensitive. The effects of acid rain on trees and forests has been an area of particular interest, and also controversy, since trees are subject to many natural stresses, including diseases, pests and climatic variations such as hard winters and high winds, as well as other forms of pollution (see Innes, 1992 for a comprehensive review). Hence, distinguishing the effects of acid rain from other potential causes of deteriorating tree health is by no means easy. The symptoms commonly associated with poor tree health which may be due, at least in part, to acid rain include thinning of crowns, shedding of leaves and needles, decreased resistance to drought, disease and frost, and direct damage to needles, leaves and bark.

A systematic monitoring programme of damage to European forests has been carried out since 1986. The programme measures defoliation according to five classes. In 1992, 22.2 per cent of all broadleaved species were

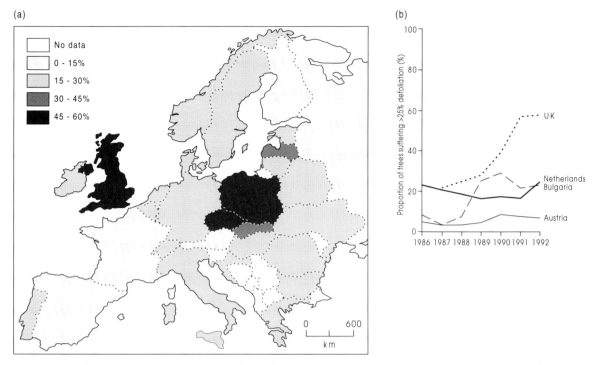

FIGURE 9.4 *Forest damage in Europe (from data in Elvingson, 1993,* Acid News *5: 8–9): (a) proportion of trees suffering >25 per cent defoliation by country (the data refer to all species except in Ireland and Russia where they refer to conifers only); (b) time series of defoliated trees in selected countries*

suffering at least 25 per cent defoliation, with 24.3 per cent of conifers suffering the same level of damage. This represents a marked increase in damage to broadleaved trees since 1986, but little change for conifers. The intensity of defoliation for all species by country is shown in Fig. 9.4a. Trends through the period 1986–92 are by no means obvious or constant for all countries (Fig. 9.4b), and country-scale data obscure regional variations, but overall there has been an increase in damage to Europe's forests over that period. The causes of damage include drought, wind, frost and disease, but everything points to air pollution as a contributing factor in those countries where the critical loads for many pollutants

are being markedly exceeded (Elvingson, 1993).

Such impacts can have knock-on effects for faunal species that depend upon forests for food and breeding. In the case of birds, Goriup (1989) suggests that while some species may face population declines and/or range contractions, others may benefit from the superabundance of dead and decaying standing timber. Some forest-dwelling birds may be affected more directly by the acidification of soils. The reproductive capacity of great tits and other species has been reduced in European forests as snail populations have declined on soils with falling pH, leading to a calcium deficiency in birds, which in turn, is manifested in a number of egg-shell defects (Graveland *et al.*, 1994).

Human health

The affects of acid rain pollutants on human health occur directly through inhalation, and indirectly through such exposure routes as drinking water and food contamination. Most air pollutants directly affect respiratory and cardiovascular systems, and high levels of SO_2 have been associated with increased mortality, morbidity and impaired pulmonary function, although distinguishing between the effects of SO_2 and other pollutants such as suspended particulate matter is not always easy. The best-documented effects relate to strong acids in aerosol form (i.e. sulphuric acid, H_2SO_4 and ammonium bisulphate, NH_4HSO_4) and the areas of major concern with respect to inhalation of these aerosols are exacerbation of the effects of asthma and the risk of bronchitis in all exposed persons. Inhalation of highly water-soluble acidic vapours such as SO_2 can also cause breathing difficulties, particularly in asthmatics.

The release of heavy-metal cations, such as cadmium, nickel, lead, manganese and mercury from soils into watercourses can eventually reach human populations through contaminated drinking water or fish. Such toxic heavy metals have been associated with a number of health problems including brain damage and bone disorders. Acidified water can also release copper and lead from water supply systems with similar adverse effects on human health. High levels of lead in drinking water, for example, have been linked to Alzheimer's disease, a common cause of senile dementia.

Materials

Acid deposition has been associated with damage to building materials, works of art and a range of other materials. In all cases, the presence of water is very important, either as a medium for transport and/or in its role in accelerating a vast array of chemical and electrochemical reactions and in creating mechanical stresses through cycles of solution and crystallisation.

Acid rain can accelerate the corrosion of metals and the erosion of stone. Sulphur compounds are the most deleterious in this respect, although rain made acidic to below a pH of 5.6 by CO_2, SO_2 or NO_x can dissolve limestone. Iron and its alloys, which are widely used in construction, are the among the most vulnerable construction materials. The metals are attacked by corrosion, particularly by SO_2. Atmospheric corrosion of iron and steel reached rates of up to 170 μm/year in the most heavily polluted areas during the 1930s, while rates in present-day Britain for plain carbon steels range from 20–100 μm/year (Lloyd and Butlin, 1992).

Limestone, marble and dolomitic and calcareous sandstones are vulnerable to SO_2 pollution since in the presence of moisture, dry deposition of SO_2 reacts with calcium carbonate to form gypsum, which is soluble and easily washed off the stone surface. Alternatively, layers of gypsum can blister and flake off when subjected to temperature variations. Soluble sulphates of chloride and other salts can also crystallise and expand inside stonework, causing cracking and crumbling of the stone surface. Such processes have caused particularly noticeable damage to statues in many urban and industrial areas (Fig. 9.5). The results of one study, in which weathering rates were measured on more than 8000 century-old marble tombstones across North America, suggest that SO_2 pollution was probably responsible for more deterioration than other weathering processes (Meierding, 1993). Upper stone faces in heavily polluted localities with mean SO_2 concentrations of 350 g/m^3 receded at a mean rate >3 mm/100 years due to granular disintegration induced

FIGURE 9.5 *Acid rain damage to a statue in one of the courtyards of the Louvre in central Paris, France*

by the growth of gypsum crystals between calcite grains.

Although modern glass is little affected by SO_2, medieval glass which contains less silica and higher levels of potassium and calcium can be weakened by hygroscopic sulphates which remove ions of potassium, calcium and sodium. Paper is affected by pollutants of SO_2 and NO_2 because these are absorbed, making the paper brittle, and fabrics such as cotton and linen respond to SO_2 in a similar way. Damage to materials such as nylon and rubber have also been attributed to SO_2 as well as low-level ozone.

COMBATING THE EFFECTS OF ACID RAIN

Working on the premise that prevention is better than cure, the most obvious method for combating the effects of acid rain is to reduce the emissions of sulphur and nitrogen oxides from polluting sources. There are numerous pathways that can be taken towards achieving this aim. Reducing emissions from power stations can be approached in a number of ways. Energy conservation is the simplest and arguably the most sensible method of reducing emissions from the burning of fossil fuel, while increasing the use of other energy sources also helps, although such alternatives also come with an environmental price (see Chapter 17). Alternatively, there are many existing technologies available which can reduce acid rain pollutant emissions from power stations and other industrial sources. These include:

- fuel desulphurisation, which removes sulphur from coal before burning;
- fluidised bed technology, which reduces the SO_2 emissions during combustion; and
- flue gas desulphurisation, which involves removing sulphur gases before they are released into the air.

Similar approaches can also be taken to reduce emissions of NO_x from power stations, while catalytic converters and lean-burn engines have been applied to NO_x emissions from motor vehicle engines (see Chapter 15).

Political aspects of emissions reduction

Many of these actions are costly in economic terms, however, and the uncertainties that have surrounded some of the suggested effects of acid rain have been the reason for some governments' unwillingness to initiate expensive acid rain reduction programmes.

Britain has been a case in point here. The Scandinavian countries were successful in persuading the UN Economic Commission for Europe to set up a Convention on Long-Range Transboundary Air Pollution (CLRTAP) which was formally signed by thirty-five states in 1979. Its members include all European countries, the USA and Canada. However, when the Convention's protocol on sulphur emissions reduction was adopted in 1985 – the '30% Club', so-called because the members agreed that their sulphur emissions in 1993 would be at least 30 per cent less than their 1980 levels – Britain, amongst others, refused to join. The case is an interesting one in the light that it shines upon the interactions between politics and science.

In the early 1980s, Britain became increasingly isolated politically in its reticence to take the issue of transboundary transportation of acid rain pollutants seriously enough to take preventative action. Initially, an influential ally was found in the former West Germany, but domestic pressure turned the Germans in favour of sulphur reductions following widespread fears over the deteriorating health of German forests: so-called 'waldsterben'. While other European countries, particularly those convinced that their ecology was being adversely affected by pollution from their neighbours, were clamouring for concerted action, Britain's response was to call for more research (Dudley, 1986). The British Government's logic was understandable in that action to reduce emissions would cost a great deal of money, and therefore they had to be convinced that the money was being spent correctly. Evidence which was convincing enough for some countries was not convincing enough for others. Other countries that refused to join the 30% Club, such as Spain and Poland, also did so primarily for economic reasons. In 1986, however, Britain did concede the need to reduce SO_2 emissions and began a planned introduction of flue gas desulphurisation equipment to a number of large British power stations.

Irrespective of the political tensions engendered by the 30% Club, its approach had inadequacies, both absolutely and relatively. In absolute terms, many believed that reductions need to be much greater to make any significant improvements, while it was also recognised that reductions needed to be more specifically targeted at the large emitters and at the worst-affected environments. Hence, the critical load idea was adopted for the new sulphur emission protocol, signed in 1994, to replace the 30% Club. The new protocol marks a significant milestone in international pollution control since it is the first time that different targets have been set for each country. The targets take the form of maximum permissible emissions per target year which are based upon the ability of the environment to withstand pollution.

Some of the largest emitters of sulphur and nitrogen are shown in Table 9.3. This table illustrates the success of several industrialised countries in reducing sulphur dioxide emissions over the period 1980–90. Decreases have been achieved in many cases through pollution control strategies, energy conservation and fuel switching. Even greater reductions have been achieved in other countries, such as Sweden, where sulphur dioxide emissions have been cut from 507 000 t in 1980 to 102 000 t in 1992 by a combination of regulatory measures and the introduction in 1988 of a sulphur tax levied on combustion plants. The effect on atmospheric concentrations in many cities in the more-developed countries is illustrated by the trend shown for Tokyo in Fig. 9.6. Worrying trends are illustrated for China and India, where emissions increased by half over the decade, with China now the second largest emitter of sulphur dioxide after the USA, although in per capita terms China and India are still low on the comparative scale.

TABLE 9.3 *Changes in national emissions of sulphur dioxide and nitrogen oxides in selected countries, 1980–90*

Country	Annual emissions (thousand tonnes*)		Change 1980–90 (%)	Per capita 1990 (kg)
	1980	1990		
SULPHUR DIOXIDE				
Canada	4643	3800	−18	143.3
USA	23 400	21 200	−10	84.7
Poland	4100	3210	−22	83.5
Russia	12 123	10 166	−16	68.2
UK	4898	3774	−23	66.3
China	13 370	19 990	50	17.7
Japan	1600	1140	−29	9.2
India	2010	3070	53	3.7
NITROGEN OXIDES				
USA	20 400	19 800	−3	79.8
Canada	1959	1943	−1	73.3
UK	2442	2690	10	47.3
Russia	2578	3050	18	20.5
Japan	2130	1940	−9	15.7
China	4910	7370	50	6.5
India	1670	2560	53	3.1

*Note: some '1990' data are for previous years, but the 1980–90 change percentage is calculated according to the growth rates from the data available. Emissions for nitrogen oxides are given as nitrogen dioxide equivalents
Source: UNEP (1993: 44–48, Tables 1.8 and 1.9).

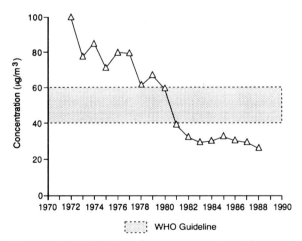

FIGURE 9.6 *Decline of mean annual sulphur dioxide concentration at Omori-Minami, an industrial city-centre site in Tokyo, 1972–88 (after UNEP/WHO, 1992)*

Recent trends for nitrogen oxides are less encouraging, with little change over the period in North America and significant increases in the UK and Russia. China and India have again increased emissions by half over the decade. Motor vehicles are a major source of nitrogen oxides and larger emissions may reflect increases in national fleets.

Environmental recovery

Even given the actual and planned reduction in acid rain emissions, the response of affected environments is not necessarily known or guaranteed. There is some evidence for fairly rapid recovery following emissions control, such as that reported for lake fauna downwind of the Sudbury metal smelters in

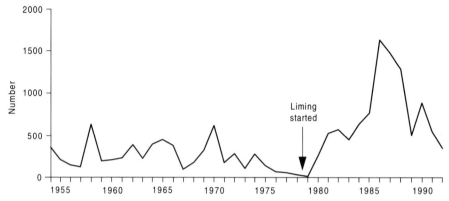

FIGURE 9.7 *Counts of upstream migrating salmon in the River Högvadsån, south-west Sweden, 1954–92 (after Fleischer et al., 1993)*

Canada after a 50 per cent reduction in sulphur emissions (Gunn and Keller, 1990), but not all organisms may respond as immediately. In many instances there is likely to be a lag time between emissions reduction and a detectable reduction in environmental effects, the so-called 'memory effect' as far as building stone weathering is concerned (Cooke, 1989).

In the case of acidified lakes, some researchers believe that it might take more than 100 years before fresh waters regain something close to their original species composition with maintained functions. This view emphasises the continued need for complementary local measures such as appropriately managed land use and the continued implementation of liming programmes designed to increase water pH. Liming is carried out in several European countries and in North America. The most comprehensive operation is that in Sweden, where a large-scale freshwater liming programme initiated in 1977 had treated about 6000 lakes up to 1992 (Fleischer *et al.*, 1993). The programme has been chemically successful and in many cases

considerable biological improvements have resulted, as the Atlantic salmon counts for a river in south-west Sweden, shown in Fig. 9.7, indicate. In this river, salmon counts rose as upstream migration increased after liming in the late 1970s; in other rivers and lakes, fish populations may require help by stocking.

FURTHER READING

Howells, G. 1990 *Acid rain and acid waters.* London, Ellis Horwood. A concise and comprehensive critical review of the acid rain issue, particularly the effects on water, soils and vegetation.

Mason, B.J. 1992 *Acid rain: its causes and effects on inland waters.* Oxford, Clarendon Press. This book focuses on acidification of lakes and streams and the effects on aquatic animals. It draws largely on experimental and modelling work conducted in the UK and Scandinavia.

Park, C.C. 1987 *Acid rain: rhetoric and reality.* London, Methuen. An analysis of acid rain problems, their scientific complexities, technical solutions and political dimensions.

10

OCEANS

Oceans cover nearly 71 per cent of the Earth's surface and contain 97 per cent of the planet's water. They play a fundamental role in the Earth's climatic system and marine ecosystems have long provided people with fish and other resources, as well as a receptacle for wastes. The perception of oceans as being infinitely vast has tended to undermine concern over their increasing exploitation by human society, despite their economic and ecological importance. Although public outcry focuses on some marine issues, such as the killing of whales and pollution from major oil spills, more pervasive threats, from other pollutants and overfishing, for example, are seldom given such attention. However, consideration of our relatively poor understanding of how oceans work, of the population dynamics of many marine biota and of the effects of certain pollutants, suggests that the arguments for caution in our use and abuse of oceans are strong.

FISHERIES

Global fish production is humankind's largest source of either wild or domestic animal protein, and is particularly important in the developing countries. The fisheries industry expanded rapidly over the last 40 years from around 22 million tonnes in 1950 to 100.3 million tonnes in 1989, falling slightly over the following 3 years. The large majority of these landings came from marine fisheries and were destined for human consumption.

As with terrestrial food crops, we are heavily reliant upon a very limited number of the 13 000 or so marine fish species. Herrings, sardines and anchovies made up 25 per cent of landings in 1989. Continuous growth in global catches throughout the 1980s was largely based on increased landings of these shoaling pelagic (free-swimming) species because most demersal (bottom-dwelling) stocks were, and still are, fully fished.

Many marine biologists suspect that the currently exploited fishery resources around the world are now close to their maximum sustainable catch limits, and many show signs of biological degradation. Overfishing of demersal species in the North Atlantic, for example, is thought to have first occurred 100 years ago in the case of North Sea plaice. The 1980s saw a significant increase in the levels of national and international controls designed to ensure the conservation and sustainable exploitation of fish stocks. The establishment of state jurisdiction zones up to 200 miles (322 km) from national coastlines, so-called Exclusive Economic Zones, has been a critically important development in facilitating this move, since 99 per cent of marine fisheries are currently taken from within this limit. But international cooperation is still needed in many cases simply because fish are no respecters of national zoning (Fig. 10.1). The North Atlantic herring, for example,

FIGURE 10.1 *European fishermen, like these in Portugal, have become subject to increasing restrictions on their activities in an effort to reduce pressure on overexploited fish stocks*

which was overfished in the North Sea in the 1960s and 1970s, spawns in British and French coastal waters, their planktonic larvae drift to shallow nursery grounds off The Netherlands, Germany and Denmark and the young herring join the adult population in feeding grounds in British and Norwegian controlled waters (Fig. 10.2). A ban on fishing between 1977 and 1982 is thought to have allowed the population to recover to pre-1960s levels.

Identifying the impacts of overfishing is, however, a complex task for fisheries managers. The location of the world's marine fisheries is governed principally by the distribution of floating plants on which they depend for food, and the most important factor determining phytoplankton production is the supply of nutrients, which is greatest in areas of upwelling. Climatic circulations can greatly alter the pattern of ocean circulation, however, and hence the pattern of fisheries. It is often difficult to distinguish between the impacts of overexploitation and the natural variability of stocks, dependent upon species biology, migratory habits, food availability and natural hydrographic factors. Even with a relatively stable and resilient stock like the North Sea herring, a fishery which has been studied as much as any other comparable marine resource, the relative role played by environmental factors in the 1970s' collapse is still not clear (Bailey and Steele, 1992). The difficulties are greater still with some other shoaling pelagic species such as anchoveta and sardine, the populations of which are notoriously difficult to assess and manage, making their fishing a high-risk business (Csirke, 1988).

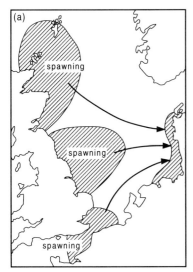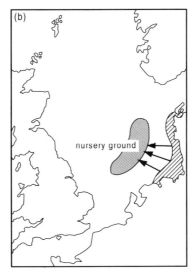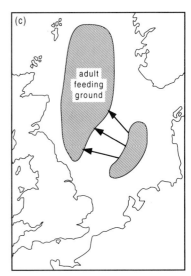

FIGURE 10.2 *Movement of North Sea herring: (a) after hatching from eggs, planktonic larvae drift from three main spawning grounds to shallow waters off Germany and Denmark; (b) young herring swim offshore to the nursery ground; (c) young herring join the adult population in their feeding ground (after Meadows and Campbell, 1988)*

FIGURE 10.3 *Catch history of the Peruvian anchoveta, 1955–89 (from data in FAO Yearbooks of Fisheries Statistics, Catchings and Landings, Rome, FAO)*

One of best-known examples of dramatically fluctuating population numbers is the Peruvian anchoveta of the southeastern Pacific. Disruption to population biomass occurs in some years due to the El Niño phenomenon, which intermittently leads to the near total collapse of coastal fisheries, a natural variability which is thought to have been exacerbated by overfishing.

The Peruvian anchoveta was the most important single fishery in the world in the 1960s, providing up to 10 million tonnes of animal protein a year. Annual catches rose from a minor 59 000 t in 1955 to peak in 1970 at just over 13 million tonnes (Fig. 10.3). Just prior to 1972, the fishery was so intense that few of the fish caught were more than 2 years old, and many biologists expressed concern about the dangers of overfishing. The 1972 El Niño event caused a reduction in recruitment and subsequent intense exploitation of the remaining stock. Continued heavy fishing after 1972 further reduced the stock. By the late 1970s, annual catches had fallen to less than a million tonnes and directed commercial fishing was abandoned. In 1984, after another intense El Niño event in 1982/83, the landed catch was just 94 000 t. Stocks appear to have collapsed due to the combination of short-term fluctuations caused by El Niño events combined with intense fishing pressures (Caviedes and Fik, 1992). The landed tonnage, however, was growing again in the late 1980s.

Similarly enormous differences between maximum and minimum catches occur for other shoaling pelagic species: the Japanese sardine, for example, appears to be affected by changes in the Kuroshio current that last for periods of one or more decades. Natural variability may also have contributed to the dramatic collapse of the California sardine fishery, which boomed in the 1930s and 1940s but had almost completely disappeared by 1950. Once thought to be a classic example of the effects of overfishing, work by Soutar (1967) has thrown up an alternative explanation. The 2000-year history of species abundance reconstructed from fish scales found in sea bed sediments suggests that irregularities in the abundance of sardine in this area were common over the period, with decades of almost total absence not unusual.

The spasmodic population dynamics of the Peruvian anchoveta, Japanese sardine and California sardine exhibit one of four types of fish population dynamics described by Caddy and Gulland (1983):

- steady-state
- cyclical
- irregular
- spasmodic.

While the human impact upon steady-state species is relatively easy to detect and management of their exploitation fairly straightforward, irregular and spasmodic stocks are particularly difficult to manage and do not lend themselves to current ideas of sustainable yields on a year-to-year basis.

Southern Ocean

Ignorance of the workings of the marine ecosystem makes life difficult for management of ocean resources, and no more so is this the case than in the Southern Ocean surrounding Antarctica. The Southern Ocean has been the site of several phases of species depletion caused by overexploitation. In the nineteenth century, seals were pushed to the edge of extinction for their fur and oil. They were followed by the great whales in the first two-thirds of this century (see below), and in the 1970s and 1980s Southern Ocean finfish stocks were heavily exploited with the marbled Antarctic rock cod and mackerel icefish showing particular signs of decline. Most recently, attention has turned to Antarctic krill, the most abundant living resource of the Southern Ocean. Commercial fishing of the small shrimp-like crustacean began in the late 1970s, but annual catches have not risen above the peak of over half a million tonnes in 1980–81 (Fig. 10.4).

The reason for the relatively modest krill catches is the establishment of a limit on harvesting by the Commission for the Conservation of Antarctic Marine Living Resources (CCAMLR), an international body set up in 1981 to manage the sustainable use of Antarctic marine living resources in the light of the devastation of seal and whale populations and the depletion of some finfish. Since krill are at the centre of the Antarctic ecosystem, providing the major food source for many larger animals such as seals, fish, squid, seabirds and baleen whales, there was

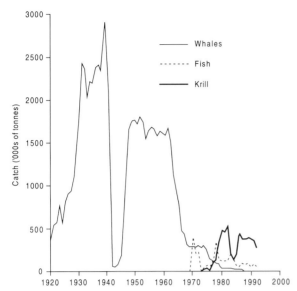

FIGURE 10.4 *Annual catch of whales, finfish and krill in the Southern Ocean, 1920–92 (after Nicol and de la Mare, 1993)*

a fear that overharvesting of krill might lead to the collapse of the entire ecosystem.

The CCAMLR's job is severely hampered by a basic lack of knowledge and information on the workings of the Antarctic ecosystem in general, and of krill in particular. While most estimates of a sustainable krill harvest are well in excess of the current estimated annual sustainable catch of all other species from the world's oceans (100 million tonnes), making it a potentially huge resource, these estimates are very rough and are based on some rather large assumptions. As Nicol and de la Mare (1993: 38) put it: 'Biologists do not have any reliable estimates of how many krill there are, nor a sense of exactly where they live and how their local distribution is governed. We do not know where they spend the winter, nor the age structure of the population'.

WHALING

Although whales have been hunted for thousands of years, modern commercial whaling for oil, and in time also for meat, is traced back to the eleventh or twelfth centuries when groups of Basque villagers began to hunt right whales from small rowing boats in the Bay of Biscay. Whaling had become big business by the eighteenth and nineteenth centuries, and large whaling fleets from many nations, including England, North America, Japan and The Netherlands, made enormous profits from their sorties.

Commercial whaling has focused on the largest, most profitable species. These are the so-called 'great whales', of which there are ten species: the blue, fin, humpback, right, bowhead (or Greenland), Bryde's, sei, minke, grey and sperm whales. The history of commercial whaling has seen intensive catches of a species until it has been driven to the brink of extinction, at which point the whalers shifted their attention to a different species in a new area of ocean. The populations of most of the great whale species are now so low as to be effectively extinct from a commercial viewpoint; about 2 million whales have been killed this century alone, and several species could ultimately disappear altogether.

The Greenland bowhead whale, for example, was first exploited in 1610 and reduced to the brink of extinction by the mid-nineteenth century. The Arctic bowhead whales, discovered in the mid-nineteenth century and quickly exploited, were almost totally depleted by 1914. As many as 30 000 sperm whales were killed in the Atlantic during peak years in the eighteenth and nineteenth centuries, and by the 1920s the Atlantic sperm whale was practically extinct.

It was early in the twentieth century that whalers began to exploit the previously untouched waters of the Southern Ocean where several species go to feed on krill. The catch rates were further increased with the introduction of factory ships in the 1920s. The humpback was the first species targeted in the Antarctic waters, but as its numbers declined,

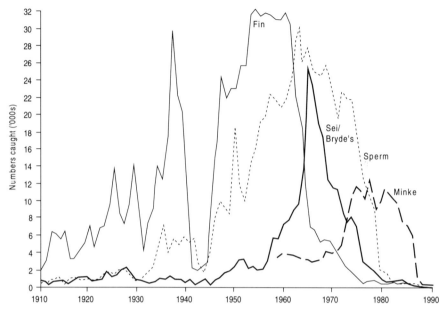

Figure 10.5 *World annual whale catch, 1910–91 (from data in UNEP, 1987 and 1993)*

the blue whale became the main quarry. An estimated 250 000 blue whales were quickly killed, leaving around 1000 today. In the second half of this century it was the turn of the fin and sperm whales. Annual catches of fin whales regularly topped 30 000 in the 1950s (Fig. 10.5).

Protecting whales

There have been a number of attempts to control the whaling industry over the years. The earliest effort was the Convention for the Regulation of Whaling, drawn up in 1931. But the convention had little beneficial effect. One of its regulatory instruments, the so-called blue whale unit, effectively lumped all great whales together, so that one unit was made up of successively greater numbers of smaller species: two fin, three humpback, five sei, and so on. Countries were spurred by the profit motive to fulfil their quotas with a few large species rather than more smaller ones, resulting in a rush on the blue whales. In 1946, this

convention was replaced by the Whaling Convention, which established the International Whaling Commission (IWC) to discuss and adopt regulations. The IWC has had a chequered history, overseeing as it did the all-time peak in whaling, in the 1960–61 season, when 64 000 whales were caught, but after increasing pressure from outside, the IWC declared a moratorium on all commercial whaling, beginning in 1985.

The ban has not been a complete success, since several countries have not abided by it, claiming that whaling should be continued for scientific purposes. Japan, Norway, and Iceland have been some of the most reluctant nations to comply with the ban. Conservationists have argued that effective non-lethal techniques for studying whales have been developed, which do away with the need to butcher the object of the scientists' enquiries. It may come as no surprise to learn that the meat and oil of the whales killed for scientific research is still processed and sold in the normal way. At the time of writing,

Norway and Japan continue to defy the ban. Conservationists have also voiced fears for the future viability of the moratorium based upon Japan's powerful position as an aid donor. It is not inconceivable that Japan could use aid as a lever to persuade developing-country IWC members to back the Japanese position.

Various attempts have been made outside the IWC to dissuade the countries who continue to kill whales. Amendments in US fisheries law designed to bolster IWC regulations have been used by conservation organisations. The amendments enable the US Government to take action against countries that diminish the effectiveness of the IWC, by blocking imports of their fisheries products or by denying them access to fish within US waters. In the mid-1980s, a group of conservation bodies took legal action against the Commerce and State departments because they feared that sanctions would not be imposed against Japan, who was planning a large sperm whale hunt after the species had received official protection. The case eventually reached the US Supreme Court in 1986. The conservationists lost, and in the meantime Japan had caught their 400 sperm whales.

Other actions by conservation organisations have proved more successful. Greenpeace estimated that a boycott of products from Iceland's fishing industry in 1988–89 cost Iceland £30 million in lost orders. This action contributed to Iceland's abandonment of scientific whaling in late 1989. In 1993, Greenpeace conducted a similar boycott campaign against Norwegian products. The conduct of this campaign saw a new development in such action when Greenpeace asked its UK members to shop at Safeway stores on a particular weekend since Safeway had come out strongly against Norway's resumption of commercial whaling.

The most recent approach to persuading the few nations who still catch whales to stop their whaling was strongly advocated to IWC delegates in 1994. Organised whale watching tours were pushed vigorously as a viable and sensible alternative way of 'using' these marine mammals. Whale watching is clearly a more sustainable method of exploiting whales than simply killing them, and the income generated from whale watching 'ecotourists' can make the organisation of these tours very profitable in economic terms (Fig. 10.6).

Commercial whale watching began in the mid-1950s in southern California where migrating grey whales were the attraction. Since then it has become a major tourist industry. In 1993, about 4 million people went on whale watching trips worldwide. Global ticket sales totalled £44 million in 1992, and with the addition of accommodation, food and other essentials the total revenue generated reached nearly £200 million (Carwardine, 1994).

In the areas of the world where whale watching is well organised, ships carrying ecotourists abide by strict regulations designed to cause minimal disturbance to the subjects of their searches. Engines must be cut within a certain distance and no vessel is allowed to approach too close. Many of the trips are made on research ships (non-lethal research, that is), with the income from carrying tourists helping to finance the research itself. In some parts of the world, old whaling vessels themselves are used, a development which offsets one of the arguments proposed in favour of continuing the killing of whales: its importance for employment.

Whale-watching tours have even started in Japan, the first setting out from Tokyo for the Ogasawara Islands in April 1988. Despite stern opposition from both government officials and whalers, Japan now has one of the world's fastest growing whale-watching industries. In 1992, nearly 20 000 people paid over £0.5 million to join these Japanese tours. If the trend continues, it will not be long before live whales in Japanese waters are worth more in economic terms than dead ones. It is a trend which the IWC hopes to

FIGURE 10.6 *A boat-load of whale watchers following two humpback whales in Massachusetts Bay off the north-east coast of the USA. Whale watching is being promoted as a viable alternative to whaling (courtesy of Mark Carwardine)*

encourage with its backing of whale watching as a sensible, more humane way of treating these creatures.

POLLUTION

The wide variety of sources, causes and effects of human-induced pollution of the marine environment is indicated in Table 10.1, while an indication of the varying persistence of marine pollutants is given in Fig. 10.7. Since land-based sources tend to be dominant, the worst effects of these pollutants are concentrated in coastal waters adjacent to areas with large population densities, and in similarly placed regional seas where the mixing of waters is limited. The open ocean, by contrast, is to date little affected. In this chapter, the focus is on pollution and its effects outside the coastal zone, while coastal pollution is covered in Chapter 11, although in practice

the distinction is somewhat arbitrary since coastal pollutants also reach the wider context of the high seas and oceans.

The Group of Experts on the Scientific Aspects of Marine Pollution (GESAMP), an international group sponsored by the UN, noted the planet-wide distribution of some pollutants, but gave the open oceans a clean bill of health in their most recent report:

> The open ocean is still relatively clean. Low levels of lead, synthetic organic compounds and artificial radionuclides, though widely detectable, are biologically insignificant. Oil slicks and litter are common along sea lanes, but are, at present, of minor consequence to communities of organisms living in open-ocean waters.
>
> (GESAMP, 1990: 1)

Others, however, warn that the lack of ecological studies in the open ocean cloud our understanding of the effects of pollutants

TABLE 10.1 *Primary causes and effects of marine pollution*

Type	Primary source/cause	Effect
Nutrients	Runoff approximately half sewage, half from upland forestry, farming, other land uses; also nitrogen oxides from power plants, cars, and so on	Feed algal blooms in coastal waters. Decomposing algae depletes water of oxygen, killing other marine life. Can spur toxic algal blooms (red tides) releasing toxicants into the water that can kill fish and poison people
Sediments	Runoff from mining, forestry, farming, other land uses; coastal mining and dredging	Cloud water. Impede photosynthesis below surface waters. Clog gills of fish. Smother and bury coastal ecosystems. Carry toxicants and excess nutrients
Pathogens	Sewage; livestock	Contaminate coastal swimming areas and seafood, spreading cholera, typhoid, and other diseases
Persistent toxicants (e.g. PCBs, DDT, heavy metals)	Industrial discharge; wastewater from cities; pesticides from farms, forests, home use, and so on; seepage from landfills	Poison or cause disease in coastal marine life. Contaminate seafood. Fat-soluble toxicants that bioaccumulate in predators
Oil	46% runoff from cars, heavy machinery, industry, other land-based sources; 32%, oil tanker operations and other shipping; 13%, accidents at sea; also offshore oil drilling and natural seepage	Low-level contamination can kill larvae and cause disease in marine life. Oil slicks kill marine life, especially in coastal habitats. Tar balls from coagulated oil litter beaches and coastal habitat
Introduced species	Several thousand species in transit every day in ballast water; also from canals linking bodies of water and fishery enhancement projects	Outcompete native species and reduce marine biological diversity. Introduce new marine diseases. Associated with increased incidence of red tides and other algal blooms
Plastics	Fishing nets; cargo and cruise ships; beach litter; wastes from plastics industry and landfills	Discarded fishing gear continue to catch fish. Other plastic debris entangles marine life or is mistaken for food. Litters beaches and coasts. May persist for 200–400 years

Source: Weber (1994: 47, Table 3.3)

there. Weber (1994) suggests that there is reason to believe that contaminants may have an effect disproprotional to their concentration, since chemical pollutants tend to amass in surface waters where larvae, eggs and micro-organisms concentrate: 'heavy metals can be 10 to 100 or more times more concentrated near the surface than in the waters below, and pesticide residues can be millions of times stronger' (Weber, 1994: 50).

Shipping and atmospheric transport are the main sources of pollution to the open oceans.

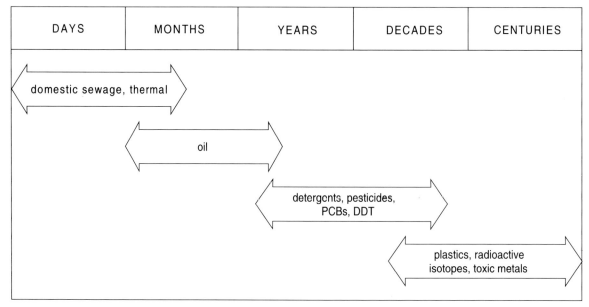

DAYS	MONTHS	YEARS	DECADES	CENTURIES

FIGURE 10.7 *Persistence of pollutants in the marine environment (after Meadows and Campbell, 1988)*

Inputs from shipping arise mainly from operational activities and from deliberate discharges. Oil is the most obvious pollutant involved, but biocides from anti-fouling paints leach into ocean water along shipping routes, and recently the accumulation of plastic debris from ships has been on the increase. Dumping and accidents also contribute.

As Table 10.2 shows, the number of oil spills which occur during normal operations is roughly equal to those from accidents, although accidental discharges tend to involve larger quantities of oil. International maritime regulations governing the operational inputs of oil to the oceans have been responsible, in part, for the decline in the number of major spills (>700 t) during the last 20 years. Individual accidental discharges do, however, have significant local impacts. The blow-out at the Ixtoc 1 drilling platform off the Mexican coast in 1979, for example, continued for 10 months and released more than 400 000 t of oil, producing a slick which spread across the Gulf of Mexico and contaminated beaches in Texas.

The dumping of wastes into the oceans is controlled by the London Dumping Convention adopted in 1972, which has been largely responsible for the gradual decrease in the amounts of industrial wastes and sewage sludge dumped at sea since it came into force. A similar agreement for the north-east Atlantic, including the North Sea and the Baltic, is the Oslo Convention, adopted in 1974. Dumping of industrial and hazardous wastes in the North Sea has been banned since the end of 1989, and is scheduled to cease in other marine waters by the end of 1995 under the London Dumping Convention. In the North Sea, under the Oslo Convention, sewage sludge is currently disposed of only by Ireland and the UK but is due to stop completely by 1998.

The Oslo Convention members also agreed, in 1988, to phase out the incineration of wastes at sea by 1994, and the London

TABLE 10.2 *Marine oil spills by cause and magnitude over the period 1974–92*

	Number of spills by magnitude (t)			
	<7	7–700	>700	Total
Operations				
Loading/discharging	2708	254	13	2975
Bunkering	540	23		563
Other operations	1143	42		1185
Accidents				
Collisions	134	177	71	382
Groundings	207	159	86	452
Hull failure	523	58	29	610
Fires and explosions	142	10	21	173
Other	2181	157	43	2381
Total	7578	880	263	8721

Source: UNEP (1993: 367)

Dumping Convention contracting parties have agreed, in principle, to discourage this activity. The Oslo decision was not based upon observed harmful effects but on conviction that adequate means would be available for land disposal. GESAMP suggest that on the available evidence:

> [T]he environmental implications of incineration at sea cannot be regarded as more significant than those of incineration on land. There are limitations and risks associated with either choice. In practice, the choice will depend on a careful analysis of technical, social and political as well as environmental factors.
>
> (GESAMP, 1990: 18).

The oceans are an attractive place to dispose of wastes due to their vast diluting capacity, but the long residence times of compounds within them does imply that any contaminants will be removed only very slowly (Jickells *et al.*, 1990). Indeed, the dilute and disperse philosophy becomes questionable for persistent pollutants which take very long periods to become harmless (e.g. radionuclides, heavy metals, certain pesticide residues) and which can be accumulated up the food chain. The dumping of high- and medium-level radioactive wastes at sea does not occur and the dumping of low-level packaged radioactive waste at sea was voluntarily banned in 1982. During the period 1949–82, the UK and the USA made most use of the twenty-six sites used in the Atlantic and twenty-one sites in the Pacific. The total amount dumped was less than 0.1 per cent of the amount that reached the oceans due to atmospheric nuclear weapons testing between 1954 and 1962. This, in turn, is only 1 per cent of the amount found naturally in the oceans. However, the mix of radionuclides involved, their rates of decay and the effects upon biological life are different in each case, and GESAMP (1990: 50) warns that 'dumping cannot be considered safe just because releases of radionuclides are small compared with the natural incidence of radionuclides in the environment'.

A similar statement might be applied to the disposal of low-level liquid waste into the oceans, from nuclear reactors and fuel reprocessing plants, particularly since the discharges are made from point sources. In some parts of the world, these discharges have been made continuously over several decades; the nuclear facilities at Sellafield in northwestern England, for example, have been discharging low-level liquid radioactive waste into the Irish Sea since the early 1950s. Such discharges are carefully monitored and limits are set with the aim of maintaining bioaccumulation in organisms well below international standards for various radionuclides. Nevertheless, discharges of nuclear waste to marine and other environments continues to be a source of considerable controversy (see Chapter 17). It is also worth noting that nuclear material for disposal is in many cases derived from military activities, and information on this source of radioactive waste to the oceans, as with other disposal options, is on the whole not available.

Black Sea

The ecosystem of the Black Sea and the adjoining Sea of Azov has been deteriorating since the late 1960s as a result of pollution, principally from land-based sources. The result has been eutrophication and contamination by pathogenic microbes and toxic chemicals which has, in turn, had severe repercussions for fisheries and the sea's amenity value (Mee, 1992).

A major part of the Black Sea is now critically eutrophic following a large increase in nutrient loads entering the sea over the last 25 years. Much of the increase has come from increased loads in rivers, reflecting the widespread use of phosphate detergents and agricultural intensification (see Fig. 12.1). The River Danube alone delivered 60 000 t of total phosphorus and 340 000 t of total inorganic nitrogen per year in the early 1990s. In recent

years, the volumes of nutrients from coastal communities' sewage has probably become comparable to the riverine loads. The northwestern portions of the sea are particularly affected, and have been changed from a diverse ecosystem supporting highly productive fisheries to a eutrophic plankton culture leading to a marine environment which is unsuitable for most higher organisms. The effects of nutrient inflows have also been traced to formerly pristine sites on the central and eastern coasts where macrobenthic species have declined significantly since the 1960s (Fig. 10.8).

A number of other factors have contributed to the catastrophic changes wrought on the Black Sea. Withdrawal of water following dam construction on many rivers in the Black Sea watershed has resulted in a substantial net decrease in runoff reaching the sea (Table 10.3). The resulting increase in salinity has had a serious affect on anadromous fish catches in the Sea of Azov, which fell by 95 per cent between 1950 and 1975. Data on chemical and microbiological pollutants are scarce but heavy metals, pesticides, radionuclides and oil are all thought to have suppressed biota. Oil spills, increased trawl

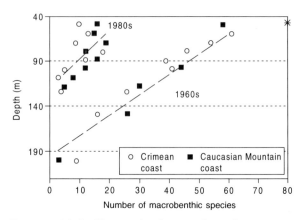

FIGURE 10.8 *Change in the number of macrobenthic species as a function of depth in the Black Sea (after Zaika, 1990)*

TABLE 10.3 *Reduction in discharge of some of the rivers flowing into the Black Sea*

River	Natural flow (km²/year)	Reduction (%)		
		1971–75	1981–85	1991–2000*
Don	27.9	19	27	43
Kuban	13.4	39	49	65
Dnieper	53.5	24	52	71
Dneister	9.3	20	40	62

*Estimated
Source: Tolmazin (1985)

fishing and invasion by opportunistic settler species have also contributed to the worsening situation. From the socio-economic perspective, the Black Sea fishing industry, which supported some 2 million fishermen and their dependents, has suffered an almost total collapse and the attractions of the sea's beaches have been severely depleted.

Steps towards dealing with the ecological problems have been made with the signing of a legal Convention for the Protection of the Black Sea and the adoption of a plan of action in 1992. There is still a need for basic data for input into designing the very large investments which will be needed to control pollution, and there is no doubt that rehabilitation will take many years.

Other regional seas

Like the Black Sea, many other regional seas are showing signs of degradation, in many cases with land-based sources being most prevalent. In the Mediterranean, for example, at least half of the nutrient, mercury, lead, chromium, zinc, pesticide, and radioactive pollutants enter from rivers draining areas outside the immediate coastal zone (Grenon and Batisse, 1989). Projected increases in the Mediterranean's population indicate that the pollution loads entering the sea will continue to increase, with the greatest rises in population likely to occur in the poorer North African and Middle Eastern countries which border the sea.

In contrast to the relatively well-developed international agreements governing marine sources of pollution, such as the London Dumping Convention and the Oslo Convention, agreements to limit problems which emanate from coastal landmasses are conspicuous by their absence. Some progress has been made in this direction, however, in the form of a series of action plans for regional seas, coordinated by UNEP (Table 10.4).

TABLE 10.4 *UNEP Regional Seas Programme*

Regional sea area	Action plan adopted
Mediterranean	February 1975
Gulf	April 1978
West/Central Africa	March 1981
South-east Pacific	November 1981
Red Sea	February 1982
Caribbean	April 1981
Eastern Africa	June 1985
South Pacific	March 1982
East Asia	October 1981
Black Sea	April 1992
South Asia	In preparation
North-west Pacific	In preparation

The Mediterranean was the first regional sea for which such a plan was formulated. The future well-being of the Mediterranean depends principally on land use planning, urban management and pollution prevention, and these are central aspects of the plan, but the difficulties of organising successful cooperation between the nation states involved – twenty-one border the Mediterranean – have been serious. There is no doubt that international cooperation is vital for the management of threats to the marine environment, and thus such initiatives need to succeed.

FURTHER READING

Clark, R.B. 1992 *Marine pollution*, 3rd edn. Oxford, Clarendon Press. A comprehensive study of the causes and consequences of marine pollution by pollution type.

Cuyvers, L. 1984 *Ocean uses and their regulation.* New York, Wiley. This book covers food, minerals, waste disposal, transport, energy and the law of the sea.

GESAMP (Group of Experts on the Scientific Aspects of Marine Pollution) 1990 *The state of the marine environment*. Oxford, Blackwell Scientific. A comprehensive assessment of current and future marine pollution threats.

Gulland, J.A. (ed.) 1988 *Fish population dynamics*, 2nd edn. Chichester, Wiley. A collection of papers on our developing knowledge of natural variations in the oceans' fish stocks.

11

COASTAL PROBLEMS

The coastal zone, which can be defined as the region between the seaward margin of the continental shelf and the inland limit of the coastal plain, is among the Earth's most biologically productive regions and the zone with the greatest human population. It also embraces a wide variety of landscape and ecosystem types, including barrier islands, beaches, deltas, estuaries, mangroves, rocky coasts, salt marshes and seagrass beds. Coral reefs and coastal wetlands are some of the most diverse and productive ecosystems, and 90 per cent of the world's marine fish catch, measured by weight, reproduces in coastal areas (FAO, 1991). More than 50 per cent of the world's population lived in coastal regions in 1970, most in large urban centres. These numbers have risen in recent years, and in 1990 it was estimated that about 60 per cent of humanity, about 3 billion people, lived in the coastal zone, and two-thirds of the world's cities with populations of 2.5 million or more are near estuaries. Within the next 20–30 years the population of this zone is expected to double (IUCN/UNEP/WWF, 1991).

This continuing increase in density of human occupation will inevitably put further pressures on coastal resources, and as Walker (1990: 276) puts it:

> [V]irtually every aspect of the coastal zone, whether it be space, climate, beauty, animal or plant life, minerals, or water, must be considered a resource. Thus everything humans do in the coastal zone (and many of the things they do outside it) alters the resource base.

PHYSICAL CHANGES

The natural operation of the processes of erosion and sedimentation affects human use of coasts by constantly changing the configuration of coastlines, effects which are superimposed upon the larger-scale changes in sea level relative to land, due to crustal movements and changes in the overall volume of marine waters. Sedimentation by the Büyük Menderes River into the Gulf of Miletus on Turkey's Aegean coast over the last 2400 years, for example, has today left the ancient port of Haracleia 30 km from the coast (Bird, 1985). Conversely, erosion of Britain's Humberside coastline between Bridlington and Spurn Head has removed a 3-km strip of coastline, with the loss of thirty villages, since Roman times (Fig. 11.1). These natural processes which affect the physical make-up of coastlines have also been modified in numerous ways by human populations. In Japan, for example, more than 40 per cent of the country's 32 000-km coastline is influenced by human action, through land reclamation, urban and port facilities or

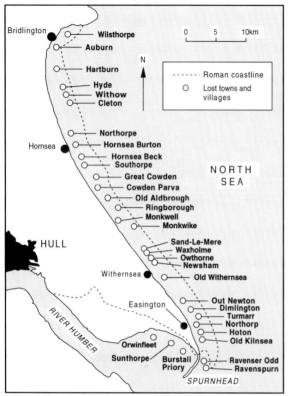

FIGURE 11.1 *Extent of coastal erosion since Roman times in the Humberside region of north-eastern England (after Willson, 1902 and Murray, 1994)*

by shoreline protection associated with lines of communication (Koike, 1985).

Deliberate efforts to control the sediment budget on certain portions of coasts are often carried out in an attempt to maintain a beach which is useful for recreation and/or coastal protection. However, interference with any portion of a coast, whether deliberate or inadvertent, will inevitably have impacts elsewhere through its effect on the coastal processes of erosion and deposition of unconsolidated sediments such as beach sand. Hence, the effects of groynes built for beach protection are often similar to those caused by other structures, such as breakwaters and jetties, built for alternative purposes. The

frequent outcome is illustrated in Fig. 11.2, which shows significant modification to Kuta beach and erosion damage to a resort area on the Indonesian island of Bali following the construction of the Denpasar airfield which extends into the ocean. Likewise, the completion in 1907 of two moles, to protect Lagos harbour from sedimentation, has resulted in a mean erosion rate at Victoria beach to the west of 6 m/year and a mean sedimentation rate at Lighthouse beach to the east of 8 m/year (UNEP/UNESCO/UN-DIESA, 1985). Similar effects can be caused by the construction of a sea-wall to protect a cliff, which may result in depletion of a beach elsewhere that was fed by the eroding cliff.

Other activities in the coastal zone can have equally significant effects on beaches, not least the removal of sands and gravel for construction purposes. In the northern region of Grand Comore in the Indian Ocean, whole beaches are disappearing due to collection of sand for cement making (UNEP, 1984).

It is not just activities in the coastal zone itself which have an impact there. River outlets also supply sediment to the coasts, so that human modification of river regimes can have a downstream impact. Changes in land use, such as cultivation, can increase sediment loads, as in the case of many of Taiwan's rivers and the Huang Ho River in China, which transports an order of magnitude more sediment than it did prior to widespread cultivation of the loess plateau (Milliman, 1990). Conversely, numerous major rivers have had their sediment budgets severely depleted, commonly by the construction of dams. The Colorado River and the Nile no longer deliver any sediment to the coast; the effects in the latter case are illustrated in Fig. 13.7. The absolute reduction in discharge due to dam construction can also have severe repercussions for coastlines in inland seas. The completion of a dam to facilitate agricultural, industrial and municipal withdrawals across the mouth of the Kora-Bogaz-Gol embayment,

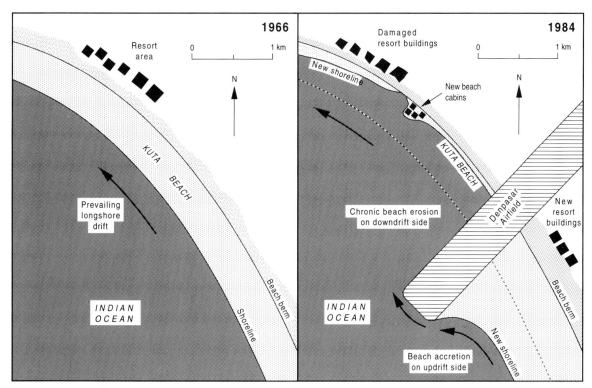

FIGURE 11.2 *Changes to Kuta beach on the Indonesian island of Bali caused by the construction of an airport runway out to sea (after Nunn, 1994)*

an area of high evaporation on the Turkmenistan coast of the Caspian Sea, is thought to have contributed to a 2.7 m decline in sea-level in the Caspian between 1930 and 1977, exposing more than 50 000 km² of former sea bed. However, a subsequent rise of about 1.5 m by 1990 (Fig. 11.3), suggests that climatic variations influencing fluvial discharge into the Caspian also contributed.

The risk from marine floods and other natural hazards has also inspired many modifications to coastlines around the world. In western Europe, the storm surge hazards posed to the low-lying coastlines of the North Sea have been combated by a number of engineering solutions. In The Netherlands, the Zuider Zee scheme (1927–32) is an early example which involved a 32-km-long, 19-m-high dam, closing off the vast estuary of the Zuider Zee. In addition to flood protection,

closure of the estuary also facilitated land reclamation and improvements to freshwater resource management.

In 1953, when North Sea surge floods drowned 1853 people and seriously damaged 50 000 buildings in The Netherlands, the Delta Project was set in motion to protect the country's southwestern coastal area. It provides for the closing off of some of the estuaries discharging into the North Sea – the Rhine, Maas and Scheldt – and three closures have been built: the Veerse Meer, Haringvliet and Grevelingen. The project has effectively shortened the Dutch coastline by 700 km.

A similar threat of marine flooding to London in south-east England, where the dangers from high tides and storm surges are exacerbated by land subsidence, led to the construction of the Thames Barrier, completed in 1983. Such schemes are not without their

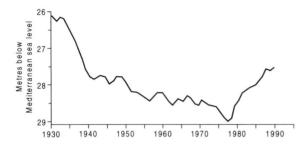

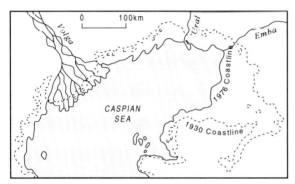

FIGURE 11.3 *Variations in sea level and northern coastline changes of the Caspian Sea (after Tolba and El-Kholy, 1992; Bird, 1985)*

problems, however. The 25-km barrier dike built across the Gulf of Finland to protect St Petersburg from storm-induced tidal surges, which is currently just over half built, has hampered water circulation in the diked-off portion, turning it into a stagnant settling basin for effluent. The resulting algal blooms and noxious odours have a deleterious effect on fishing and coastal recreation, while siltation behind the dike and ice build-up along it also threaten the port's maritime economy. Worries over similar effects behind tidal power barrages have also been expressed. A further question mark over the efficiency of these and other coastal flood protection schemes is currently posed by the potential for rapid global sea-level rise associated with global warming (see Chapter 8).

POLLUTION

Pollution of the coastal zone comes from activities on the coast, in coastal waters and from inland, in the latter case usually arriving at the coastal zone via rivers. Pollutants can be divided into non-conservative types which are eventually assimilated into the biota by processes such as biodegradation and dissipation, and conservative pollutants which are non-biodegradable and are not readily dissipated, and thus tend to accumulate in marine biological systems (Table 11.1, see also Fig. 10.7).

The wide range of human activities that contribute pollutants to coastlines, and the pathways by which they reach the coast (see also Table 10.1), are illustrated in an issue which is causing considerable concern: the increasing discharges of nutrients to coastal environments – a problem that is envisaged to constitute a worldwide issue in the next few decades. Human activities have increased nutrient flows to the coast in numerous ways:

- forest clearing;
- destruction of riverine swamps and wetlands;
- application of large amounts of synthetic fertilisers;
- industrial use of large amounts of nitrogen and phosphorus;
- addition of phosphates to detergents;
- production of large livestock populations;
- high-temperature fossil fuel combustion, adding large amounts of nitrogen to acid rain; and
- expansion of human populations in coastal areas.

The extent of the increase is indicated for a part of the North American seaboard by Nixon (1993), who used a complex mix of historical data to estimate the changes in nutrient inputs to the Narragansett and Mount Hope bays (Fig. 11.4). Up to around

TABLE 11.1 *Classification of marine pollutants*

NON-CONSERVATIVE

- Degradable wastes (including sewage, slurry, wastes from food processing, brewing, distilling, pulp and paper and chemical industries, oil spillages)
- Fertilisers (from agriculture)
- Dissipating wastes (principally heat from power station and industrial cooling discharges, but also acids, alkalis and cyanide from industries which can be important very locally)

CONSERVATIVE

- Particulates (including mining wastes and inert plastics)
- Persistent wastes (heavy metals such as mercury, lead, copper and zinc; halogenated hydrocarbons such as DDT and other chlorinated hydrocarbon pesticides, and polychlorinated biphenyls; radionuclides)

Source: after Clark (1992)

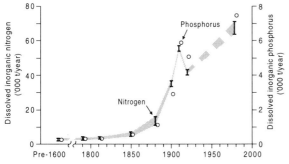

FIGURE 11.4 *Inorganic nitrogen and phosphorus input to Narrangansett and Mount Hope bays, north-east USA, pre-seventeenth to late twentieth centuries (after Nixon, 1993)*

1800, the inflow of nutrients from the ocean was five to ten times larger than inputs from the land and the atmosphere, but today about five times more inorganic nitrogen enters Narragansett Bay from the land and

atmosphere than from the coastal ocean. The particularly dramatic rise which occurred over a 30-year period from 1880 to 1910 was a response to the proliferation of public water and sewage systems through much of the urban areas of the northeastern USA at this time.

Excessive nutrient flows to coastal environments cause oxygen depletion, reduced diversity, increased toxic and other undesirable phytoplankton blooms, and generally negative changes in population structure and dynamics. Hence, algal blooms have been associated with catastrophic mass deaths of marine mammals, fish and invertebrates and the emergence of new disease pathogens, such as a new form of cholera present in several Asian nations, which threatens to become the agent of an eighth cholera pandemic (Levins *et al.*, 1994).

All kinds of urban and industrial activities in the coastal zone contribute pollution (Fig. 11.5), the effects of which can become particularly critical in areas where the mixing of coastal water with the open ocean is relatively slow, as in bays, estuaries and enclosed seas (e.g. see the account of the Black Sea in Chapter 10). For example, in dryland coastal cities where fresh water is produced by desalination plants, the discharge of chlorine back into coastal waters may represent a severe pollution problem, as in Kuwait where about 17 t of chlorine are dumped each day (Khordagui, 1992).

Petroleum hydrocarbons are a particular hazard to coastal ecology, chiefly at oil terminals but also from accidental spillages. Habitats such as mangroves and marshes are especially vulnerable to oil spills, although wildlife is often the most conspicuous victim. Generally, younger individuals are more sensitive than adults, and crustaceans are more sensitive than fish. The release of 36 000 t of oil into Prince William Sound in Alaska from the grounded supertanker *Exxon Valdez* in March 1989 resulted in the death of 36 000

FIGURE 11.5 *An unusual form of coastal pollution in Pakistan where a 10-km stretch of sands at Gadani is used for ship-breaking*

seabirds, 1000 sea otters and 153 bald eagles in the following 6 months (Maki, 1991). The loss of these animals from intermediate and top trophic levels has had many indirect effects at lower levels, and recovery of most biotic communities to their pre-spill levels could take up to 25 years (Shaw, 1992).

Heavy metals and organochlorine compounds have long been recognised as among the most deleterious contaminants to biota in coastal and estuarine waters, and a large proportion of these pollutants arrive at coasts in rivers draining wide areas outside the coastal zone. A study of the Mediterranean found that more than half the mercury, chromium, lead, zinc and pesticides (as well as a similar level of nutrients and radioactive materials) came from such rivers (UNEP, 1989a).

Since heavy metals and organochlorine compounds cause problems even in very low concentrations, and the measurement of such small quantities is technically difficult, many studies of these coastal pollutants focus on concentrations in marine organisms. Mussels and oysters have been widely used because as filter feeders they bioaccumulate contaminants. Several investigations of long-term trends in pollutants have measured concentrations in the feathers of seabirds since old specimens can often be found in museums. Mercury contamination measured in herring gulls and common terns from the German North Sea coast, for example, indicates that the rivers Elbe and Rhine are the major sources of the pollutant. Mercury flows peaked at three times 1880 levels after the Second World War when heavy metals from ignition devices in munitions were released in large quantities (Thompson *et al.*, 1993). Significant declines in the last two decades reflect measures to reduce river-borne pollution, however, and the trend will probably continue.

HABITAT DESTRUCTION

Coastal habitats, particularly mangroves, salt marshes, seagrass beds and coral reefs, are subject to many types of pressures from human activities. Direct destruction of these habitats for urban, industrial and recreational growth, as well as for aquaculture, combined with overexploitation of their resources and the indirect effects of other activities, such as pollution and sedimentation or erosion, are causing rapid losses all over the world. Rates of habitat loss are difficult to assess, but the Group of Experts on the Scientific Aspects of Marine Pollution reported: 'If unchecked, this trend will lead to global deterioration in the quality and productivity of the marine environment' (GESAMP, 1990: 1). The sections below detail the situation for two coastal habitats in the tropics which have come under substantial pressures in recent times.

Coral reefs

Coral reefs, tropical marine ecosystems found in shallow waters (generally <30 m) and largely restricted to the areas between 30° N and 30° S, rank among the most biologically productive and diverse of all natural ecosystems. They cover less than 0.2 per cent of the ocean floor, yet support 25 per cent of marine species. Their high productivity is a function of their efficient biological recycling, high retention of nutrients and their structure, which provides a habitat for a vast array of organisms. From the economic viewpoint, coral reefs protect the coastline from waves and storm surges, prevent erosion, and contribute to the formation of sandy beaches and sheltered harbours. They play a crucial role in fisheries, by providing nutrients and breeding grounds, and provide a source of many raw materials such as coral for jewellery and building materials. Most recently, their potential for tourism has been realised – one which plays an important role in the economies of many Caribbean island states particularly.

The last 20 years or so has been marked by a growing concern for the future of the world's coral reefs. Reefs have always been subject to natural disturbances such as hurricanes, storms, predators, diseases and fluctuations in sea level, but in recent times they have also been subject to increasing pressures from human action. Island and coastal societies have long used coral reefs for subsistence purposes, as sources of food and craft materials, but the development of commercial fisheries, rapidly increasing populations still dependent upon a subsistence lifestyle, the growth of coastal ports and urban areas, increases in soil loss due to deforestation and poor land use, and most recently the exponential growth of coastal tourism, have combined to put unprecedented pressures on reefs from both direct and indirect impacts (IUCN, 1988).

The wide range of human impacts upon coral reefs is indicated in Table 11.2 with examples from the Pacific. They can be divided roughly into direct impacts caused by overexploitation of reefs (overcollecting, damaging fishing techniques, and careless or poorly supervised recreational use), and indirect impacts resulting from other activities (siltation following land clearance, damage caused by coastal developments and pollution, and that caused by military activities).

In many cases reefs are subject to a combination of impacts. In the Philippines, where only 30 per cent of remaining reefs are considered to be in good or excellent condition, siltation due to forest denudation, blast fishing, and the use of sodium cyanide by tropical fish collectors, are the main causes of coral decline. Chronic overfishing has also been a constant hazard to reefs in Jamaica since the 1960s, and this stress has combined with natural hurricane damage, disease and algal blooms to reduce coral cover around the island's coastline from about 50 per cent in the late 1970s to less than 5 per cent in the mid-1990s (Hughes, 1994).

The changing nature of human threats to the coral reefs of Kaneohe Bay, on the east coast of the Hawaiian island of Oahu, is shown in Fig. 11.6. Dredging and filling by the military between 1938 and 1950 significantly

TABLE 11.2 *Human-induced threats to coral reefs with selected examples from the Pacific Ocean*

Threat	Example
OVER-COLLECTING	
Fish	Futuna Island, France
Giant clams	Kadavu and islands, Fiji
Pearl oysters	Suwarrow Atoll, Cook Islands
Coral	Vanuatu
FISHING METHODS	
Dynamiting	Belau, USA
Breakage	Vava'u Group, Tonga ("tu'afeo")
Poison	Uvea Island, France
RECREATIONAL USE	
Tourism	Heron Island, Great Barrier Reef, Australia
Scuba diving	Hong Kong
Anchor damage	Molokini Islat, Hawaii, USA
SILTATION DUE TO EROSION FOLLOWING LAND CLEARANCE	
Fuelwood collection	Upolu Island, Western Samoa
Deforestation	Ishigakishima, Yaeyama-retto, Japan
COASTAL DEVELOPMENT	
Causeway construction	Canton Atoll, Kiribati
Sand mining	Moorea, French Polynesia
Roads and housing	Kenting National Park, Taiwan
Dredging	Johnston Island, Hawaii, USA
POLLUTION	
Oil spillage	Easter Island, Chile (1983)
Pesticide spillage	Nukunonu Atoll, New Zealand (1969)
Urban/industrial	Hong Kong
Thermal	Northwestern Guam, USA
Sewage	Micronesia
MILITARY	
Nuclear testing	Bikini Atoll, Marshall Islands (1946–58)
Conventional bombing	Kwajalein Atoll, Marshall Islands (1944)

Source: compiled from sources quoted in Nunn (1990); IUCN (1988)

increased problems of turbidity and sedimentation in most of the bay and caused direct damage in southern parts where coral was used as fill to create new land. In 1970, while northern reefs were recovering from this impact and colonising some of the dredged zones, two new sewage outlets caused further reef decline in the southern portion. Closure of these sewage outlets in 1977–78 allowed a rapid recovery in the central part of the bay and a more prolonged recovery in the south. The new sewage outlet outside the bay has not damaged coral, due to the high level of mixing and flushing at the site.

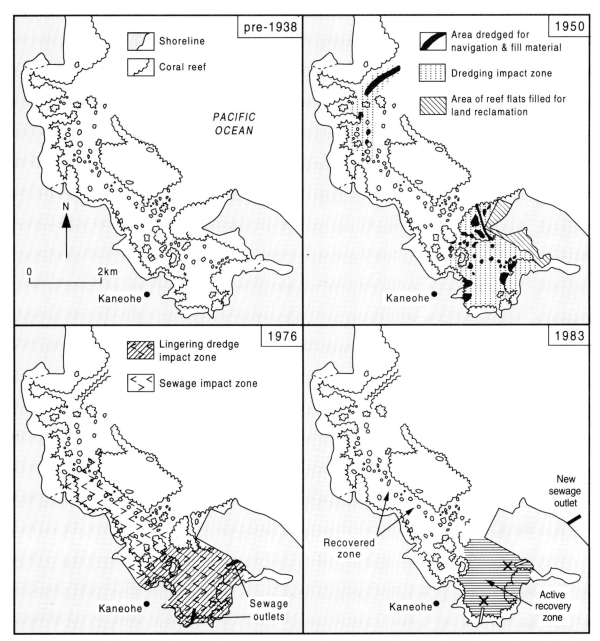

FIGURE 11.6 *Effects of dredging, filling and sewage discharge in Kaneohe Bay, Oahu, Hawaii, USA (after Maragos, J.E. 1993, Impact of coastal construction on coral reefs in the US-affiliated Pacific Islands.* Coastal Management *21: 235–69. Reproduced by permission of Taylor & Francis)*

Given the importance of coral reefs to many local human communities, the economic knock-on effects of damage can also be signif-

icant. Reefs in Rosario Coral Reef National Park in Colombia were severely affected after the Canal del Dique, leading to the coast, was redredged and reopened in 1986. The resulting cover of sediment, which deprived symbiotic algae in the coral of the sunlight needed to photosynthesise, led to a mortality rate of 55 per cent in 1989. Loss of elkhorn and staghorn corals, in particular, has left the coastline susceptible to wave attack, and the resultant erosion now threatens homes and port facilities. As the reef has declined, the local fish catch has dropped due to the disruption of the food chain, and the diving industry for tourists has suffered as sedimentation spreads. High nutrient levels in the canal water have also led to eutrophication and the spread of brown algae across areas of the park.

Although the anthropogenic impact is unmistakable in the above examples of coral reef stress, two other recent issues surrounding reef degradation are not so clear-cut. These are the population explosion of the predatory crown-of-thorns starfish (*Acanthaster planci*) since the 1960s, and the coral bleaching phenomenon since the 1980s (Salvat, 1992).

Crown-of-thorns populations exploded at the end of the 1950s and throughout the 1960s in many parts of the Indian and western Pacific oceans, causing widespread damage to reefs. No less than 90 per cent of the fringing reefs of Guam and 14 per cent of the Great Barrier Reef were destroyed, the dead corals being replaced by an algae and sea urchin dominated community. Initially the finger of suspicion for the outbreaks was pointed at human action, through the destruction of starfish predators and the effects of pollution on larvae. Subsequently, however, researchers have viewed the phenomenon as part of essentially natural cyclical events.

Similar uncertainty surrounds the widely reported phenomenon of coral bleaching which has resulted in the widespread mortality of reefs, particularly in the Caribbean. While bleaching commonly occurs naturally when sea surface temperatures rise during events such as El Niño, a worldwide overview concluded that bleaching events were much more usual in the 1980s (Williams and Bunkley-Williams, 1990), leading to suggestions of links with global warming.

The uncertainty surrounding these two issues of coral reef vitality highlights the general lack of a historical perspective on the environmental parameters affecting reefs and their population dynamics. A heightened realisation of the important role of corals in the marine ecosystem, their high biodiversity and their importance to local communities should help to turn concern at their degradation into action towards managing these key habitats more sustainably. Such management is needed most urgently for those atoll island states whose future is threatened by the altered environmental dynamics of a warmer world (see Chapter 8). In response to global sea-level rise, some reefs appear to 'keep-up' with rising water levels, others 'catch-up' after a lag period by rapid vertical accretion of coral, while others are terminally affected by initial drowning and effectively 'give-up' (Neumann and Macintyre, 1985).

Mangroves

Mangrove forests are made up of salt-adapted evergreen trees, and are found in intertidal zones of tropical and subtropical latitudes. These diverse and productive wetland ecosystems are under threat from human action in many parts of the world, despite their importance as breeding and feeding grounds for many species of fish and crustaceans and their role in protecting coasts from erosion, among the many other functions of wetlands in general (see Table 12.4). Although estimates of mangrove area in the literature vary considerably, due to differing survey methods and definitions, Table 11.3 shows a number of countries in Asia and Africa where mangroves have declined significantly from their pre-agricultural extent. Increasing population

TABLE 11.3 *Extent of mangroves and major causes of destruction, selected Asian and African countries, late 1980s*

Country	Extent (km²)	Pre-agricultural extent (km²)	Major causes of destruction
India	3100	12 600	Agriculture, urban development
Philippines	1000	4500	Charcoal production and fuelwood, fish ponds
Singapore	<5	75	Urban/industrial development
Viet Nam	1600	3800	Herbicide spraying during Viet Nam War, agriculture and fish ponds
Côte d'Ivoire	29	1600	Fuelwood
Ghana	3	2100	Charcoal production and fuelwood, salt extraction, wood for construction
Guinea	3000	4000	Agriculture, fuelwood
Guinea-Bissau	2400	3100	Agriculture
Sierra Leone	1000	6800	Agriculture, fuelwood, wood for construction

Sources: WRI (1986); World Bank (1989); WRI (1990); Collins *et al.* (1991); Sayer *et al.* (1992); Jagtap *et al.* (1993)

numbers and a general lack of protection and management appear to be the common denominators behind this destruction. Conversion to agriculture, fish ponds, urban/industrial uses, and destruction for fuelwood or charcoal production are the main causes.

Some of the largest losses along the African coastline have been in the west of the continent. The most serious threat to mangroves in Guinea, Guinea-Bissau and Sierra Leone, formerly some of the most extensive areas of this habitat in West Africa, is clearance and conversion of the land to rice farming, although exploitation for fuelwood also plays an important part. Estimates for Guinea suggest that about 0.25 million tonnes of fuelwood per year are taken from the mangrove forests (Table 11.4). Clearance for fuelwood has been the major factor behind the almost total disappearance of mangroves in Ghana and Côte d'Ivoire (Sayer *et al.*, 1992).

TABLE 11.4 *Use of mangrove fuelwood in Guinea*

Use	Amount (t/year)
Urban	54 000
Rural	52 000
Fish smoking	58 000
Salt making	93 000
Total	257 000

Source: Bertrand (1993)

In Asia, as Fig. 11.7 indicates, the clearance of mangrove forests for agriculture has been a progressive process over the last century or so in the delta regions of the Irrawaddy and Ganges rivers, while in the Philippines, where more than 75 per cent of the mangrove cover has been lost in less than 70 years, cutting for fuelwood and charcoal production has been superseded as the most serious

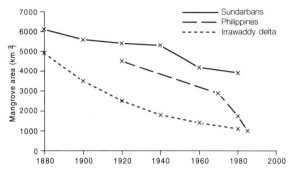

FIGURE 11.7 *Decline of mangroves in the Sundarbans (Bangladesh and India), the Philippines and the Irrawaddy delta (Myanmar) (from data in Richards, 1990; World Bank, 1989; Collins* et al.*, 1991)*

cause by clearing for the establishment of fish ponds (see Chapter 5). Fish ponds grew to cover an area of 2050 km² in the Philippines in 1987 from 890 km² in 1952 (World Bank, 1989). Destruction of Indonesia's estimated 44 000 km² of mangroves began more recently, being little affected before 1975. Since then, however, conversion to fish ponds has been widely practised in Sulawesi, Java and Sumatra, and large areas have been cleared on many islands to produce wood chips which are exported to Japan for cellulose and paper making.

The mangrove forests of the Ganges and Brahmaputra deltas, known as the Sundarbans, remain the largest mangrove ecosystem in the world, and parts of this area have been protected as a Reserved Forest since 1875. It is a key wildlife habitat for a large number of bird, mammal, reptile and amphibian species and provides the last stronghold of the royal Bengal tiger. About half a million people also depend directly on the Sundarbans for their livelihood. New areas have been planted on offshore islands as part of the management plan, both for fuelwood harvesting and coastal protection against cyclonic storm surges, but the area is under increasing pressure from Bangladesh's

growing population. A mass replanting programme has also been undertaken in postwar Viet Nam, successfully re-establishing many thousands of hectares destroyed by spraying with the herbicide Agent Orange, but large areas still remain damaged and there has been virtually no natural regrowth after the war.

Local populations use mangrove forests for a wide variety of their resources, including wood (for fuel, construction materials and stakes), leaves for fodder, fruits for human consumption, the rich mud which is used as manure, tannin extracted from barks, leaf extracts for medicinal purposes, and captive fisheries of prawns, shrimps and fish (Fig. 11.8). The questionable wisdom of destroying this diverse ecosystem is illustrated by data presented for Thailand by Christensen (1983), which show that when converted to rice paddies, a hectare of former mangrove yields US$168 a year, but an undamaged hectare farmed for shrimp produces US$206 a year and up to ten times that amount if carefully managed. This is quite apart from the mangrove's role in coastal protection, a particularly important one in the storm-prone, low-lying coastal zones where they are found.

A similar cost–benefit analysis has been made for the 8700 ha Sarawak Mangroves Forest Reserve in Malaysian Borneo by Bennett and Reynolds (1993), who calculate that the area is worth about US$25 million a year in forestry, marine fisheries and tourism revenues, as well as providing up to 3000 jobs in fisheries alone. Most of this value would be lost with conversion to an alternative form of land use, which would greatly reduce levels of revenue. Conversion to aquaculture ponds or oil palm plantations, the greatest threats, would also necessitate highly expensive engineering works to prevent coastal erosion, flooding and other damage, as well as causing immeasurable loss in terms of the reserve's wildlife importance.

FIGURE 11.8 *Captive fishing in the mangroves of the Gulf of Guayaquil, Ecuador. The area has also become the centre of a shrimp farming industry, one of the country's major sources of foreign exchange. Since the mid-1960s, about 20 per cent of mangroves in the gulf have been cleared for shrimp production*

MANAGEMENT AND CONSERVATION

Management of the coastal zone is necessary both to limit damaging activities and to protect coastal resources, as well as to restrict development in areas prone to natural hazards such as hurricanes, tsunamis, subsidence and inundation. On the national scale, Costa Rica has one of the world's most comprehensive and ambitious shoreland restriction programmes (Sorensen and McCreary, 1990) which covers a jurisdictional area 200 m wide in the marine and terrestrial zone. The law divides this zone into two components: the *zona pública* and the *zona restringida* (restricted zone). The *zona pública* extends inland 50 m from the mean high tide level or the inland limit of wetlands and the upstream limit of estuaries as defined by salt or tidal influence, and the *zona restringida* covers the remaining 150 m inland. The *zona pública* is devoted to public use and access, and commercial development is generally prohibited unless it is for enterprises which

need to be in a coastal location, such as port installations. In the *zona restringida*, development is controlled by a permit and concessions system based on a detailed regulation plan formulated by local government. A concession is a development right on a specific area of land for a particular use over a fixed time period.

The importance of adequately managing coastal areas from ecological and economic viewpoints, which are often intimately related, can be illustrated at a regional level by Mauritania's coastal Parc National du Banc d'Arguin. The park covers an area of more than 1 million hectares and includes extensive tidal mudflats and extremely rich offshore fishing grounds. The site is internationally significant as the most important coastal wetland in Africa through its role as a breeding and wintering ground for millions of waterfowl: it hosts the largest concentration of wintering waders in the world. It is also a vital spawning ground and nursery for the fish which are Mauritania's main foreign currency earner. The area became a National Park in 1976, and has subsequently been listed under both the Ramsar and World Heritage Conventions. National and international conservation efforts focus on maintaining and strengthening the wise use of the mullet fishery by the Imraguen fishermen who reside in the park and use traditional, sustainable harvesting methods in which they collaborate symbiotically with wild dolphins to catch grey mullet. There is also a need to keep commercial fishing boats from entering the area, while investigation is also being made into the possibilities of developing ecotourism to increase the use of the park's resources.

In other coastal areas, however, management problems are made more complex by large human population densities and the consequent heavy pressures put on resources. The links between coastal problems and pollution sources outside the coastal zone also make integrated management over the entire watershed necessary, given the importance of rivers in contributing to pollution. The initiatives taken to improve the quality of Chesapeake Bay, the largest estuary in North America, are widely considered to be a good model for dealing with such complex pollution problems of coastal zones. The Chesapeake Bay Program, adopted by the six states and the District of Columbia which together have jurisdiction over all of the watershed, aims to reduce nutrient flows into the bay largely through subsidies for improved agricultural practices, reduction of urban runoff and treament of sewage (USEPA, 1983). Although these measures have had some success in reducing pollutants, further efforts are deemed necessary and the complex political situation highlights the possibility of conflict between polluter and polluted: 'it is not yet certain whether upstream landowners and governments are willing to restore the bay by restricting their own activities when downstream neighbors pay for the impacts of those activities' (WRI, 1992: 189).

Many of the management decisions made at any level need to balance conflicting interests and must do so with the cost-effectiveness of particular strategies in mind. These issues are illustrated in the small area of Britain shown in Fig. 11.1, where erosion continues to affect existing economic activity. Many farmers along the North Sea coast of Humberside are losing their fields at a rate of up to 2 m/year. The local council is sympathetic to calls for coastal protection schemes, but such projects are costly in financial terms and the priority given to individual farms is likely to be lower than that given to the gas terminals at Easington which handle 25 per cent of Britain's gas supplies. An added complicating factor is the worry that sediment eroded from the North Sea coast is transported up the Humber estuary and deposited on the river banks. Halting the erosion could therefore increase the flood risk at Hull. The need for an integrated management plan for the coastal zone, which includes a thorough appreciation of marine and terrestrial sediment dynamics, is therefore paramount.

FURTHER READING

Bird, E.F.C. 1985 *Coastline changes: a global review.* Chichester, Wiley. Many case studies of physical changes to coastlines on a country-by-country basis.

Carter, R.W.G. 1988 *Coastal environments.* London, Academic Press. An introduction to the physical and ecological nature of coastal environments and their management.

Hinrichsen, D 1990 *Our common seas: coasts in crisis.* London, Earthscan. This book documents growing pressures on coastal environments worldwide and provides case studies of local conservation successes.

Viles, H.A. and Spencer, T. 1995 *Coastal problems.* London, Edward Arnold. A comprehensive review of all types of coastal problems, their origins and management.

12

RIVERS, LAKES AND WETLANDS

Fresh water plays an integral role in the functioning of all environments and societies. Since its earliest inception, human society has seen freshwater bodies as a vital resource and entire ancient 'hydraulic civilisations' developed on certain rivers, notably those of the Tigris–Euphrates, the Nile and the Indus. Society is no less reliant on this fundamental natural resource today. Fresh water from lakes, rivers and wetlands is utilised for municipal, agricultural, industrial, recreational and power generation uses; it forms a convenient medium for transport and a sink for wastes. Society's use of fresh water, its availability and quality, have thrown up numerous issues of controversy and these issues will become more acute as global water withdrawals continue to increase. Estimates of the gross global withdrawals for the year 2000, for example, are 5200 km^3, a nine-fold increase over the 1900 level, and this increase continues to accelerate (Shiklomanov, 1993). Global data mask regional differences, however, which are essentially functions of climatic influences. All countries suffer from periodic excesses of fresh water in the form of floods (see Chapter 20), while others also experience perennial shortages, in some cases due to a complete absence of permanent rivers (e.g. Malta and Saudi Arabia).

An idea of the wide variety of ways in which human activity can adversely affect freshwater ecosystems is given by noting the effects upon freshwater fish, a biological indicator of ecosystem health (Table 12.1). This chapter deals with surface freshwater ecosystems: rivers, lakes and freshwater wetlands, although some of the effects of activities such as deforestation, urbanisation and mining on the quality and quantity of freshwater bodies are covered elsewhere (see Chapters 3, 14 and 18). Wetlands found in coastal regions, such as mangroves and estuaries, are covered in Chapter 11, and issues surrounding underground sources of water are covered in Chapters 5 and 14.

WATER POLLUTION

Sources of water pollution can be traced to all sorts of human activity, including agriculture, irrigation, industry, urbanisation and mining. Most forms of water pollution can be classified into three categories:

- excess nutrients from sewage and soil erosion,
- pathogens from sewage,
- heavy metals and synthetic organic compounds from industry, mining and agriculture.

TABLE 12.1 *Summary of main pressures facing freshwater fish and their habitats in temperate areas*

Danger	Effects
Industrial and domestic	Pollution, elimination of effluents stocks, blocking of migratory species
Acid deposition	Elimination of stocks in poorly buffered areas
Land use (farming and forestry)	Eutrophication, acidification, sedimentation
Eutrophication	Algal blooms, deoxygenation, changes in species
Industrial development (including roads)	Sedimentation, obstructions, transfer of species
Warm water discharge	Deoxygenation, temperature gradients
River obstruction (dams)	Blocking of migratory species
Infilling, drainage and canalisation	Loss of habitat, shelter, food supply
Water abstraction	Loss of habitat and spawning grounds
Fluctuating water levels (reservoirs)	Loss of habitat, spawning and food supply
Fish farming	Eutrophication, introductions, diseases, genetic changes
Angling and fishery management	Elimination by piscicides, introductions
Commercial fishing	Overfishing, genetic changes
Introduction of new species	Elimination of native species, diseases, parasites

Source: Maitland (1991)

Three other forms should also be mentioned:

- thermal pollution from power generation and industrial plants,
- radioactive substances,
- turbidity problems caused by increased sediment loads or decreased water flow.

Organic liquid wastes can be broken down by bacteria and other micro-organisms in the presence of oxygen, and the burden of organics to be decomposed is measured by the biochemical oxygen demand (BOD). Liquid organic wastes include sewage, many industrial wastes (particularly from industries processing agricultural products) and runoff which picks up organic wastes from land. As a river's dissolved oxygen decreases, with increasing loads of organic wastes, so fish and aquatic plant life suffer and may eventually die. Heavy volumes of organic wastes can overload a riverine system to the point at which all dissolved oxygen is exhausted.

Pollution of water due to temperature increase, so-called 'thermal pollution', also reduces its dissolved oxygen content. This occurs in two ways. An increase in water temperature decreases the solubility of oxygen on the one hand, and increases the rate of oxidation, thereby imposing a faster oxygen demand on a smaller content, on the other. Thermal pollution also has a number of more direct adverse effects on river ecology, including a general increase in undesirable forms of algae and reduced reproduction and growth of some species of fish (Langford, 1990).

Inorganic liquid wastes become dangerous when not adequately diluted. Even in very small concentrations, however, some heavy metals (such as cadmium, lead and mercury) are particularly dangerous and can bioaccumulate up the food chain, ultimately damaging human health.

Pathogens from human waste spread disease and represent the most widespread contamination of water. Water-related diseases can be classified into those that are water-borne (e.g. diarrhoea, cholera and polio); those that are related to a lack of personal cleanliness (e.g. trachoma and typhoid); and those that are related to water as a habitat for certain disease vectors (e.g. schistosomiasis, malaria and onchocerciasis).

Data for freshwater quality are generally sparse, cover limited periods, and are subject to changes in analytical methods and the movement of gauging stations, but a coordinated effort at worldwide monitoring has been made within the framework of UNEP's Global Environmental Monitoring System, or GEMS (UNEP/WHO, 1988). The network, launched in 1977, comprises 240 river stations, forty-three lake stations and sixty-one groundwater stations.

Meybeck *et al.* (1989) indicate how particular sources of water pollution have ebbed and flowed with time in the highly industrialised countries. The growth of urban areas during the Industrial Revolution created the first wave of serious water pollution, from domestic sources, around the turn of the century. These have been superseded by industrial pollutants, and during the second half of the century by nutrient pollution (particularly from agricultural sources) and micro-organisms towards the end of the 1990s. A similar pattern of peaks in serious pollution problems is being experienced in the rapidly industrialising countries, but compressed into the postwar decades, and it is here that some of the most serious water pollution problems are being faced today. The other regions where water pollution has reached crisis proportions are in the countries of the former Soviet bloc. As with many other pollution issues, it is often the poorer sectors of society that suffer most from the detrimental effects of poor water quality.

As Meybeck *et al.*'s (1989) analysis suggests, most sources of water pollution are well known and methods for reducing them or their detrimental effects have been devised. Most of today's continuing water pollution problem areas are due to a lack of political will, poor coordination between governments and/or a lack of funds.

RIVERS

River pollution

The World Bank (1992a) found from a sample of GEMS monitoring sites, classified according to the development status of countries, that through the 1980s, high-income countries had seen some overall improvement in river water quality, middle-income countries had on average shown no change, and low-income countries experienced continued deterioration.

The major sources of river pollution in the USA are monitored by the Environmental Protection Agency (EPA). In 1988, based on a 30 per cent sample of US river miles, no less than 55 per cent of total river length was deemed to be impaired by pollution from agricultural sources. Municipal pollution impaired 16 per cent of river length, mining 13 per cent and habitat modification 13 per cent (USEPA, 1990). Other sources of pollution, all of which impaired less than 10 per cent of total river length, were storm sewers/runoff, silviculture, industrial, construction, land disposal and combined sewers. As far as causes of pollution are concerned, siltation, nutrients, pathogens, organic enrichment, metals and pesticides were the major culprits.

Agriculture is the leading non-point source of pollutants such as sediments, pesticides and nutrients, particularly nitrogen and phosphorus. The growing use of chemical fertilisers, fuelled by increasing demands for food, has led to a steady rise in the concentration of nitrates in many world rivers and groundwaters in recent decades. It has been estimated that less than 10 per cent of rivers in Europe, for example, are pristine with regard to nitrate concentrations (Meybeck, 1982). While nitrate pollution of waters, linked to cancers of the digestive tract and the so-called 'blue baby syndrome', do not currently pose a serious threat to public health, the continued rapid rise in concentrations may be a cause for concern in the future. The startling rise in nitrate concentration in the Hungarian Danube indicated in Fig. 12.1, which mirrors curves in fertiliser usage, is fairly representative of many European rivers. The guideline for drinking water quality in Hungary is 20 mg NO_3/l. Data on pesticide contamination of rivers are not sufficient to give a clear picture of the situation even in some of the better-monitored parts of the world, but rising levels of usage give a reasonable cause for concern, particularly in the developing countries where many pesticides now banned in the developed world are still used (see Chapter 5).

Riverine pollution from sewage is certainly an acute problem facing many of the world's poorer countries where sewage treatment is inadequate or non-existent, but even in some of the richer nations the level of treatment is still far from satisfactory. In 1990, the proportion of national population served by primary, secondary or tertiary facilities was just 21 per cent in Portugal and 42 per cent in Japan, for example (OECD, 1993).

The combined effects of various pollutants have had serious impacts in many world rivers. In Metro Manila, capital of the Philippines, for example, just one in ten households was served by a wastewater disposal system in the mid-1980s and more than half of the country's industry is located in the city. All Metro Manila's rivers are biologically dead and virtually unnavigable due to pollution and heavy siltation (World Bank, 1989).

Similarly dramatic pollution black-spots have been recorded in many of the countries of eastern Europe. In Poland, where biologically 75 per cent of the rivers were dead in 1988 (Carter, 1993b), the River Vistula, once known as the Queen of Rivers, is now a potent symbol of environmental degradation. Most of Poland's large towns and much of its industry are located along the banks of the Vistula, and its waters provide the country's most important source of drinking and industrial water and a sink for domestic and industrial wastes. Concentrations of pollutants in the Vistula are high and rising. Great increases in eutrophication and organic pollutants were recorded in the 1960s, and although the rate of increase slowed for several pollutants in the 1980s, nitrogen and phosphorus are still rising

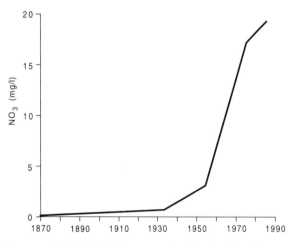

FIGURE 12.1 *Nitrate concentration in the River Danube in Hungary, 1870–1985 (after Hinrichsen, D. and Láng, I. 1993, in Carter, F.W. and Turnock, D. (eds)* Environmental problems in Eastern Europe. *Reproduced by permission of Routledge)*

TABLE 12.2 *Water quality in the Mississippi*

Dissolved concentration (µg/l)	1970	1975	1980	1985	1991
Lead	18.00	1.50	0.42	4.90	1.00
Cadmium	18.00	0.36	1.40	1.00	1.00
Chromium	18.00	0.36	2.50	1.00	1.80
Copper	98.00	4.10	4.10	5.70	3.80

Source: data from OECD (1993)

at alarming rates (Kajak, 1992). About 70 per cent of the sewage entering the river is untreated and the rest is treated unsatisfactorily. The most important pollutants are nitrogen, phosphorus, phenols, NaCl and SO_4. NaCl comes mostly from coal and sulphur mines, nitrogen and phosphorus from sewage as well as dispersed sources, mostly agricultural. Carter (1993b) also notes that the mercury content of the Vistula below Cracow in 1980 was more than 200 times the permitted norm and that the Vistula is now so poisoned or potentially corrosive that stretches are considered unusable even for factory coolant systems, let alone drinking water.

In places where the economic power and political will are available, however, concerted efforts have been made to improve river water quality with some notable successes. Twenty-one years of records on the Mississippi River, for example, have highlighted the decline in dissolved concentrations of several heavy metals (Table 12.2) and similar success in reducing contamination has been achieved in Europe's Rhine.

In terms of organic wastes, the history of the River Thames provides a clear example of the reversibility of many sorry stories of riverine pollution. The quality of the Thames declined in the nineteenth century as London grew and the flushing water closet became widely used, discharging into the river. Five

cholera epidemics occurred between 1830 and 1871 and during the long, dry summer of 1858, the so-called 'Year of the Great Stink', Parliament had to be abandoned on some days because of the stench. Such a direct impact on the nation's legislators engendered positive action, however, and conditions had improved by the 1890s, when oxygen levels were up to 25 per cent saturation (Fig. 12.2) with the introduction of sewage treatment plants. During the first half of the twentieth century, however, sewage treatment and storage did not keep pace with growing population, and dissolved oxygen reached

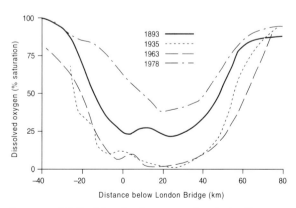

FIGURE 12.2 *Dissolved oxygen levels in the River Thames estuary in 4 years between 1893 and 1978 (after Wood, L.B. 1982,* The restoration of the tidal Thames. *Reproduced by permission of IOP Publishing Limited)*

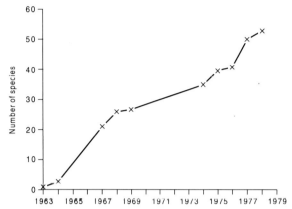

FIGURE 12.3 *Increase of fish diversity in the River Thames estuary as monitored by collections from the West Thurrock power station cooling water intake, 1963–78 (from data in Ellis, 1989)*

zero at 20 km downstream of London Bridge during many summers.

Following tighter controls on effluent and improved treatment facilities introduced in the postwar years, improvement was gradual, with 1976–77 generally taken as the time when water quality reached satisfactory levels. The general improvement in the quality of the Thames water is indicated by the increased diversity of fish monitored at the water intake of power stations (Fig. 12.3). Much publicity accompanied the landing of the first salmon caught in the Thames since 1833 off the West Thurrock screen in 1974.

River management

River modification and management are rooted in society's view of rivers both as hazards to be ameliorated and resources to be maximised. There are numerous ways in which people have modified rivers: by constructing dams, building levées, widening, deepening and straightening channels (Table 12.3; see also Chapters 13 and 20). The beginnings of direct river modification in the UK, for example, can be traced back to the first century AD with fish ponds and water mills, and changes to facilitate transportation and to effect land drainage (Sheail, 1988). Severe pollution in the mid-nineteenth century led to an organised system of river management. A marked east–west rainfall gradient, combined with an uneven population distribution (a high proportion live in eastern and southeastern England where the rainfall is low at 600–700 mm/year) means that many rivers in

TABLE 12.3 *Selected methods of river channelisation and their US and UK terminologies*

Method	US term	UK equivalent
Increase channel capacity by manipulating width and/or depth	Widening/deepening	Resectioning
Increase velocity of flow by steepening gradient	Straightening	Realigning
Raise channel banks to confine floodwaters	Diking	Embanking
Methods to control bank erosion	Bank stabilisation	Bank protection
Remove obstructions from a watercourse to decrease resistance and thus increase velocity of flow	Clearing and snagging	Pioneer tree clearance; control of aquatic plants; dredging of sediments; urban clearing

Source: after Brookes (1985)

central and southern England, and many urban rivers throughout the UK, are overexploited. Today, most towns and cities rely on interbasin water transfers and virtually all the major rivers in the UK are regulated directly or indirectly by mainstream impoundments, interbasin transfers, pumped storage reservoirs, or groundwater abstractions (Petts, 1988).

Ideally, river management should pursue four objectives (Mellquist, 1992):

- balancing between users' interests,
- optimization of resource use,
- inclusion of environmental interests and those of the general public when exploiting water resources,
- cleaning up after past abuses.

In practice, these objectives can be conflicting, and the relative weight given to each by decision-makers is affected by a wide range of influences which include economic, political and environmental considerations. River management is no different from any other natural environmental management issue in that it involves compromises. Some conservationists, for example, argue that river regulation and environmental conservation are intrinsically incompatible since regulation modifies the natural environment in which original communities of organisms became established (e.g. Hellawell, 1988). Indeed, in some cases the ecological requirements of organisms are destroyed or modified beyond the limits of adaptations and the organisms are unable to survive.

Other conservationists, however, adopt another view of the situation. Moore and Driver (1989) point out that in England and Wales there are more than 500 water supply reservoirs, with a total water surface of 20 000 ha, much of which has been flooded this century. Waterfowl, amphibious and aquatic wildlife have all benefitted from this change, one which is especially important after so

much wetland has been drained and ploughed for increased agricultural production. The general importance of reservoirs to wildlife conservation is indicated by the designation of 174 reservoirs as Sites of Special Scientific Interest. Modern reservoirs include landscaping and conservation in their planning, construction and operation, while protection from recreational activity, a use of resources which is potentially damaging to wildlife, is an essential part of management at these sites.

The importance of ecological good health in river systems has been increasingly recognised in recent times, both from the purely moral standpoint and from the point of view of human society's use of their resources. There are numerous ways in which damage done in the past can be rectified, many of which are not new in themselves. In 1215, the Magna Carta demanded the removal of numerous weirs along the River Thames so that migratory Atlantic salmon and sea trout could pass upstream to their spawning grounds. There was, however, little response to this edict nor to similar statutes issued in the fifteenth century. Currently, the Thames Water Authority plans to install fish passes on navigation weirs as part of a salmon rehabilitation programme initiated in 1979 (Mann, 1988). Extensive damage to migratory salmonid fish stocks has also occurred in Finland, caused by the dredging of rivers and brooks both to facilitate boat traffic and, increasingly this century, for timber-floating. Timber-floating has now almost completely ceased and Finnish water legislation obliges water authorities to make good any damage caused by dredging. The restoration of rapids and their restocking have been increasingly used to rehabilitate damage caused by dredging (Jutila, 1992).

A critical final point to make with regard to river management, however, is the simple fact that management of the channel alone is usually insufficient. It is important to realise that rivers are an integral part of the landscapes through which they flow. Rivers affect the land in their drainage basins and vice versa. Hence, management of human activities in the entire basin, embodied in 'catchment management plans', is the sensible way forward, as Burt (1993) points out in the UK context.

LAKES

Lacustrine degradation

As with rivers, human use of fresh water from lakes has led to numerous impacts. In some cases, entire lakes have dried up as rivers feeding them have been diverted for other purposes. Owens Lake in California is a case in point. Levels began to drop in the second half of the nineteenth century due to offtakes for irrigated agriculture, but an accelerated decline in water levels occurred with the construction of a 360-km water export system by the Los Angeles Department of Water during the first 20 years of this century. The lake, which was 7.6 m deep in 1912, had disappeared by 1930. In its place, the dry bed of lacustrine sediments covering 220 km^2 is a source of frequent dust storms which are hazardous to local highway and aviation traffic, often exceed California State standards for atmospheric particulates and have increased morbidity among people suffering from emphysema, asthma and chronic bronchitis (Reinking et al., 1975). From 1930 to 1969 the Los Angeles system was extended northwards to tap streams flowing into the saline Mono Lake, the level of which had dropped by 14 m between 1941 and 1981, threatening the numerous species endemic to the closed basin lake.

Similar offtakes, in this case for agriculture, have had dramatic consequences for the Aral Sea. Diversion of water from the Amu Darya and Syr Darya rivers to irrigate plantations,

predominantly growing cotton, has had severe impacts on the Aral Sea since 1960. Expansion of the irrigated area in the former Soviet region of Central Asia, from 2.9 million hectares in 1950 to about 7.2 million hectares by the late 1980s, was spurred by Moscow's desire to be self-sufficient in cotton. As a result, inflow to the lake from the two rivers, the source of 90 per cent of its water, was virtually halted in the 1980s. Since 1960, the lake's surface area has shrunk by about 50 per cent (Fig. 12.4), its volume has dropped by two-thirds and salinity levels have risen more than three-fold and now stand close to the salinity of sea water in the open ocean. As a consequence, the Aral Sea fishing industry, which landed 40 000 t in the early 1960s, has been decimated: virtually all native fish species have disappeared. The exposure of large areas of previously submerged sediments has become a dust bowl from which an estimated 43 million tonnes of saline material is deposited on surrounding cropland, with suspected detrimental effects on soil fertility up to several hundred kilometres from the sea coast. The ecological plight of the Aral Sea was not the result of ignorance or lack of forethought, since anecdotal or informal projections of disaster were made in some Soviet quarters well before the situation became serious, but seemingly it is the result of an unspecified cost–benefit analysis in which environmental and health impacts were given little consideration (Glantz et al., 1993).

The introduction of non-native species to lakes is another way in which, often unwittingly, the human impact has caused serious ecological change. Introducing fish species to lakes to provide employment and additional sources of nutrition was a common aspect of economic development projects during the 1950s and 1960s. Predatory species such as largemouth, bass and black crappie, introduced into the oligotrophic Lake Atitlan in Guatemala, succeeded in wiping out many of the smaller fishes which had been previously

(a)

(b)

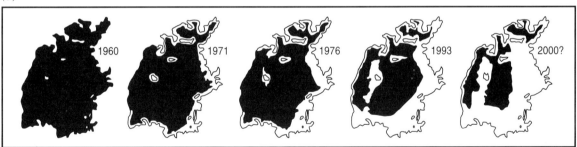

FIGURE 12.4 *(a) Irrigated areas in Central Asia (after Pryde, 1991) and (b) changes in the surface area of the Aral Sea (Glantz* et al., *1993,* Global Environmental Changes *3: 174–98. By permission of the publishers, Butterworth-Heinemann, Ltd. ©)*

used by indigenous lakeside people. Similar dramatic effects on the pelagic food web resulted from the introduction of the Amazonian peacock bass to Gatun Lake in the Panama Canal. One serious consequence of this predator's elimination of the lake's smaller native fishes was a dramatic increase in malaria-carrying mosquitos, since minnows which had previously occupied the shallower lakeside waters had fed off mosquito larvae (Fernando, 1991)

The ecological changes forced by exotic introductions should not necessarily always be seen as bad. In the East African Lake

Victoria, the predatory Nile perch, introduced in the early 1960s, has consumed to very low numbers (some say to extinction) most of the lake's endemic openwater Haplochromine species. However, the Nile perch has significantly boosted the economic output from the lake, with annual fish landings rising from about 40 000 t to 500 000 t over the 20 years to 1992, and the loss of some species may be the price that has to be paid (Viner, 1992). If this view is adopted, however, it is sad to note that the Nile perch is already being exploited at its maximum sustainable yield, and with more processing plants planned it looks set to become overexploited, a fate that afflicts virtually all such new commercial fisheries. Lack of finance prevents the countries concerned from supporting adequate fishery regulatory bodies. One of the main lessons to be learned from Lake Victoria's experience, therefore, is that if a radical change is forced upon an ecosystem, the human component of it requires proper management to keep in step with that forced change.

Another range of problems common to many lakes are those stemming from the introduction of plant species, particularly waterweeds. The water hyacinth is a frequent culprit, growing very fast, covering the water surface (so reducing light and oxygen for other plants and for fish), increasing evaporation, blocking waterways and presenting a serious threat to hydroelectrical turbines. It has become a major problem in Lake Victoria in the 1990s.

There are numerous examples of lakes being degraded inadvertently, many of which are due to lacustrine sensitivity to pollutants. Some examples of problems caused by pollution, particularly acidification from industrial and mining activity, are covered in Chapters 9 and 18. One of the world's most pervasive water pollution problems, and one which is widespread in all continents, is the eutrophication of standing water bodies (Ryding and Rast, 1989). Eutrophication is a natural phenomenon in lakes, brought on by the gradual accumulation of organic material through geological history. But human activity can accelerate the process, so-called 'cultural eutrophication', by enriching surface waters with nutrients, particularly phosphorus and nitrogen. Such increased nutrient concentrations in lakes have been attributed to wastewater discharge, runoff of fertilisers from agricultural land and changes in land use which increase runoff.

Eutrophication is an important water quality issue which causes a range of practical problems. These include the impairment of the following: drinking water quality, fisheries and water volume or flow. The main causes of these problems include algal blooms, macrophyte and littoral algal growth, altered thermal conditions, turbidity and low dissolved solids. Health problems range from minor skin irritations to bilharzia, schistosomiasis and diarrhoea.

Monitoring of Lac Léman (Lake Geneva) indicates how the build-up of nutrient pollutants can be rapid (Fig. 12.5). Water quality deteriorated from its relatively clean state during the 1950s as concentrations of phosphorus increased, primarily from point-source discharges, to reach a critical stage in

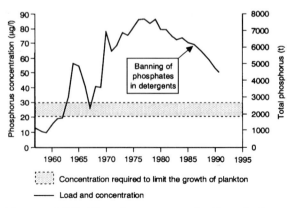

FIGURE 12.5 *Phosphorus concentrations in Lac Léman, Switzerland, 1957–91 (after CIPEL, 1992)*

the late 1970s. The situation has improved, however, following the introduction of tertiary wastewater treatment plants and a ban on the use of phosphates in detergents in 1986 (Rapin *et al.*, 1989).

Lake management

The restoration and protection of lakes presents all sorts of problems for policy-makers and environmental managers, and the management experience of the North American Great Lakes illustrates the difficulties well. A major restoration effort has been in progress throughout the Great Lakes Basin since the US and Canadian governments committed themselves to the Great Lakes Water Quality Agreement in 1978. The agreement adopted an ecosystem approach, which incorporates the political and economic interests of the 'institutional ecosystem' as well as elements of the natural ecosystem, after two previous management strategies had failed to address fully the dangers faced by the lakes (Hartig and Vallentyne, 1989). Current efforts are focused on reducing pollution entering the lakes, incorporating the 'critical loads' approach adopted in studies of acid rain pollution (see Chapter 9). Serious pollutants include volatile organic compounds (VOCs), heavy metals and industrial and agricultural chemicals which come from both point and non-point sources such as municipal and industrial effluent, rainfall and snowmelt. Special and immediate attention is being given to forty-three severely degraded so-called Areas of Concern.

The importance of integrating the ecological needs with those of the lake's users, which are, of course, intimately related, is most acute in lakes which support high population densities. However, the balance of interests also depends upon the nature of the lake and the type of overall regulatory body involved. The example of Lake Baikal in eastern Siberia, the deepest and volumetrically the largest fresh-water lake in the world and home to 800 species of plant and 1550 species of animal life, most of which are endemic to its particularly pure waters, illustrates the point well. Theoretically, the state ownership and centralised decision-making of the former USSR was the ideal structure for successfully integrating ecological and user interests on such a unique freshwater body. In practice, however, as numerous examples of environmental problems quoted in this chapter and elsewhere in this book illustrate, the ecological side of the equation was too often given inadequate consideration.

Efforts to protect Lake Baikal against pollution, a struggle which dates from the mid-1960s, was the first major issue in the rise of public environmental awareness in the former USSR (Pryde, 1972). Although there have been small logging and industrial processing operations on the shores of the lake for several decades, concern was raised at the proposal to establish two large wood processing plants. Great debates occurred over the possible effects of effluents from the plants and the effects of soil erosion resulting from deforestation of the lakeshore to supply the plants. Public outcry succeeded in forcing the introduction of new measures to protect the lake's waters, including a ban on logging on surrounding steep slopes, the establishment of several protected areas around the lake, and the eventual conversion of one of the processing plants to a less-damaging manufacturing role. Nevertheless, the success of these measures is still to be evaluated and the real fate of Lake Baikal is still in the balance (Pryde, 1991).

WETLANDS

The term 'wetland' covers a multitude of different landscape types, located in every major climatic zone. They form the overlap between dry terrestrial ecosystems and permanently inundated aquatic ecosystems such as rivers, lakes or seas.

It is only in the last 30 years or so that wetlands, which currently cover about 6 per cent of the Earth's land surface, have been considered as anything more than worthless wastelands, only fit for drainage, dredging and infilling. But an increasing knowledge of these ecosystems in academic and environmentalist circles has yielded the realisation that wetlands are an important component of the global biosphere, and that the alarming rate at which they are being destroyed is a cause for concern.

Wetlands perform some key natural functions and provide a wealth of direct and indirect benefits to human societies (Table 12.4). Wetlands form an important link in the hydrological cycle, acting as temporary water stores. This function helps to mitigate river floods downstream, protects coastlines from destructive erosion, and recharges aquifers. Chemically, wetlands act like giant water filters, trapping and recycling nutrients and other residues. This role is partly a function of the very high biological productivity of wetlands, amongst the highest of any world ecosystem (see Table 1.1). The importance of this function is indicated by the fact that the 6.4 per cent global wetland area is estimated to contribute 24 per cent of global terrestrial primary productivity (Williams, 1990a). As such, wetlands provide habitats for a wide variety of plants and animals. All these functions provide benefits to human societies, from the direct resource potential provided by products such as fisheries and fuelwood, to the ecosytem value in terms of hydrology and productivity, up to a value on the global level in terms of the role of wetlands in atmospheric processes and general life-support systems (Odum, 1979).

Wetland destruction

Estimates of the global area of wetlands drained vary considerably, but a total of 1.6 million km^2 by 1985 is considered by Williams (1990a) to be a reasonable figure. Nearly three-quarters of this total has occurred in the temperate world, and the prime motivation has been to provide more land to grow food. The wetlands of Europe have been subject to human modification for more than 1000 years. In The Netherlands, where particularly active reclamation and drainage periods occurred in the seventeenth, nineteenth and twentieth centuries, more than half of the national land area is now made up of reclaimed wetlands (Fig. 12.6), and Dutch drainage engineers have hired out their skills to neighbouring countries for more than 300 years.

In the USA, where wetlands are very largely concentrated in the east of the country, with major areas around the Great Lakes, on

TABLE 12.4 *Wetland functions*

Hydrology
Flood control
Groundwater recharge/discharge
Shoreline anchorage and protection

Water quality
Wastewater treatment
Toxic substances
Nutrients

Food-chain support/cycling
Primary production
Decomposition
Nutrient export
Nutrient utilisation

Habitat
Invertebrates
Fisheries
Mammals
Birds

Socio-economic
Consumptive use (e.g. food, fuel)
Non-consumptive use (e.g. aesthetic, recreational, archaeological)

Source: after Sather and Smith (1984)

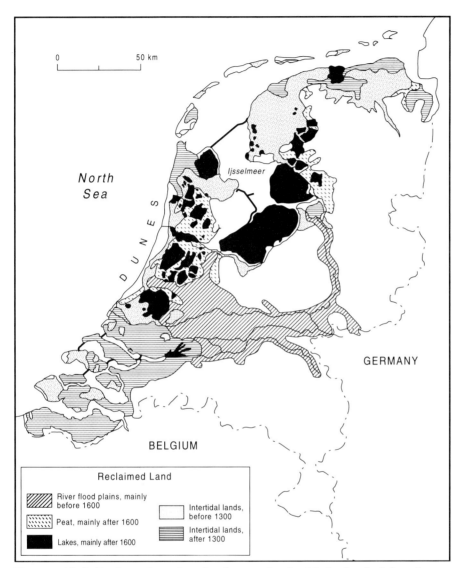

FIGURE 12.6 *Reclamation of wetland types in the Netherlands (after de Jong and Wiggens, 1983)*

the lower reaches of the Mississippi River, and on the Gulf of Mexico and Atlantic seaboard south of Chesapeake Bay, about half of the pre-settlement area of wetlands had been lost by 1975 (Williams, 1990b). Agriculture has been the major beneficiary from the drainage operations in the USA, while urban and suburban development, dredging and mining account for much of the rest.

Although the rich soils of former wetland areas often provide fertile agricultural lands, conversion to cropland is not always successful. Drainage of the extensive marshlands around the Pripyat River in Belarus during the postwar Soviet period, for example, was largely unsuccessful. The low productivity of reclaimed areas necessitated the application of large amounts of chemical fertilisers, and

FIGURE 12.7 *Mexico City's last remaining area of wetland at Xochimilco which is in danger of drying up due to the continuing need for water from the world's largest city*

some former wetland zones also became prone to wildfires (Pryde, 1991).

There have been numerous other reasons for wetland reclamation. Urban sprawl has been a major cause in many areas. Mexico City, for example, was surrounded by five shallow lakes when the Spanish first arrived in 1519, but as the city has expanded since then, all but one small wetland area have been desiccated for their water and land. The remaining area at Xochimilco is threatened by the city's declining water table (Fig. 12.7). On the national scale, in Albania, where about 9 per cent of the country was drained in the period 1946–83, reclamation was as much driven by the fight against malaria as the desire to improve food output, although most of the former marshes were converted to irrigated agriculture (Hall, 1993). In southern Sudan, the Jonglei Canal Project was designed to utilise the large proportion of the White Nile's discharge which currently feeds the Sudd Swamps. The 360-km artificial channel,

which aims to bypass and effectively drain the swamps, will provide much-needed additional water to the dryland countries of Sudan and Egypt if it is ever completed – the project has been hindered by civil war in the region. Plans to utilise the water resources of wetlands also threaten other sites, such as the inland Okavango delta in semi-arid northern Botswana. The general well-being of the Okavango is also under increasing pressure from cattle encroaching into the lush grazing of the delta, a threat currently held at bay by a 300 km cordon fence, but the lure of resources other than water and land has proved to be a serious cause of wetland degradation in many areas.

The fuel resources harboured in wetlands represent perhaps the most important reason in this respect. In many developing countries fuelwood collection is a prime cause of degradation. It is the major threat to wetlands in Central America, as it is in many coastal mangrove swamps in Africa and Asia (see

Chapter 11). In more temperate latitudes, peat is the most important organic fuel of the wetland habitat. Hand-cutting of peat has been practised in rural communities in northern Europe for centuries and continues today in remote parts of Ireland, Scotland, Scandinavia and Russia, but since the 1950s, peat has been cut from much larger areas with machines. Worldwide production reached 100 million tonnes in the late 1970s (Kivinen and Pakarinen, 1981), most for fuel but with exploitation for horticultural purposes particularly important in some countries. Russia is by far the world's largest producer, with Ireland in second place, deriving 15 per cent of energy consumption from peat burning. In Ireland, 94 per cent of raised bogs and 86 per cent of blanket bogs have been lost, inspiring urgent conservation measures for remaining areas. At present rates of exploitation, all unprotected raised bogs will be lost by 1997, and all unprotected blanket bogs will be lost early in the next century (Irish Peatland Conservation Council, 1992).

Wetland protection

Realisation of the value of wetlands to human and animal populations has prompted moves to protect this threatened landscape in recent decades, although in many cases conservationists still face a difficult task in protecting wetlands from destructive development projects. In the case of the Göksu Delta on the south coast of Turkey, threatened by a proposed tourist complex, airport and shrimp farm, the Turkish Society for the Protection of Nature enlisted international help in publicising the danger to the delta and successfully lobbied the Turkish Government, which in 1990 declared the delta a Protected Special Area. The Göksu Delta is a key European site for migrant and wintering birds, holding important populations of no less than twelve globally threatened species, and its beaches are among the main nesting sites for the two Mediterranean species of sea turtle. Following the designation of the delta as a protected area, studies are underway to develop a management plan for the area.

The most comprehensive national efforts towards wetland conservation have been made in the USA, where numerous state and federal laws have a bearing on wetland protection. An attempt is also being made to reverse the loss with the Wetlands Reserve Program, established in 1990, which seeks to restore 405 000 ha of privately owned freshwater wetlands that have been previously drained and converted to cropland.

On the international front, one of the earliest attempts to protect a major world biome was the Ramsar Convention, designed to protect wetlands of international importance. Formally known as the Convention on Wetlands of International Importance especially as Waterfowl Habitats, it was adopted by eighteen countries attending its first conference at Ramsar, Iran in 1971. The convention aims to stem the progressive encroachment on, and loss of, wetlands, and represents one of the first attempts to impose external obligations on the land use decisions of independent states. Each contracting country's main obligation is to designate at least one wetland of international importance for inclusion in a formal list. As of January 1994, this Ramsar List had 641 sites covering 37 million hectares in seventy-four contracting countries. However, many important wetland sites are still not listed and the Convention lacks any legal powers, relying on persuasion and moral pressure to achieve its aims. It is in the developing countries of the world where most concern over the future of wetlands is focused. The rising resource needs of rapidly growing populations, particularly to expand food production, means that many unique wetlands remain under considerable pressure. As with rivers and lakes, sensible wetland management can only be undertaken in these circumstances by incorporating all activities and interest groups present in a particular drainage basin.

FURTHER READING

Boon, P.J., Calow, P. and Petts, G.E. (eds) 1992 *River conservation and management*. Chichester, Wiley. A collection of twenty-nine papers on conservation, recovery, rehabilitation and legislative protection.

Gleick, P.H. (ed.) 1993 *Water in crisis: a guide to the world's fresh water resources*. New York, Oxford University Press. A collection of articles on key issues plus more than 200 tables of data on freshwater resources.

Mason, C.F. 1991 *Biology of fresh water pollution*, 2nd edn. London, Longman. A good introduction to the causes and consequences of river and lake pollution, and techniques for pollution abatement.

Williams, M. (ed.) 1990 *Wetlands :a threatened landscape*. Institute of British Geographers Special Publication 25. Oxford, Blackwell. A comprehensive appraisal of the world's wetlands from both physical and human perspectives. The book focuses on the nature of wetlands, the effects of human impacts and strategies for their management.

13

BIG DAMS

Middleton (1995)

People have constructed dams to harness water resources for at least 5000 years. The first dams were used to control floods and to supply water for irrigation and domestic purposes. Later, the energy of rivers was harnessed behind dams to power primary industries directly, and more recently still, hydroelectricity has been generated using water held in reservoirs, allowing the impoundment and regulation of river flow. The modern era of big dams dates from the 1930s and began in the USA with the construction of the 221-m Hoover Dam on the Colorado River. But in the last 40 years there has been a marked escalation in the rate and scale of construction of big dams all over the world, made possible by advances in earth-moving and concrete technology. Initially, these dams were for hydroelectricity genera-tion, and subsequently for multiple purposes, primarily power, irrigation, domestic and industrial water supply and flood control. Some rivers have been intensively manipu-lated in this way, North America's River Columbia, for example, has, since the mid-nineteenth century, become the site for no less than nineteen major dams and more than sixty smaller ones, making it the world's largest generator of hydroelectricity (Lee, 1989).

The most active phase of large dam construction, defined by the International Commission on Large Dams as structures above 15 m in height, was in the 1950s, 1960s and early 1970s, when 373 dams were completed each year. At the end of 1960, there were 7408 such dams registered. By 1986, the total had reached 36 562. Of those constructed, 64 per cent were in Asia, with no less than half of the world total in China (WRI, 1987; 1992). Excluding China, most of the new structures have been built in the temperate zone, but tropical and subtropical countries such as Brazil, Mexico, India, Thailand, Indonesia, Zimbabwe, Nigeria, Côte d'Ivoire and Venezuela have also become prominent in dam construction. Seventy-nine per cent of these dams are less than 30 m in height, and only 4 per cent exceed 60 m.

COSTS - HYDRO

THE BENEFITS OF BIG DAMS

There is no doubt that many big dam schemes have been successful in achieving their primary objectives. Egypt's Aswan High Dam, completed in 1970, illustrates some of the benefits of big dam construction. Hydroelectricity generated by the dam, a renewable energy source which does not produce any harmful atmospheric pollution, is cheap to operate after the initially high capital costs of dam construction and saves on the purchase of fossil fuels from abroad. The Aswan High Dam generates about 20 per cent of Egypt's electricity.

The natural discharge of the Nile is subject to wide seasonal variations, with about 80 per

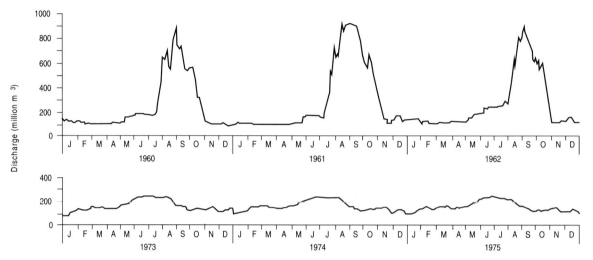

FIGURE 13.1 *Daily discharge regime of the River Nile at Aswan, before and after construction of the High Dam (after Beaumont et al., 1988)*

cent of the annual total received during the flood season from August to October, and marked high and low flows depending upon climatic conditions in the main catchment area in the Ethiopian highlands. The dam allows management of the flow of the Nile's discharge, evening out the annual flow below the dam (Fig. 13.1) and protecting against floods and droughts. Management of the Nile's flow has also had benefits for navigation and tourism, resulting from the stability of water levels in the river's course and navigation channels. Irrigation water for cropland is also provided by the dam's reservoir storage, which has allowed 400 000 ha of cropland to convert from seasonal to perennial irrigation and the expansion of agriculture onto 490 000 ha of new land, a particularly important aspect for a largely hyper-arid country with just 3 per cent of its national area suitable for cultivation (Abu-Zeid, 1989).

Large dams are often seen as symbols of economic advancement and national prestige for many developing nations (Petts, 1984), but the huge initial capital outlay needed for construction often means that agendas are set

to varying extents by foreign interests. A key element in the financing of Ghana's Akosombo Dam, completed in 1965, was the sale of cheap electricity to the Volta Aluminium Company, a consortium with two US-owned companies that produces aluminium from imported alumina, despite the fact that Ghana has considerable reserves of alumina of her own. The construction of the Cahora Bassa Dam in Mozambique (Fig. 13.2) in the 1970s, when the country was still a Portuguese colony, was also largely catering to outside interests. Most of the electricity is sold to South African industry, and part of the original reason for flooding the 250 km-long Lake Cahora Bassa was to establish a physical barrier against Frelimo guerillas seeking independence. The plan for the dam also envisaged the settlement of up to 1 million white farmers in the region, who, it was thought, would fight to protect their new lands.

Within the country, too, concerns have often been raised about the main benefits of a new dam being directed towards urban areas. Wali (1988) points out that although the Bayano Hydroelectric Complex in Panama provides

FIGURE 13.2 *The dam on the River Zambezi at Cahora Bassa in northern Mozambique. Just 10 per cent of the dam's electricity generation capacity would be sufficient for all Mozambique's needs, but the major transmission lines built by the Portuguese serve South Africa, and much of Mozambique is not connected to the dam's supplies*

30 per cent of the country's electricity, no less than 83 per cent of national production is consumed in Panama City and Colón, so that the dam is reinforcing the concentration of wealth in the urban areas.

Hence, it is clear that the undoubted benefits of big dams are not always gained solely by the country where the dam is located, and that within the country concerned, the demands of urban populations can outweigh those of rural areas. Many of the drawbacks of such structures, however, are borne by the rural people of the country concerned. Despite the success of many big dams in achieving their main economic aims, their construction and associated reservoirs create significant changes in the pre-existing environment, and many of these changes have proved to be detrimental. It is the negative side of environmental impacts that have pushed the issue of big dams to a prominent position in the eyes of environmentalists and many other interest groups.

ENVIRONMENTAL IMPACTS OF BIG DAMS

The environmental impacts of big dams and their associated reservoirs are numerous, and Goodland (1990) has outlined the main areas that they influence (Table 13.1).

TABLE 13.1 *Areas of influence of dam and reservoir projects*

1. The catchments contributing to the reservoir or project area and the area below the dam to the estuary, coastal zone and offshore
2. All ancillary aspects of the project such as power transmission corridors, pipelines, canals, tunnels, relocation and access roads, borrow and disposal areas and construction camps, as well as unplanned developments stimulated by the project (e.g. logging or shifting cultivation along access roads)
3. Off-site areas required for resettlement or compensatory tracts
4. The airshed, such as where air pollution may enter or leave the area of influence
5. Migratory routes of humans, wildlife or fish, particularly where they relate to public health, economics, or environmental conservation

Source: Goodland (1990)

The temporal aspect of environmental impacts within a certain area is also important. The river basin itself can be thought of as a system which will respond to a major change, such as the construction of a dam, in many different ways and on a variety of timescales. While the creation of a reservoir creates an immediate environmental change, the permanent inundation of an area not previously covered in water, the resulting changes in other aspects of the river basin, such as floral and faunal communities, and soil erosion, will take a longer time to re-adjust to the new conditions.

The range of environmental impacts consequent upon dam construction, and their effects on human communities, can be considered under three headings which reflect the broad spatial regions associated with any dam project: the dam and its reservoir; the upstream area; and the downstream area.

The dam and its reservoir

The creation of a reservoir results in the loss of resources in the land area inundated. Flooding behind the Balbina Dam north of Manaus, Brazil has destroyed much of a centre of plant endemism, for example. In some cases the loss of wilderness areas threatened by new dam projects has raised considerable debate, both nationally and internationally. A case in point was the Nam Choan Dam Project on the Kwae Yai River in western Thailand, first proposed in 1982. The proposed reservoir lay largely within the Thung Yai Wildlife Sanctuary, one of the largest remaining relatively undisturbed forest areas in Thailand, containing all six of the nation's endangered mammal species. Debate over the destructive impact of the project resulted in it being shelved indefinitely in 1988 (Dixon *et al.*, 1989).

Some resources, such as trees for timber or fuelwood, can be taken from the reservoir site prior to inundation, although this is not always economically feasible in remote regions. There are dangers inherent in not removing them, however. Anaerobic decomposition of submerged forests produces hydrogen sulphide which is toxic to fish and corrodes metal that comes into contact with the water. Corrosion of turbines in Surinam's Brakopondo reservoir has been a serious problem. In a similar vein, decomposition of organic matter by bacteria in the La Grande 2 reservoir in Quebec, Canada has released large quantities of mercury by methylation. Mercury has bioaccumulated in reservoir fish tissue to levels often exceeding the Canadian standard for edible fish of 0.5 mg/kg (Harper, 1992).

Cultural property may also be lost by the creation of a reservoir – twenty-four archaeological sites dating from 70–1000 AD were inundated by the Tucuruí Dam reservoir in Brazil, for example – although in some cases such property is deemed important enough to

FIGURE 13.3 *One of the temples moved from Philae Island which became submerged following completion of Egypt's Aswan High Dam (courtesy of Charles Toomer)*

be preserved. Lake Nasser submerged some ancient Egyptian monuments but major ones – including the temples of Abu Simbel, Kalabsha and Philae – were moved to higher ground prior to flooding (Fig. 13.3).

Big dams often necessitate resettlement programmes if there are inhabitants of the area to be inundated, and the numbers of people involved can be very large. Some of the biggest projects in this respect have been in China. The Sanmen Gorge Project on the Huang Ho River involved moving 300 000 people and the proposed Three Gorges Dam on the Yangtze River may involve the displacement of up to 1.2 million people.

Some indication of the trade-off between land lost, people displaced and power generated is indicated in Table 13.2 for a selection of big dam projects.

For people who are displaced, the move can be a traumatic one. The resettlement of 57 000 members of the Tonga tribe from the area of the Kariba Dam on the Zambezi illustrates some of the adverse effects for the people concerned. Obeng (1978) describes the culture shock suffered in moving to very different communities and environments. Drawn-out conflicts over land tenure resulted between the new settlers and previous residents, and since the resettlement area was drier than the Tongan homelands, problems with planting and the timing of harvests were faced. Deprived of fish and riverbank rodents which traditionally supplemented their cultivated diet, the Tongas faced severe food shortages. When the government sent food aid to relieve the suffering, the food distribution centres became transmission sites for trypanosomiasis.

Development following the construction of big dams can also act as a pull for migrants, bringing associated problems of pressure on local resources. The influx of migrants to the Aswan area has led to an increase in population from 280 000 in 1960 to more than 1 million by the late 1980s, mainly due to the increase in job opportunities (Abu-Zeid, 1989).

Over the longer term, other effects of reservoir inundation become evident. The alteration of the environment can have significant impacts on local health conditions. In some cases these can be beneficial. Onchocerciasis or river blindness, for example, a disease which is common in Africa, is caused by a small worm transmitted by a species of blackfly. The blackflies breed in fast-running, well-oxygenated waters and dam construction can reduce the number of breeding sites by flooding rapids upstream. This has been the case in Ghana's Akosombo and Nigeria's Kainji dams (Worthington, 1978), although the flies may

TABLE 13.2 *Hydropower generated per hectare inundated, and number of people displaced for selected big dam projects*

Project and country	Approx. rated capacity (MW)	Normal area of reservoir (ha)	Kilowatts per hectare	People relocated
Pehuenche (Chile)	500	400	1250	
Guavio (Colombia)	1600	1500	1067	
Itaipu (Brazil and Paraguay)	12 600	135 000	93	8000 families
Sayanogorsk (Russia)	6400	80 000	80	
Churchill Falls (Canada)	5225	66 500	79	
Tarbela (Pakistan)	1750	24 300	72	86 000
Grand Coulee (USA)	2025	32 400	63	
Tucuruí (Brazil)	6480	216 000	30	30 000
Keban (Turkey)	1360	67 500	20	30 000
Three Gorges (China)*	13 000	110 000	12	1 200 000
Batang Ai (Sarawak, Borneo)	92	8500	11	3000
Cahora Bassa (Mozambique)	2075	266 000	8	25 000
Aswan High Dam (Egypt)	2100	40 000	5	100 000
BHA (Panama)	150	35 000	4	4 000
Kariba (Zimbabwe and Zambia)	1500	510 000	3	50 000
Akosombo (Ghana)	833	848 200	0.9	80 000
Brokopondo (Surinam)	30	150 000	0.2	5000

*Under construction, 50 000 people already relocated
Sources: Barrow (1981); Goldsmith and Hildyard (1984); Wali (1988); Dixon *et al.* (1989); Goodland (1990); Gleick (1993)

find alternative breeding sites in new tributary streams.

Malaria, conversely, is likely to increase as a result of water impoundment, since the mosquitos which transmit the disease breed in standing waters. Local malaria incidence has increased around Tucuruí, Brazil, although management by fluctuating water levels and stranding larvae can help as in the USA's Tennessee Valley Authority water management complex.

Schistosomiasis, also known as bilharzia, a very debilitating though rarely fatal disease which is widespread throughout the Third World, is transmitted in a different way: by parasitic larvae that infect a certain aquatic snail species as the intermediate host. Incidence of schistosomiasis was considerably increased by the construction of the Akosombo Dam, with infection rates among 5–19-year-old children rising from 15 per cent to 90 per cent within 4 years of its completion (Worthington, 1978). Similar figures have been reported from other large dams, such as Kariba in Zambia (Hira, 1969).

Other biological consequences of large reservoirs include the rapid spread of water-weeds that cause hazards to navigation and a number of secondary impacts, notably water losses through evapotranspiration. Water-fern appeared in Lake Kariba 6 months after the dam was closed and after 2

TABLE 13.3 *Rates of sedimentation in some Chinese reservoirs*

Reservoir	River	Total sediment deposited (million m³)	Period of record (years)	Storage lost (%)
Sanmenxia	Huang Ho	3391	7.5	35
Qingtongxia	Huang Ho	527	5	84
Yanguoxia	Huang Ho	150	4	68
Liujiaxia	Huang Ho	522	8	11
Danjiangkou	Hanshui	625	15	4
Guanting	Yongdinghe	553	24	24
Hongshan	Laohe	440	15	17
Gangnan	Hutuohe	185	17	12
Xingqiao	Hongliuhe	156	14	71

Source: Biswas (1990)

years had covered 10 per cent of the 420-km² lake area. More dramatic still was the spread of water hyacinth on Surinam's Brokopondo reservoir, which covered 50 per cent of the lake's surface within 2 years (UNEP, 1989b). Similar serious difficulties have been encoutered at Aswan and Pa Mong in Viet Nam.

New reservoirs also have effects on geomorphological and, in some cases, tectonic processes. The trapping of sediment is a particularly important aspect of reservoir impoundment. The siltation of reservoirs has a number of knock-on effects downstream of the dam (see below), but it can also seriously affect the useful life of the dam itself. Some examples of sedimentation rates in Chinese reservoirs are shown in Table 13.3. An extreme example of rapid sedimentation behind a dam is provided by China's Sanmenxia reservoir. River impoundment began in 1960, but within just 7.5 years of operation the reservoir had lost 35 per cent of its total storage capacity of 9700 million m³ due to sedimentation (UNEP, 1989b).

A wide range of techniques is available for reservoir desiltation, the cost of which needs to be budgeted for. Tolouie *et al.* (1993) document the case of the Sefid-Rud reservoir in northwestern Iran which lost over 30 per cent of its storage capacity in the first 17 years after construction. Desiltation successfully restored about seven per cent of total capacity in seven years (Fig. 13.4), but the reservoir had to be emptied during the non-irrigation season to enable sediment flushing. Emptying the reservoir released a highly erosive flow downstream of the dam and hydroelectricity generation was prevented during the operation.

Local heightening of water tables following reservoir impoundment can have deleterious affects on new irrigation schemes through waterlogging and salinisation. Waterlogging is an occasional problem around the Kuban reservoir on the River Kuban, near Krasnodar in southern Russia, when the reservoir is filled above its maximum normal level to aid navigation and benefit rice cultivation. The result has been the ruin of over 100 000 ha of crops, and water damage to 130 communities, including 27 000 homes, 150 km of roads and even the Krasnodar airport (Pryde, 1991). Local changes in groundwater conditions

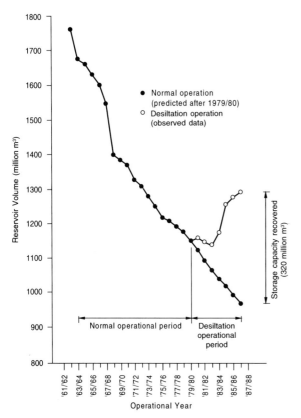

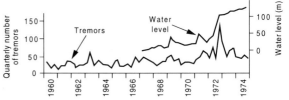

FIGURE 13.5 *Number of earth tremors at the Nurek Dam, Tajikistan, 1960–74 (after Soboleva, O.V. and Mamadaliev, U.A. 1976, The influence of Nurek Reservoir on local earthquake activity. Engineering Geology 10: 293–305. Reproduced by permission of Elsevier Science)*

FIGURE 13.4 *Variation of storage capacity in the Sefid-Rud reservoir, Iran during normal and desiltation operation (after Tolouie et al., 1993)*

have also affected slope stability, causing landslides around some reservoirs. Water displaced by a landslip at the Vaiont Dam in Italy in 1963 overtopped the dam, killing more than 2000 people in the resulting disaster (Kiersch, 1965).

The sheer size of some reservoirs can also create new geomorphological processes. Artifical lakes behind dams on the Volga River are so large that storms can produce ocean-like waves which easily erode the fine wind-blown soils lining the shores. When this process undercuts trees, or creates shoals, navigational hazards result (Pryde, 1991).

The stress changes on crustal rocks induced by huge volumes of water impounded behind

major dams have been suspected of inducing earthquakes in some regions. Nurek Dam on the Vakhish River in central Tajikistan is the best-documented example of a large dam, in this case a 315-m-high earth dam, causing seismic activity. Filling of the dam, located in a thrust-faulted setting, began in 1967 and substantial increases in water level were mirrored by significant increases in earthquakes per quarter (3 months) during the first 8 years of the dam's lifetime (Soboleva and Mamadaliev, 1976; see Fig. 13.5).

The reservoirs behind the Hoover Dam in the USA and Canada's Manic 3 have also induced local seismic activity, although earthquake incidents suspected to have been caused by other big dams, such as those at the Konya Dam near Bombay in India, Egypt's Aswan High Dam, and at the Kurobe Dam on Honshu Island, Japan are unlikely to be due to reservoir-induced stresses (Meade, 1991).

The creation of new water bodies with large surface areas is thought by many to affect local climate. Thanh and Tam (1990) suggest that Lake Volta has shifted the peak rainfall season in central Ghana from October to July/August, for example, but few monitoring programmes have proved such effects conclusively, to date. Changes in the local temperature regime have, however, been observed at the 45 000 ha Rybinsk reservoir north of Moscow in Russia, where the frost-free period has been extended by 5–15 days per year on average in an area of influence that extends for 10 km around the reservoir's shoreline (D'Yakanov and Reteyum, 1965). Evaporation from reservoir surfaces may affect local humidity and the incidence of fog has been observed to rise in some areas.

The upstream area

A variety of upstream impacts can be induced or exacerbated by big dam projects. Some of these, in turn, may impact the dam project itself. Notable in this respect is the improved access to previously remote areas. Deforestation in the watershed above the Ambukloo Dam in the Philippines has led to sedimentation of the reservoir, reducing its useful life from 60 years to 32 years (UNEP, 1989b). Conversely, afforestation of catchments above dams has been carried out in many areas specifically to limit sediment accumulation in reservoirs. In the UK, for example, many water authorities have bought land in upper catchments to plant new forests.

The downstream area

Downstream of a reservoir, the hydrological regime of a river is modified. Discharge, velocity, water quality and thermal characteristics are all affected leading to changes in geomorphology, flora and fauna, both on the river itself and in estuarine and marine environments.

The trapping of sediment behind dams leads to reduced loads in the river downstream. The effects of construction of the Danjiangkou Dam on the sediment load of the Han River, a tributary of the Yangtze, are shown in Fig. 13.6. The resulting flow downstream of the dam is highly erosive, with degradation of the bed and banks observed 480 km below the dam at Xiantao

(Chien, 1985). Similar effects on the River Nile have been noted, downstream of the Aswan High Dam, and the lack of silt arriving at the Nile delta has had effects on coastal erosion, salinisation through marine intrusion and a decline in the eastern Mediterranean sardine catch (Abu-Zeid, 1989). Figure 13.7 illustrates some of the changes that have occurred in the northern Nile delta in the last 200 years, due both to the effects of barrages and dams on the Nile and associated water use, as well as the spread of irrigated farmland and the natural rise in sea level which can be expected to accelerate with global warming. Expected further changes due to these factors are also shown. To some extent the loss of fisheries off the Nile delta has been offset by a new fishing industry in Lake Nasser, which has provided employment to 7000 fishermen. Downstream changes in salinity due to construction of the Cahora Bassa Dam in Mozambique are also threatening mangrove forests at the mouth of the Zambezi. Mangroves provide the breeding grounds for prawn and shrimp, a major source of foreign currency, but strategic water release from the dam could be used to offset the possible deleterious effects on shrimp and prawn catches (Gammelsrød, 1992). The absolute reduction in volumes of flow following dam construction also affects the ecology of downstream seas, as the example of the Black Sea illustrates well (see Chapter 10).

Dams can also affect marine and lake fish populations through the barrier they create which effectively cuts off access to spawning grounds. This effect has been evident on salmon and aloses in the River Garonne and its tributaries in southwestern France since the Middle Ages (Décamps and Fortuné, 1991). In the twentieth century, decline in the landed catches of Caspian Sea sturgeon, a source of caviar, from 40 000 t early this century to just 11 000 t in the 1970s, is attributable primarily to large hydroelectric dams on the Volga and the consequent loss of spawning grounds.

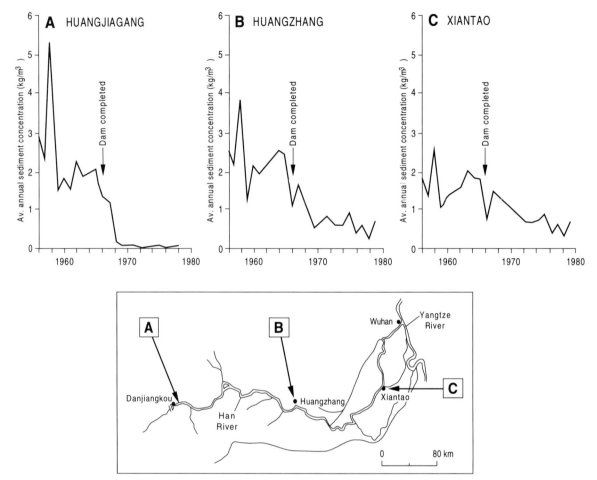

FIGURE 13.6 *Effect of construction of the Danjiangkou Dam on sediment loads of the Han River, China (after Chien, 1985)*

However, catches had largely recovered to pre-dam levels in the 1980s with the establishment of new sturgeon farms on the Caspian shores (Pryde, 1991).

POLITICAL IMPACTS OF BIG DAMS

Increasing public awareness of the environmental and social implications of big dams has generated some heated debates in recent years, in some cases leading to the shelving of construction plans. The Nam Choan Dam Project in Thailand has been mentioned in this respect. Another example is the Tasmanian Franklin River Project which was stopped on environmental grounds in 1983. India's Narmada Dam and the Chinese Three Gorges Project have also come under severe criticism over their anticipated impacts. In such cases, the obvious benefits of dam construction must be carefully weighed against the costs measured in environmental and social terms, and the potential impacts predicted and ameliorated by sensible planning (Goodland, 1986). There is little doubt that many of the

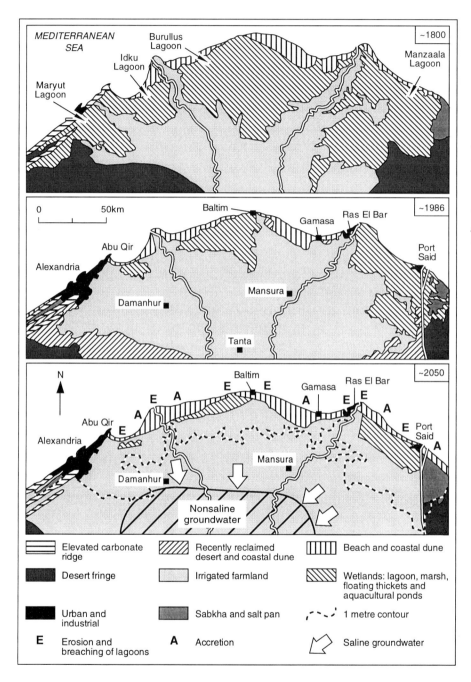

FIGURE 13.7 *Nile delta changes (reprinted with permission from Stanley, D.J. and Warne, A.G. 1993. Nile delta: recent geological evolution and human impact. Science 260: 628–34. Copyright 1993 American Association for the Advancement of Science)*

adverse impacts of dams can be reduced greatly by good planning and anticipation, and aid agencies that finance such projects now require an Environmental Impact Assessment before approval of funding is given. Progress has also been made in the widening scope given to consultation prior to dam construction (Table 13.4). Nevertheless, a sensible operating schedule is also a key factor – many of the problems caused by dams are

TABLE 13.4 *Broadening of consultation in the design of big dams*

Design team	Approximate era
Engineers	Pre-Second World War
Engineers + economists	Post-Second World War
As above + environmentalists	Late 1980s
As above + affected people	Early 1990s
As above + NGOs	Late 1990s
As above + national consensus	Early 2000s?

Source: after Goodland *et al.* (1993b)

the result of operators aiming to maximise water use, through releases for hydroelectricity generation and irrigation, for example, to such an extent that other concerns are given too little consideration.

The building of dams on rivers flowing through more than one country brings international political considerations onto the agenda of big dam issues. Such considerations are particularly pertinent in dryland regions where rivers represent a high percentage of water availability to many countries. The main issues at stake here are those of water availability and quality.

In several international river basins, peaceful cooperation over the use of waters has been achieved through international agreement. One such agreement, between the USA and Mexico over use of the Rio Grande, was signed in 1944 and is operated by the International Boundary and Water Commission. This body ensures equal allocation of the annual average flow between the two countries.

In other international basins, such as the Tigris–Euphrates in the Middle East, the lack of agreement represents a significant potential for conflict. While there is currently a water surplus in this region, the scale of planned developments raises some concern (Agnew and Anderson, 1992). Turkey's Southeastern Anatolian Project, a regional development scheme on the headwaters of the two rivers,

centres on twenty-two dams. In early 1990, when filling of the Ataturk Dam reservoir commenced, stemming the flow of the Euphrates, immediate alarm was expressed by Syria and Iraq, despite the fact that governments in both countries had been alerted and discharge before the cut-off had been enhanced in compensation. Syria and Iraq nearly went to war when Syria was filling its Euphrates Dam. Full development of the Southeastern Anatolian Project could reduce the flow of the Euphrates by as much as 60 per cent, which could severely jeopardise Syrian and Iraqi agriculture downstream. The three Tigris–Euphrates riparians have tried to reach agreements over the water use from these two rivers, and the need for such an agreement is becoming ever more pressing.

FURTHER READING

Cummings, B.J. 1990 *Dam the rivers, damn the people*. London, Earthscan. The story of two major hydroelectric developments in the Amazon and the conflicts between them and local environments and populations.

Goldsmith, E. and Hildyard, N. 1984 (Vol. 1), 1986 (Vol. 2) *The social and environmental effects of large dams*. Wadebridge, Cornwall, Wadebridge Ecological Centre. Volume 1 presents arguments against water development based on large dams, and Volume 2 contains thirty-one detailed case studies of big dam projects. These books have a very definite bias against the development of big dams (see also Trussell below).

Petts, G.E. 1984 *Impounded rivers – perspectives for ecological management*. Chichester, Wiley. A comprehensive appraisal of the ecological effects of dams and their reservoirs on world rivers.

Trussell, D. (ed.) 1992 *The social and environmental effects of large dams*, Vol. 3. Wadebridge, Cornwall, Wadebridge Ecological Centre. The third volume in this series reviews the very wide literature on large dams by topic. As with the other volumes, the aim throughout is to present a strong case against big dams.

14

URBAN ENVIRONMENTS

Large numbers of people have lived in close proximity to each other in cities for thousands of years. The first urban cultures began to develop about 5000 years ago in Egypt, Mesopotamia and India, but the size of cities and their geographical distribution expanded dramatically after the Industrial Revolution in the present millennium. The growth rates of cities in recent decades has been unprecedented. In 1970, four world cities had a population of more than 10 million people; by the year 2000 there will be twenty-four such cities. While there were thirty-five cities of greater than 3 million people in 1970, by the end of the century that total will reach eighty-five, by which time 47 per cent of the world's population will be living in urban areas (UN, 1989a). Fig. 14.1 indicates the growth of twenty so-called megacities. Many cities in the developed world, such as New York and London, will have grown little in the last 30–50 years of the present century, but cities in the industrialising world show remarkable growth over the same period. Estimates indicate that the populations of Mexico City, São Paulo, Karachi and Seoul will have grown by more than 800 per cent in the second half of the twentieth century. The urban area of Mexico City, which is probably already the world's largest conurbation, expanded from 27.5 km² in 1900 to cover 1250 km² in 1990.

The phenomenal growth of some cities, and the high concentrations of people they represent (the urban density of Mexico City in 1980 was 14 082 people/km²), has created some acute environmental problems both outside and within the city limits.

Cities represent a completely artificial environment; they absorb vast quantities of

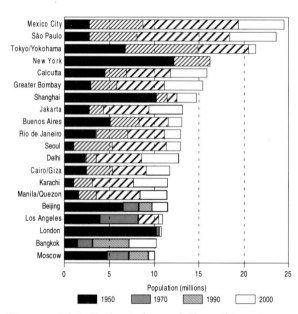

FIGURE 14.1 *Estimated population of twenty megacities in 1950, 1970 and 1990, and projections for the year 2000 (from data in UN, 1989a)*

resources from surrounding areas and create high concentrations of wastes to be disposed of. The degree to which cities impinge on their hinterlands is indicated by a few examples. About 10 per cent of prime agricultural land has been lost to urbanisation in Egypt. The twentieth century growth of São Paulo was fuelled by the expansion of coffee plantations in south-east Brazil which reduced the forest cover of São Paulo State from 81 per cent in 1860 to 6 per cent in the late 1980s (Monteiro, 1989). The demand for water in Tehran spurred the construction of a series of dams and canals in the early decades of this century, to bring water 50 km from the River Karaj to the west, reducing the water available for rural agriculture. By the 1970s, supplies were again running low, so water was diverted more than 75 km from the River Lar to the north-east (Beaumont *et al.*, 1988). In Rio de Janeiro's Guanabarra Bay, pollution from two oil refineries, two ports, 6000 industries, twelve shipyards, sixteen oil terminals, sewage and garbage dumps has reduced commercial fishing by 90 per cent, mangrove cover by 90 per cent, led to outbreaks of water-borne diseases such as infectious hepatitis and typhoid, and is silting the bay by 81 cm/100 years (Kreimer *et al.*, 1993).

The acute environmental problems that occur within many cities, particularly in the developing world, their underlying reasons, and the scale of the clean-up task faced by urban authorities are well summarised in the case of Manila Metro, capital of the Philippines. A 1990 population of 8 million is projected to rise to 13 million by the year 2000. All the city's rivers are biologically dead. Each day, 2000 t of solid waste is left uncollected, to be burnt, thrown into waterways or moulder on the ground. Much of the garbage which is collected is dumped on 'Smokey Mountain', a 23-ha open tip which represents a severe health hazard to the 20 000 people who reside on its fringes and earn a living by scavenging from the dump (Jimenez and Velasquez, 1989).

About 65 per cent of the country's 1500 recognised industrial enterprises are located in the Manila Metro area, and only one-third to one-half of them are thought to comply with minimal air and water pollution emission standards. One million vehicles, more than half country's total, operate in the Manila Metro area. Just half of these vehicles are thought to meet even minimal emission standards. The annual cost to the economy due to congestion alone is estimated to be more than US$50 million, which is low by the standards of other Asian capitals, while the economic burden of air pollution may be an order of magnitude higher (Table 14.1).

TABLE 14.1 *Some estimates of the annual cost of congestion and air pollution in selected Asian cities*

City	Cost (US$ million/year)	
	Congestion	Air pollution
Bangkok Metropolitan Area (1989)	272	380–580
Bangkok Metropolitan Region (1993)	400	1300–3100
Seoul	154	—
Manila	51	—
Jakarta	68	400–800

Source: after Brandon and Ramankutty (1993)

The basic cause of Manila Metro's severe environmental problems is that 8 million people are using infrastructure, much of which dates from the US colonial period, estimated to be adequate for about 2 million people, at most. A large proportion of the solid and liquid wastes are simply inaccessible for collection by virtually any means due to the density of squatter settlements, inappropriate collection systems and the simple lack of services such as septic tank desludging. The problems of physical infrastructure are exacerbated by the goverment's inability to stop polluters, largely a function of serious understaffing at the metropolitan regulation agency (World Bank, 1989).

SURFACE WATER RESOURCES

One of the most important environmental issues that stems from urban modifications to the hydrological cycle is that of poor water quality. Runoff from developing urban areas is usually choked with sediment during construction phases, when soil surfaces are stripped of vegetation, and a finished urban zone greatly increases runoff due to widespread impermeable city surfaces of tarmac and concrete, and networks of storm drains and sewers. This drastically modified urban drainage network feeds large amounts of urban waste products into rivers and ultimately into oceans.

Many rivers that flow through urban areas are biologically dead, thanks to heavy pollution. Hardoy *et al.* (1992: 73) sum up the state of urban rivers in developing countries as follows: 'Most rivers in Third World cities are literally large open sewers'. They go on to point out that of India's 3119 towns and cities, only 209 have partial sewage treatment facilities and just eight have full facilities. India's Jamuna River, for example, contains 7500 coliform organisms per 100 ml of water on entering New Delhi, a figure which rises to 24

million coliform organisms per 100 ml after flowing through the city. For comparison, the WHO guidelines for such microbiological pollution are <10 coliform organisms per 100 ml for drinking water and <1000 per 100 ml for irrigation purposes. Industrial effluents combine with this domestic source of riverine pollution to make urban rivers the most polluted freshwater sources on Earth.

All the rivers flowing through Jakarta, Indonesia, are heavily polluted from numerous, mostly untreated, discharge sources: household drains and ditches, overflows and leaks from septic tanks, commercial buildings, and industries. Water-related diseases such as typhoid, diarrhoea and cholera increase in frequency downstream across the metropolitan area (Hardoy *et al.*, 1992). Untreated sewage and discharge from 20 000 classified water-polluting industries which feed into Bangkok's canal system have created a distinct sag in the dissolved oxygen profile of the Chao Phraya where the canals feed the river (Phantumvanit and Liengcharernsit, 1989). Although the example of the Thames at London (see Fig. 12.2) shows how such near-anaerobic river conditions can be improved, neither the money nor the political will are currently as forthcoming in Thailand.

The local hydrological impact of the Saudi capital, Riyadh, provides a very contrasting example to the depressing catalogue of riverine disaster areas typically associated with large, rapidly growing cities. Discharge of Riyadh's wastewater feeds the Riyadh River, which scarcely existed 20 years ago, but now flows throughout the year down what was the seasonal Wadi Hanifa. The water, which is originally derived from desalinated Gulf sea water, is partially treated before being released to flow down the steep-sided wadi and enter open countryside, eventually disappearing 70 km from Riyadh. The new flow has created an attractive valley lined by tamarisk trees and phragmites which is becoming an important recreational site for Riyadh's 2.3

million population. Beyond the wadi, significant irrigated agriculture has grown up, drawing on the groundwater around the river. This unique new feature is, however, under some threat from needs to further recycle the much-needed water resource (Meynell, 1993).

Different types of environmental problems are encountered in permafrost areas where surface water and soil moisture is frozen for much, and in some places all, of the year. Frozen rivers and lakes mean that many of the uses such water bodies are commonly put to at more equable latitudes, such as sewage and other waste disposal, are not always available. The low temperatures characteristic of such regions also means that biological degradation of wastes proceeds at much slower rates than those elsewhere. Hence, the impacts of pollution in permafrost areas tend to be more long-lasting than in other environments.

The nature of the permafrost environment also presents numerous environmental challenges to the construction and operation of settlements, challenges which have been encountered in urban developments associated with the exploitation of hydrocarbons and other resources in Alaska, northern Canada and northern Russia. Disturbance of the permafrost thermal equilibrium during construction can cause the development of thermokarst – irregular, hummocky ground. The heaving and subsidence caused can disrupt building foundations and damage

pipelines, roads, railtracks and airstrips. Terrain evaluation prior to development is now an important procedure in the development of these zones, following expensive past mistakes. Four main engineering responses to such problems have been developed: permafrost can be neglected, eliminated, preserved, or structures can be designed to take expected movements into account (Johnston, 1981). Preservation of the thermal equilibrium is achieved in numerous ways, such as by insulating the permafrost with vegetation mats or gravel blankets, and ventilating the underside of structures which generate heat (e.g. buildings and pipelines).

GROUNDWATER

The water needs of urban population and industry is often supplemented by pumping from groundwater, and pollution of this source is another problem of increasing concern in many large cities. Seepage from the improper use and disposal of heavy metals, synthetic chemicals and other hazardous wastes such as sewage, is a principal origin of groundwater pollution. Some of the major pollutants involved in a selection of major cities are shown in Table 14.2. The quantity of such compounds reaching groundwater from waste dumps in Latin America, for example, is thought to be doubling every 15 years

TABLE 14.2 *Some examples of urban groundwater pollution*

City/region	Country	Major pollutants
Merida	Mexico	Bacteria
Milwaukee	USA	Cl, SO_4, bacteria
Birmingham	UK	Majors, metals, B, P, Si, CN, organics
Narbonne	France	SO_4, NO_3
Cairo	Egypt	NO_3, majors, metals
	Bermuda	micro-organisms, Cl, NO_3

Source: after Lerner and Tellam (1993: 324, Table 1)

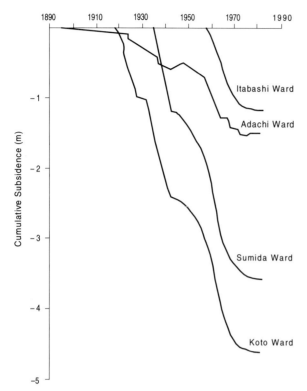

FIGURE 14.2 *Ground subsidence in the Ward area of Tokyo Metropolis (after TMG, 1985)*

(World Bank, 1992a). Aquifers do not have the self-cleansing capacity of rivers and, once polluted, are difficult and costly to clean.

A frequent outcome of overusing groundwater is a lowering of water-table levels and consequent ground subsidence. In Mexico City, use of subterranean aquifers for more than 100 years has caused subsidence of up to 9 m in some central areas (Schteingart, 1989), greatly increasing the flood hazard in the city and threatening the stability of some older buildings, notably the sixteenth century cathedral.

Marked subsidence episodes in Tokyo have mirrored phases of economic and industrial growth. The Tokyo Metropolitan Government suggests that ground subsidence began in the city as economic activity grew after the First World War and came to a halt for some years

in the 1940s following the destruction of industries in the Second World War (TMG, 1985; see Fig. 14.2). Renewed industrial activity during the Korean War accelerated the process once more, but the trend was slowed in the 1960s with the introduction of pumping regulations, and subsidence has been virtually halted since these regulations were strengthened in 1972.

In other coastal cities, depletion of aquifers has created problems of seawater intrusion. Overpumping of groundwater in the Tel Aviv urban area depleted groundwater levels to below sea level over an area of 60 km² in the 1950s, requiring a programme of freshwater injection along a line of wells parallel to the coast in an attempt to redress the saltwater/freshwater balance. The programme was successful, effectively stabilising the aquifer and preventing saltwater intrusion.

A similar pattern of events occurred in Brooklyn, New York City, although here no attempt was made to prevent seawater intrusion. By 1947, pumping of the increasingly saline groundwater had ceased and all freshwater supplies were provided by surface sources. Cessation of pumping gradually allowed the water table, which had been reduced to about 11 m below sea level, to rise again. During the half century of pumping however, deep basements, building foundations and subways had been sunk, and these were subject to flooding as the groundwater levels rose, necessitating expensive remedial measures.

Rising groundwater levels have become a critical problem for many 'post-industrial' cities as manufacturing industries have given way to service industries which are much less demanding of water, and legislation has been introduced to control subsidence problems. In London, where loss of water from aged pipes is an additional reason for groundwater levels rising, the period of change from a generally falling to a rising water table occurred in the late 1970s. The potential effects upon the fabric of London's urban environment are now being assessed, with particular interest being shown by the insurance industry. A report issued by the Construction Industry Research and Information Association estimated that the cost of pumping to maintain groundwater levels below the level at which serious damage would occur was up to £30 million (CIRIA, 1989).

The rising groundwater problem has also been reported from many Middle Eastern cities where rainfall is commonly low, potential evaporation high, and natural recharge small and sporadic. Inadvertent artificial recharge from leaking potable supplies, sewerage systems and irrigation schemes has caused widespread and costly damage to structures and services and represents a significant hazard to public health (George, 1992).

THE URBAN ATMOSPHERE

Urban areas have a diverse catalogue of effects on local elements of climate which are well documented (e.g. Landsberg, 1981), but the most serious environmental issue pertaining to the urban atmosphere is that of quality. The principal sources of air pollution in urban areas are derived from the combustion of fossil fuels for domestic heating, for power generation, in motor vehicles, in industrial processes and in the disposal of solid wastes by incineration. These sources emit a variety of pollutants, the most common of which have long been sulphur dioxide (SO_2), oxides of nitrogen (NO and NO_2, collectively known as NO_x), carbon monoxide (CO), suspended particulate matter (SPM) and lead (Pb). Ozone (O_3), another 'traditional' air pollutant associated with urban areas and the main constituent of photochemical smog, is not emitted directly by combustion, but is formed photochemically in the lower atmosphere

from NO_x and volatile organic compounds (VOCs) in the presence of sunlight. Sources of the VOCs include road traffic, the production and use of organic chemicals such as solvents and the use of oil and natural gas.

These atmospheric pollutants affect human health, directly through inhalation, and indirectly through such exposure routes as drinking water and food contamination. Most traditional air pollutants directly affect respiratory and cardiovascular systems. For example, CO has a high affinity for haemoglobin and is able to displace oxygen in the blood, leading to cardiovascular and neurobehavioural effects. High levels of SO_2 and SPM have been associated with increased mortality, morbidity and impaired pulmonary function, and O_3 is known to affect the respiratory system and irritate the eyes, nose and throat and to cause headaches. Certain sectors of the population are often at greater risk: the young, the elderly and those weakened by other debilitating ailments, including poor nutrition.

Elements of the natural and built environment can also be adversely affected. Sulphur and nitrogen oxides are principal precursors of acid deposition (see Chapter 9), SO_2, NO_2 and O_3 are phytotoxic – O_3, in particular, has been implicated in damage to crops and forests – and damage to buildings, works of art and materials such as nylon and rubber have been attributed to SO_2 and O_3.

In more recent times, these traditional urban air pollutants have been supplemented by a large number of other toxic and carcinogenic chemicals which are increasingly being detected in the atmospheres of major cities. They include heavy metals (e.g. beryllium, cadmium and mercury), trace organics (e.g. benzene, formaldehyde and vinylchloride), radionuclides (e.g. radon) and fibres (e.g. asbestos). The sources of these pollutants are diverse, including waste incinerators, sewage treatment plants, manufacturing processes, building materials and motor vehicles.

TABLE 14.3 *Cities of the former Soviet Union with the poorest air quality**

City	Main pollutants	Principal sources
ARMENIA		
Yerevan	Benzo(a)pyrene, SP, NO, NO_2	Chemical industry, power plants, non-ferrous metallurgy, transport
BELARUS		
Mogilev	Benzo(a)pyrene, CS_2, NO_2	Chemical industry, ferrous metallurgy
GEORGIA		
Kutaisy	Benzo(a)pyrene, phenol, SP	Chemical and petrochemical industries, truck plant, transport
KYRGYZSTAN		
Bishkek	Benzo(a)pyrene, formaldehyde, NO	Power plants, transport
RUSSIA		
(Asia)		
Angarsk	Benzo(a)pyrene, formaldehyde, SP	Petrochemical, medical and biological industries, power plants
Bratsk	Benzo(a)pyrene, formaldehyde, CS_2, HF	Non-ferrous metallurgy, pulp and paper mills, power plants
Kemerovo	Benzo(a)pyrene, formaldehyde, CS_2	Fertiliser production, chemical industry, non-ferrous metallurgy
Komsomol'sk-na-Amure	Benzo(a)pyrene, formaldehyde, NH_3, lead	Ferrous metallurgy, power plants, petrochemical industry
Krasnoyarsk	Benzo(a)pyrene, formaldehyde, SP	Chemical industry, non-ferrous metallurgy, construction materials production, transport
Novokuznetsk	Benzo(a)pyrene, HF, NO_2	Metallurgy, coal mining, power plants
Novosibirsk	Benzo(a)pyrene, formaldehyde, NO_2, NH_3	Transport, power plants, construction materials production
Khabarovsk	Benzo(a)pyrene, formaldehyde, phenol	Power plants, construction materials production, petrochemical industry, road and rail transport
(Europe)		
Berezniky	CS_2, H_2SO_4, NO, NO_2	Chemical industry, fertiliser production
Volzskyi	Formaldehyde, metilmercaptan, NO_2, CS_2	Petrochemical and chemical industries
Gubakha	Sterene, benzo(a)pyrene, formaldehyde	Chemical industry, power plants
Magnitogorsk	Benzo(a)pyrene, sterene, CS_2, NO_2	Ferrous metallurgy
Novocherkassk	Benzo(a)pyrene, formaldehyde, SP	Metallurgy, petrochemical industry, power plants
UKRAINE		
Kiev	Benzo(a)pyrene, formaldehyde, NH_3	Chemical and petrochemical industries, construction materials production

*Annual mean of three or more pollutants exceeded the 24-h national limit in 1990
Source: after Shahgedanova and Burt (1994)

Concentrations of these chemicals are generally low, where they are measured, but this occurs at few sites to date.

Few long-term air quality monitoring programmes have been implemented in cities, and runs of available data are often characterised by changes in the location of sample stations, but data from the former Soviet Union indicate that urban areas with high concentrations of heavy industry using outdated technology have had a 'calamitous' effect on air quality (Shahgedanova and Burt, 1994). Table 14.3 lists cities where the annual levels of three or more pollutants exceeded the 24-h national limit in 1990. Some of the commonest pollutants present in excessive concentrations are highly toxic: benzo(a)pyrene, a carcinogenic coal-tar by-product, phenol and formaldehyde.

However, it seems unlikely that the acute air pollution problems of these urban areas will receive much attention as long as financial resources are limited and more pressing national problems such as housing and food supply continue to head government priorities.

Monitoring of urban air quality has been undertaken at a global network of megacities by UNEP and the WHO since 1974. Available data indicate that while cities in the industrialised countries have made significant reductions in air pollution during the past four decades, rapidly growing urban areas in the industrialising countries pose serious threats to the millions of people who live in them. The clear distinction in urban air quality between rich and poor countries is indicated in Figure 14.3.

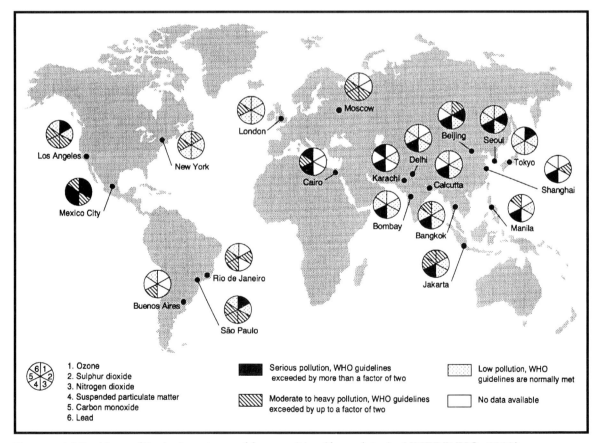

FIGURE 14.3 *Air quality in twenty world megacities (from data in UNEP/WHO, 1992)*

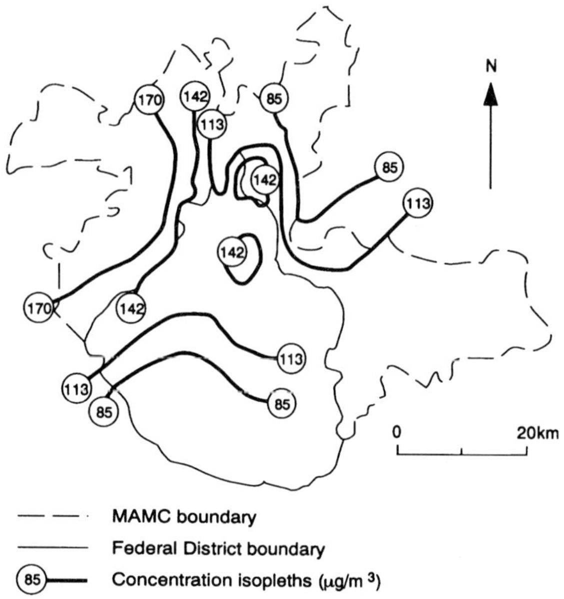

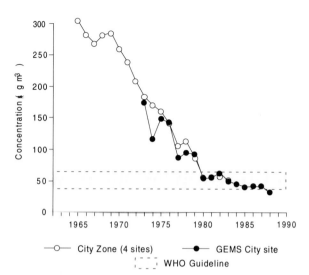

FIGURE 14.4 *Mean annual sulphur dioxide distribution in Mexico City, 1989 (after UNEP/WHO, 1992). The WHO guideline standard is 40–60 µg/m³*

FIGURE 14.5 *Mean annual sulphur dioxide concentrations in the city of London, 1965–88 (after UNEP/WHO, 1992)*

Mexico City emerges as the worst affected city, with WHO guidelines exceeded by a factor of two or more for levels of SO_2 (see Fig. 14.4), SPM, CO and O_3. Levels of Pb and NO_2 are almost as bad, exceeding WHO limits by up to two times. The city's poor air quality is exacerbated by its location, in an elevated mountain-rimmed basin where temperature inversions occur on average 20 days per month from November to March, which impairs dispersion of pollutants (Collins and Scott, 1993). Although data are sparse, no particular trends in the six pollutants monitored are discernible in Mexico City, despite the city's rapid growth. This can be attributed to use of cleaner fuels, better emission control, replacement of old industries, and technological improvements (UNEP/WHO, 1992). Older taxis, for example, are being replaced with newer models equipped with catalytic converters.

Not all efforts to control air pollution in the city have been unmitigated successes however. Concern over rising atmospheric Pb levels, which averaged 8 g/m³ in 1986 (five times the national standard), resulted in the national oil company reducing the lead content of gasoline sold in the city in September of that year. An unexpected side-effect was a dramatic increase in ozone concentrations, a result of the reaction between atmospheric oxygen and the replacement gasoline additives in ultraviolet sunlight (Ezcurra, 1990).

The severe pollution conditions observed at several megacities could have been much worse if control measures had not already been introduced. Examples include Beijing, Delhi, Seoul and Shanghai, and the need for such measures is well illustrated at Shanghai where the male lung cancer mortality rate has doubled from twenty-one to forty-four per 100 000 men from 1963 to 1985 (Rukang, 1989).

The beneficial effects of tighter legislative controls on air quality are indicated by

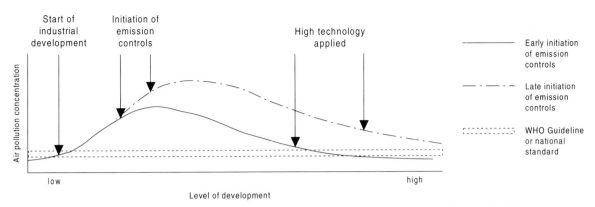

FIGURE 14.6 *Model of air quality evolution with development status (after UNEP/WHO, 1992)*

London's annual mean SO_2 concentrations, which have fallen from 300–400 g/m³ in the mid-1960s to around 20–30 g/m³ in the late 1980s (Fig. 14.5). The introduction and enforcement of 'Smoke Control Orders' under the 1956 Clean Air Act (amended in 1964 and 1968), a response to the infamous London smogs of the 1950s, is the most important factor responsible for this steady 30-year fall in ambient concentrations. Similar successes have been recorded for most of the six pollutants measured in Los Angeles, New York and Tokyo, although Los Angeles still has the most serious O_3 problem in the USA. A model for the progression of air pollution problems through time at different levels of development is shown in Fig. 14.6. Pollution rises with initial industrial development, to be brought under control through legislation on emissions. Air quality then stabilises and improves as development proceeds, to be reduced to below acceptable standards by high technology applications.

GARBAGE

The rapid, and often unauthorised, growth of urban areas has in many cases outpaced the ability of urban authorities to provide adequate facilities, such as the collection of

TABLE 14.4 *Some large cities with poor household garbage collection facilities*

City (Country)	Proportion of garbage not collected (%)
Accra (Ghana)	90
Bogotá (Colombia)	50
Dar es Salaam (Tanzania)	65
Guatemala City (Guatemala)	32
Jakarta (Indonesia)	40
Karachi (Pakistan)	65
Kampala (Uganda)	90
Manila (Philippines)	50

Source: after Hardoy et al. (1992)

household garbage. Table 14.4 illustrates the scale of the problem in some cities for which reasonable estimates are available. Many other urban areas similarly afflicted are not included due to lack of adequate information.

Although the environmental problems associated with garbage do not disappear with its collection (see Chapter 16), uncollected garbage exacerbates many of the environmental hazards covered in this chapter. It can be a serious fire hazard; it attracts pests and disease vectors, creating

Figure 14.7 *Marabou storks are a common site in Kampala, Uganda. In the countryside, these scavengers often feed on carrion and were probably first attracted to the cityscape by discarded human corpses during the times of Idi Amin and Milton Obote. They now live on garbage and rodents. Less than 10 per cent of the city's population benefits from a regular collection of household wastes*

health hazards; and local disposal by burning or dumping adds to pollution loads and clogs waterways, so increasing the dangers of flooding.

Several animal species, particularly rats, have become adapted to the urban environment by scavenging from urban refuse. Larger species, too, have been drawn to garbage bins and dumps, and are also regarded as pests, such as the urban foxes which inhabit many British cities and polar bears in Churchill, Manitoba, northern Canada. In Uganda's capital city, Kampala, carnivorous Marabou storks strut the streets like normal citizens, living off garbage and doing a useful job in controlling smaller pests (Fig. 14.7).

Some degree of waste recovery occurs in most cities. In many cities of the Third World, large numbers of residents are self-employed in the business of garbage recycling. Mexico City and Cairo are just two examples where large squatter communities live and work on official or unofficial rubbish dump sites. In the case of Cairo, the Zabbalean religious sect has cornered the market in garbage collection, scavenging and recycling, feeding edible portions to their domestic livestock and selling inorganic materials to dealers. Elsewhere, metropolitan authorities run similar programmes. In Beijing, for example, a state-run recycling scheme has been in operation since the 1950s, and in New York City, Local Law 19, brought into force in 1989, requires all residents, institutions and businesses to separate a variety of materials for collection and recycling.

HAZARDS AND CATASTROPHES

The high concentrations of people and physical infrastructure in cities make them distinctive in several ways with regard to hazards. Where money is available, cities are worth protecting because of the large financial and human investment they represent. Adequate provisions of water supply and sanitation are designed to offset the risks of disease, and other infrastructure, such as expensive flood protection schemes, protects against geophysical hazards (see Chapter 20).

In developing countries, where escalating urban growth rates and a lack of finance make such provisions inadequate, it is usually the poorest sectors of urban society that are most at risk from environmental hazards. Rapid urban growth and rising land prices have used up the most desirable and safest sites in most Third World cities, leaving increasingly hazard-prone land for poorer groups. Such hazards include the pervasive dangers of high pollution levels and the intensive dangers of industrial accidents. The accidental discharge from a pesticide production plant in Bhopal, northern India in 1984, for example, killed more than 3000 shanty-town dwellers. It was primarily caused by inadequate management and lax safety procedures.

High concentrations of poor housing are built on slopes on hillsides prone to sliding (e.g. Caracas), or in deep ravines (e.g. Guatemala City); on river banks susceptible to flooding (e.g. Delhi), and on low-lying coastlines prone to marine inundation (e.g. Rio de Janeiro). Even the destruction caused by citywide hazards, such as earthquakes, can be magnified in these unstable sites: in Guatemala, 65 per cent of deaths in the capital caused by the 1976 earthquake occurred in the badly eroded ravines around the city.

In other situations, however, the damage and loss of life caused by earthquakes can be greatest in more built-up parts of the urban environment, when buildings themselves become hazardous if they are not constructed to withstand earth tremors. The widespread failure of relatively new constructions in urban areas of Armenia in the 1988 earthquake echoed the experiences of Mexico City in 1985. Seemingly, more-sophisticated technology able to withstand tremors had not been incorporated into new buildings for reasons of cost (Krimgold, 1992). Failure of urban infrastructure following earthquakes is one of the commonest causes of damage and loss of life. The most serious earthquake disaster in the USA, in San Francisco in 1906, was largely a function of infrastructural failure. Disruption of gas distribution and service lines caused the outbreak of many fires and interrupted water distribution, so making it difficult to put the fires out.

The most critical environmental problems faced in urban areas of the developing world, however, stem from the disease hazards caused by a lack of adequate drinking water and sanitation. In 1990, at least 170 million people in urban areas worldwide lacked a source of potable water near their homes and 375 million did not have adequate sanitation (World Bank, 1992a). Table 14.5 shows some of the countries worst affected.

Water-borne diseases (e.g. diarrhoea, dysentry, cholera and guinea worm), water-hygiene diseases (eg typhoid and trachoma) and water-habitat diseases (e.g. malaria and schistosomiasis) both kill directly and debilitate sufferers to the extent that they die from other causes. Again, it is the less-well-off sectors of urban society that are most at risk. The effect of improvements to water supply and wastewater disposal on life expectancy have been clearly shown in the industrial countries, when services were improved during the nineteenth and twentieth centuries. The trend shown for three major French cities in Fig. 14.8 is typical in this respect, with life expectancy increasing from about 32 years in 1850 to about 45 years in 1900, with the timing

TABLE 14.5 *Countries with particularly poor drinking water and sanitation provision in urban areas*

Country	Access to safe drinking water (% of urban population)	Access to sanitation (% of urban population)
AFRICA		
Burkina Faso	44	35
Guinea-Bissau	18	30
Rwanda	46	45
Sao Tomé and Principe	33	8
ASIA		
Afghanistan	39	20
Bangladesh	37	37
Myanmar	38	35
Viet Nam	48	48

Source: after UNEP (1991: 225–226, Table 4.6)

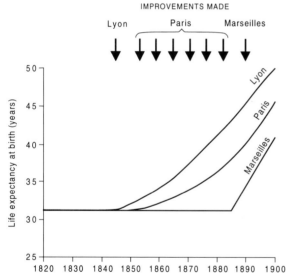

FIGURE 14.8 *Rising life expectancy with improvements in water supply and sanitation in three French cities, 1820–1900 (reprinted with permission from Briscoe, J. 1987, A role for water supply and sanitation in the child survival revolution.* Bulletin of the Pan American Health Organization *21: 92–105. Copyright Pan American Health Organization, Washington DC)*

of changes corresponding closely to improvements in water and sanitation provision.

CURITIBA – TOWARDS A SUSTAINABLE URBAN ENVIRONMENT

While the history of many developed countries' experience in dealing with urban problems offers many indicators as to how the environmental difficulties which plague so many of the developing world's major cities can be solved, the solution to the awesome environmental problems outlined in this chapter depends largely upon available finance and political will. Some of the efforts to combat these problems have been described, and, to conclude, it is useful to examine the experience of Curitiba, a city of about 1.5 million people in south-east Brazil, as a model for other cities.

Curitiba has greatly improved its urban environment with a series of innovative transport, land use and waste disposal measures. A new public transport system, involving competitively priced bus services using exclusive bus lanes, has been used as a framework for current and future urban development. The system, which carries 1.3 million passengers per day, has reduced congestion on the roads, and lowered air pollution as a typical Curitiba private vehicle now uses 30 per cent less fuel than the average for eight comparable Brazilian cities. The city also boasts one of the lowest motor accident rates per vehicle in the country, and its inhabitants enjoy a very low average transport expenditure. Land use policy has aimed to improve urban conditions, introducing pedestrian precincts and cycleways, concentrating on redeveloping existing sites rather than expanding the urban area, and expanding parks and other green spaces.

Efforts to improve the city's squatter settlements (*favelas*) have focused on garbage, with education programmes to encourage recycling through municipal collection. The problem of access to the high-density *favelas* has been tackled by offering bus tickets in exchange for household refuse brought to accessible roadsides, with a marked decrease in litter and consequent improvement in *favela* quality of life (Lowe, 1991; Rabinovitch, 1992).

FURTHER READING

Cadman, D. and Payne, G. (eds) 1990 *The living city: towards a sustainable future.* London, Routledge. A collection of reviews and case studies which focuses mainly on social and economic aspects of First and Third World urban development.

Hough, M. 1989 *City form and natural process.* London, Routledge. This thoughtful book looks at the operation of cities and their conflict with nature and argues that closer harmony with natural processes should provide an alternative basis for urban design.

Hardoy, J.E., Mitlin, D. and Satterthwaite, D. 1992 *Environmental problems in Third World cities.* London, Earthscan. A comprehensive review of environmental issues in rapidly growing cities of the developing world.

Lowe, M.D. 1991 *Shaping cities: the environmental and human dimensions.* Worldwatch Paper 105. Worldwatch Institute, Washington, DC. This pamphlet looks at past problems and future prospects for urban planning from the perspectives of quality of life and environment.

15

TRANSPORT

Movement is a basic element of the daily rhythm of life in every society, a fundamental human activity and need. Transport modes make such movement easier, whether it be a trip to the shop, or the movement of raw materials or goods from one place to another. From the earliest times, transport has been an integral part of the evolution of civilisations, and rapid transport and good communications using modern transport networks of road, rail and air are essential elements of all advanced economies, while global economic integration relies upon efficient maritime transport. Development of most parts of the less-prosperous economies is likewise intimately linked to transport facilities. Transport is becoming faster and more efficient and there is no better indicator of this trend than the relentless growth of car ownership and travel that has occurred in recent decades: in 1950 there was one car for every forty-six people worldwide; by 1970, this number had risen to one per eighteen people, and by the early 1990s there was one car per twelve people (Lowe, 1994).

Such mobility comes with an environmental price tag, and the effects of transport on the environment are wide-ranging. Some of the major issues are summarised in Table 15.1; they include air pollution and noise from road and air traffic, and marine pollution from shipping (see Chapter 10). The transport sector of economies is also a major consumer of resources, including energy, minerals and land. The burdens imposed vary greatly between different types of transport, however, and some indication of the differential impacts of transport modes is given in Table 15.2.

TABLE 15.1 *Major environmental effects of transport*

LAND
- Use and wastage of land and its associated ecosystems
- Excavation and use of minerals (e.g. gravels) for road construction
- Generation of solid waste as vehicles are withdrawn from use

BIOSPHERIC
- Introduction of immigrant species to new environments
- Barriers to migration of species.

ATMOSPHERIC
- Emissions of greenhouse gases, particulates, fuel and fuel additives
- Noise and vibrations

HYDROLOGICAL
- Contamination of surface and groundwater from surface runoff and spillages of petrol and oil and transported substances
- Modifications of hydrological regimes during construction of roads, ports, canals and airports

Source: modified after OECD (1991a)

TABLE 15.2 *Differential impact of transport modes on a number of environmental variables*

Environmental variable	Unit	Air	Car	Car with three-way catalytic converter	Rail	Bus	Bicycle	Pedestrian
Land use	m²/person	1.5	120	120	7	12	9	2
Primary energy use	g coal equivalent units/pkm	365	90	90	31	27	0	0
Carbon dioxide emissions	g/pkm	839.5	200	200	60	59	0	0
Nitrogen dioxide emissions	g/pkm	6.4	2.2	0.34	0.08	0.02	0	0
Hydrocarbons	g/pkm	1.4	1	0.15	0.02	0.08	0	0
Carbon monoxide emissions	g/pkm	8.1	8.7	1.3	0.05	0.15	0	0
Air pollution	Polluted air m³/pkm	95 000	38 000	5900	1200	3300	0	0
Accident risks	Hours of life lost/1000 pkm	1.4	11.5	11.5	0.4	1	0.2	0.01

Note: pkm = passenger kilometre. One passenger travelling 1 km is 1 pkm, ten passengers travelling 50 km equals 500 pkm. The same logic is used in freight transport where the unit of work done is the tonne kilometre (tkm)
Source: Teufel (1989)

IMPACTS ON LAND

All forms of transport consume land resources, whether for nodes such as airports, railway stations and ports, or for the route corridors of roads, railways or canals. New transport developments are often opposed by certain groups for these very reasons, because they alter the nature of the landscape in which they are placed. Concern for the impacts of new transport routes on the countryside is not a recent phenomenon. Canal and railway companies in eighteenth and nineteenth century Britain faced opposition from landowners, artists and writers on environmental grounds in the same way that motorway construction is opposed today by individuals and pressure groups intent on preserving the countryside. In many countries, however, desire by governments to improve communications often overrides other concerns, even to the detriment of areas supposedly protected from damage by developments. In the UK, the current road programme will affect no less than 161 Sites of Special Scientific Interest, one national nature reserve, three local nature reserves, five county sites of special wildlife interest and two trust reserves (Whitelegg, 1992). New motorways are often built in response to congestion problems as more cars are used, but unfortunately for those who decry the loss of landscapes to motorways, more roads can also create more traffic, spurring the demand for still further expansion of road networks.

Transport developments not only despoil landscapes, but also consume resources in their construction and operation. There are numerous historical examples of the impacts new routes have upon local resources. The introduction of wood-burning steam ships to West African rivers, for example, which occurred in the 1830s on the River Niger, heralded a significant pulse of deforestation along river banks. Similarly, when rail travel began with the line from Dakar to Saint Louis in Senegal in 1885, timber was felled on a large scale to clear the way for tracks, to make sleepers, to build bridges and to fuel steam engines. In many countries today, road building is the largest consumer of aggregates from quarries which leave permanent scars on the landscape. About 125 000 t of crushed rock are needed to construct 1 km of motorway, for example.

Impacts upon the landscape can, in turn, present hazards to the transport routes themselves, as well as other land uses in the vicinity. In permafrost areas, the importance of maintaining the thermal regime of ground ice during and after construction of routeways was learnt the hard way during the construction of the Alaska Highway in the early 1940s. Even minor disturbances to the thermal equilibrium caused permafrost to thaw, presenting problems of heaving and subsidence due to frost action, and the creation of impassable mires. Soil erosion and problems of slope failure are other hazards the highway engineer has to confront. Areas receiving high rainfalls are among the most susceptible, and slopes which are lacking in vegetation are often the most likely to fail. The major stability problems faced on steep slopes along the Kuala Lumpur–Ipoh Highway in peninsular Malaysia are shown in Table 15.3. Serious soil erosion problems can also occur when vehicles are used off roads. Such off-road driving was found by Jones et al. (1986) to be a prime cause of surface destabilisation and consequent soil erosion by wind in a Middle Eastern town, the blows of dust causing a considerable problem to the occupants of the town, particularly for the operation of machinery.

New transport routes also spawn environmental impacts indirectly through their *raison d'être* of improving access to places and resources, thereby encouraging industrial and other land uses. In Russia, for example, construction of the Trans-Siberian Railway in

TABLE 15.3 *Gradient and proportion of slopes (%) with major slope stability problems on the Kuala Lumpur–Ipoh highway*

Stability problem	Slope gradient		
	33–45 degrees	**63 degrees**	**83 degrees**
Slope failure	8	19	0
Slumps	21	31	54
Gullies	11	8	0

Source: modified after Bayfield *et al.* (1992: 80, Table 2)

FIGURE 15.1 *Heavy industry on the Trans-Siberian Railway just outside Irkutsk, Russia*

the late nineteenth century opened up the vast mineral resources of Siberia, resulting in numerous sites of environmental degradation (Fig. 15.1). Agriculturalists too often follow in the wake of new transport routes, to convert new land to food production. Cultivation only began in Australia, for example, with the arrival of European settlers, and in modern times it is generally agreed that agriculturalists are the most serious agent of tropical forest loss, their movement often closely following transport routes. The deforestation of Rondônia State in Brazil, following the arrival of the BR-364 Highway in the 1970s is a classic example (see Chapter 3).

IMPACTS ON THE BIOSPHERE

Transport both facilitates and hinders movements of flora and fauna and opens up new routes for species dispersal. The movement of people has assisted numerous 'biological invasions' of great ecological significance, both deliberate and inadvertent (e.g. Drake *et al.*, 1989; Hengeveld, 1989). History is punctuated with many examples of plants and animals deliberately transferred between regions for commercial purposes. Western Europe, for example, imported new crops such as potatoes, maize, tomatoes and tobacco from the Americas, and many of today's staple African food crops have been brought to the continent by outsiders. In other cases, however, introduced species have become such successful colonisers that their introduction has given cause for concern, necessitating attempts to reverse their rapid spread when they are considered to have reached pest status (see Chapter 5).

There are also numerous cases of unintentional dispersal of, and subsequent colonisation by, species which have 'hitched a ride' on various forms of transport. More than 120 aquatic species have been introduced into marine and estuarine systems and inland seas in this way, most as a result of canal building or through transport in ships' ballast water (Baltz, 1991). Following most of

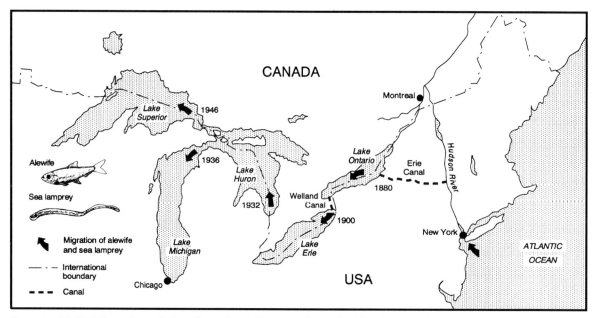

FIGURE 15.2 *Migration of alewife and sea lamprey into the North American Great Lakes*

these introductions, native fish species disappear or are greatly reduced due to competition or predation.

A classic example is the invasion of the North American Great Lakes by two species from the Atlantic, the alewife and sea lamprey, following completion of the Erie Canal (now the New York State Canal) and the Welland Canal during the nineteenth century (Fig. 15.2). As a result, the once common Atlantic salmon, lake charr, lake trout and lake herring have been severely depleted by competition for food from the alewife and predation by the sea lamprey. Since several of these greatly reduced native species are of commercial importance, the Great Lakes Fishery Commission currently spends about US$1 million per year in lamprey control, mainly through the use of chemical lampricides in nursery streams. Ironically, in Europe the sea lamprey has declined over most of its range because of river pollution and obstructions, is now extinct in many large rivers where it was formerly abundant, and is now regarded as a threatened species to be conserved (Maitland, 1991).

More recently, the Great Lakes have been invaded by another species, the zebra mussel, a small, striped native of the Caspian Sea which probably arrived in the ballast tanks of a European tanker. Within 2 years of their first appearance in 1988, zebra mussel densities had reached 700 000 individuals per square metre in parts of Lake Erie, choking out native mussels in the process. There has been a subsequent spread southwards into the Detroit, Cumberland and Tennessee rivers with potential to cause enormous damage to fisheries, dams and boats as well as devastation to the rich endemic aquatic communities (Stolzenburg, 1992).

Similar examples can be quoted for land transport. In Brazil, the house sparrow had spread 1500 km to Belém within a short time of the Belém–Brasilia highway being opened, and within 6 years had spread on to reach Marabá in 1973 (Goodland and Irwin, 1974). The sparrow is a carrier of the bugs which transmit Chagas' disease (American trypanosomiasis), a debilitating and usually fatal disease.

The dispersion of diseases is a particularly notable aspect of the transport revolution, with rapid journey times sharply reducing the natural checks upon the spread of diseases. The point can be illustrated by the case of smallpox:

[I]n Columbus's time, crossing the Atlantic was slow compared to the progression of smallpox. Since all carriers of the virus manifest symptoms of the disease, most of the infected travellers would have either become sick and died or recovered before reaching the New World. As a result, smallpox probably did not reach the Americas until several decades after Columbus's voyage.

(Levins *et al.*, 1994: 57).

Today the situation is very different; travel to almost anywhere in the world can be accomplished in a few days at most, less than the average incubation time of many disease pathogens. Modern rapid transportation can turn a local disease into a worldwide pandemic. A good example is the outbreak of cholera in South America in 1991, thought to have been introduced from China in the ballast water of a freighter which docked at a port in Peru (see Chapter 20).

Many plant diseases have also been introduced to naive populations through modern transport. An example is the devastating Dutch elm disease fungus which killed many elm trees in Britain in the 1970s. The fungus is thought to have been brought to the island on timber imported through southern ports.

In contrast to these examples of transport facilitating dispersal, the construction of new routes can also have the opposite effect by introducing a barrier to the movement of organisms. Klein (1971) describes the impact upon wild reindeer of the construction of a railway line in northern Norway which disrupted normal herd movements. The track effectively cut off access to one part of the reindeer's range, leaving a smaller area of their former range to be heavily grazed. In another example involving railways, rock tipping and sliding during construction of a line through Hell's Gate, a narrow gorge in the Fraser River canyon in western Canada, blocked the upriver spawning migration path of Pacific salmon in 1913 and 1914. Pacific salmon were a vitally important economic resource on the Canadian west coast at the time and their unusual biology was instrumental in causing a dramatic and long-lasting drop in the numbers spawning. The effects on the sockeye salmon catch from the Fraser River is estimated to be as high as US$2.2 billion over the period 1914–45 (Ellis, 1989).

AIR POLLUTION

Transport is a major source of air pollution due to its heavy dependence upon the combustion of fossil fuels, either in vehicles or at power stations. The major pollutants involved are:

- carbon dioxide
- carbon monoxide
- nitrogen oxides
- hydrocarbons
- sulphur oxides
- lead
- suspended particulate matter.

These pollutants have a wide variety of environmental impacts, being harmful to flora and fauna and detrimental to human health (see Chapter 14). In the wider perspective, carbon dioxide is a greenhouse gas and nitrogen and sulphur oxides contribute to acid rain.

Pollution emissions per passenger kilometre are greatest for air travel (Table 15.2), and since aircraft emissions are largely released into the sensitive upper atmosphere, where gases are longer-lived than at ground level, they give rise to particular concern. Emissions of nitrogen oxides and water vapour during cruise mode are perhaps the most critical in this respect. In the troposphere, nitrogen oxides contribute to ozone formation, in turn contributing to smogs at ground level, while, as a greenhouse gas, nitrogen oxides emitted directly to the troposphere may remain resident there as much as a hundred times longer than those released from terrestrial sources. Although estimates are difficult, some suggest that up to 30 per cent of aircraft emissions occur while cruising in the stratosphere, and in this region nitrogen oxides deplete ozone at a level which some researchers regard to be equivalent to the depletion caused by CFCs. Water vapour released during stratospheric flight is another area of some concern. Since the natural water vapour content of the stratosphere is low, the effect of additions could be marked, leading to increased frequency of cirrus clouds, with a possible increase in atmospheric temperature (Archer, 1993). Since the global air fleet size and movements are expected to have doubled over 1990 levels by the year 2000, these environmental impacts will be the subject of increasing concern in the years to come.

Despite the proportionally large impact of air travel, the heavy reliance on motor vehicles in most countries makes road transport the largest polluting source in the transport industry. In the UK, for example, road transport accounted for 93 per cent of passenger travel and 81 per cent of freight moved in 1990 (DoE, 1992). Concern over the environmental effects of motor vehicle pollution has led to some concerted action to reduce certain pollutants. Lead is a particular example, due to its association with reduced mental development in infants and children, inhibition of haemoglobin synthesis in red blood cells in bone marrow, and impairment of liver and kidney function. Lead's use in petrol as an anti-knock agent has been reduced considerably in many countries in recent years, with consequent reductions in concentrations in human blood. In many cases, lead-free petrol has been widely introduced, with its adoption by motorists encouraged by lower taxes. The lead content in petrol in twenty megacities of the world is shown in Table 15.4.

Karachi remains one of the worst cities for lead pollution due to Pakistan's high lead concentrations in petrol and the high concentration of traffic in the city. The number of registered vehicles more than doubled between 1980 and 1989 from 300 000 to about 650 000 at an annual growth rate of 12.5 per cent, 2.5 times higher than the population growth rate. Vehicle numbers are projected to continue climbing and to reach 1.1 million by the year 2000 (Beg, 1990). Mean concentrations of inorganic lead in Karachi were about 1–3 g/m³ with maximum values of 7–9 g/m³ in areas of heavy traffic. These values are far above the WHO annual mean guideline range of 0.5–1.0 g/m³. Beg's study showed that more than four-fifths of the lead is in respirable particles, which underlines the threat to human health.

The dramatic effects on air quality of lowering the lead content of petrol is indicated in Fig. 15.3 for Los Angeles, where in the mid-1970s atmospheric lead concentrations exceeded Federal standards virtually every day of the year but had been reduced to zero within 10 years. The importance of legislation to reduce vehicle pollution is paramount in Los Angeles, where there is almost no public transport network, so that residents have to rely on personal vehicles for virtually all transportation. This situation makes the Los Angeles Basin the area with the worst air quality in the USA. Eight million vehicles in the urban area of 12 million people represents probably the greatest number of vehicles per person in the world. While the widespread use of lead-free petrol has brought atmospheric lead levels under control in recent years, ozone concentrations still present a serious problem (Fig. 15.3), although an Air Quality Management Plan for southern California

TABLE 15.4 *Lead content of petrol used in twenty megacities*

City	Lead content of petrol (g/l)	Comments
Bangkok	0.15	
Beijing	0.4–0.8	80% unleaded
Bombay	0.15	
Buenos Aires	0.6–1.0	
Cairo	0.8	
Calcutta	0.1	
Delhi	0.18	
Jakarta	0.6–0.73	
Karachi	1.5–2.0	
London	0.15	>33% unleaded
Los Angeles	0.026	>95% unleaded
Manila	1.16	
Mexico City	0.54	
Moscow	0	No leaded fuel sold
New York	0.026	95% unleaded
Rio de Janeiro	0.45	ethanol and gasohol used
São Paulo	0.45	40% ethanol, 60% Petrol/gasohol
Seoul	0.15	
Shanghai	0.4	
Tokyo	0.15	>95% unleaded

Source: UNEP/WHO (1992: 41, Table 4.2)

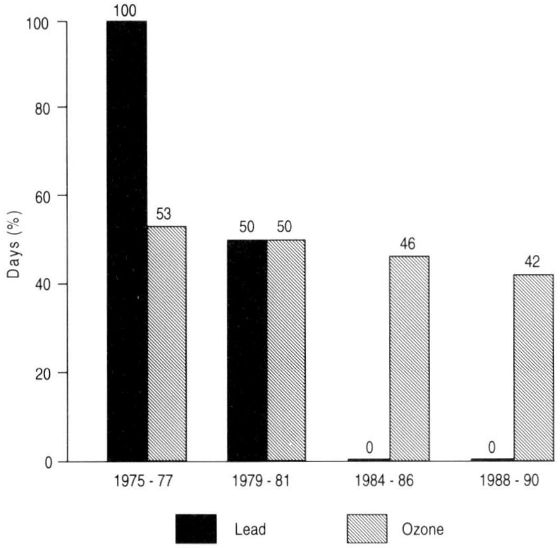

FIGURE 15.3 *Percentage of days when atmospheric levels of lead and ozone exceeded Federal standards in Los Angeles, USA, 1975–90 (after UNEP/WHO, 1992)*

approved in 1989 aims to reduce emissions of NO_x and VOCs by 80 per cent in 20 years.

Much of the success in reducing pollutants from motor vehicles in Los Angeles and elsewhere in the Western world is attributable to the increasingly widespread use of catalytic converters which were first developed in the USA in the 1950s and 1960s and fitted to US cars from the mid-1970s. The catalytic converter is a device fitted to vehicles which removes certain pollutants by a chemical reaction. The pollutants targeted in vehicle exhausts are unburnt hydrocarbons, carbon monoxide and nitrogen oxides which are converted by platinum, palladium and rhodium into carbon dioxide, water vapour and nitrogen. The ability of the three-way catalytic converter to reduce by more than 80 per cent the output from the internal combustion engine of these three pollutants, has led to its widespread adoption. From 1993, for example, all new cars sold in the European Union have had to be fitted with catalytic converters, and since these systems are only effective with lead-free petrol, their introduction will also continue to reduce the amount of lead pollution from motor vehicles.

However, even with catalytic converters fitted, cars still produce significant atmospheric pollution. Carbon dioxide, a greenhouse gas, is the most significant pollutant, produced in large quantities because cars fitted with catalytic converters tend to be less fuel efficient than those without. Other pollutants which are also still produced are emitted before the engine has warmed, a particular problem where short journeys are concerned, as with a high proportion of city car travel. Table 15.5 indicates the pollution produced by a car during its 10-year lifespan, assuming it is driven 13 000 km per year, consuming 10 litres of petrol per 100 km. The high carbon dioxide output is greater still if the manufacturing process and eventual scrapping of the car is included, bringing the total carbon dioxide produced to 59.7 t (Whitelegg, 1994).

The use of catalytic converters, and other emission-reducing technologies such as lean-burn engines, electronic fuel injection, computer-controlled spark ignition and exhaust recirculation systems, can only go so far in reducing polluting emissions. A technical fix to the vehicle engine system or an additive to fuel cannot solve the global emission problem. Indeed, recent experience indicates that these efforts do not keep pace with increasing traffic. It seems that we may have to abandon crude oil as a primary energy for transportation within three or four decades, due to dwindling supplies (Svidén, 1993). Even if this estimate is pessimistic, the end of crude oil is undoubtedly a foreseeable event, and thus provides humankind with the opportunity 'to redesign the complete energy supply and conversion chain from mining and refining to the combustion in the vehicle engines in such a way that it can meet the global ecological criteria' (Svidén, 1993: 145).

Some examples of switches to less-polluting fuels have already occurred. Diesel has some advantages over petrol. Diesel engines use around 30 per cent less fuel, hence they emit less carbon dioxide; they also emit fewer hydrocarbons and fewer nitrous oxides. Conversely, diesel engines produce larger quantities of sulphur dioxide, particulate matter, noise and smoke. Other alternative fuels are also in use in various parts of the world. In Tokyo, for example, taxis are obliged to use liquefied petroleum gas (LPG) rather than petrol because LPG, like all natural gas derivatives, contains less carbon than gasoline and hence produces less carbon dioxide, carbon monoxide and hydrocarbons, as well as less nitrous oxides, when burnt.

Other alternatives already in use include methanol and ethanol, which when burnt produce lower quantities of carbon dioxide and nitrous oxides than petrol. Methanol and ethanol also have the advantage that they can be produced renewably from biomass. At one time in the 1980s, virtually all cars in Brazil were run on either pure ethanol or a mix of ethanol and petrol, the ethanol being fermented from sugar-cane juice, although more recently, lower international oil prices and supply problems have stalled the

TABLE 15.5 *Estimated emissions of air pollutants over 10 years from a car fitted with a three-way catalytic converter*

Pollutant	Total emissions
Carbon dioxide	44.3 t
Carbon monoxide	325 kg
Nitrogen dioxide	46.8 kg
Hydrocarbons	36 kg
Sulphur dioxide	4.8 kg

Source: after Whitelegg (1994)

FIGURE 15.4 *Severe traffic congestion in Kampala, Uganda*

country's ethanol programme. A significant problem with these fuels derived from biomass is the large areas necessary for cultivation. For example, replacement of all the petrol used by cars and taxis in the UK with ethanol derived from biomass, for example, would require an area three times the size of the UK to grow the biomass (Transnet, 1990).

Overall, alternative fuels, such as methanol, diesel and gas have the potential as short-term solutions to the growing problem of pollution from transport, but unless cleaner ways of creating fuels are developed, the growing flow of traffic will wipe out these gains. One of the fuels that is a strong candidate for a cleaner replacement to those derived from crude oil is hydrogen, sometimes referred to as the ultimate clean combustible fuel since its combustion emits only water vapour and small quantities of nitrogen oxides. Electrically driven vehicles also present some hope for future development when the electricity is generated from renewables, although at present their range is limited by the need for frequent recharging, a function of battery storage capacity. Nevertheless, such technological developments to reduce pollution would still leave many of the undesirable aspects of society's heavy dependence on cars, such as congestion and sprawl, unresolved (Fig. 15.4). Hence, a wider perspective on the problem from the policy viewpoint is needed (see below).

NOISE

Road, air and rail transport is a major source of noise in both urban and rural areas. It has been estimated that about 17 per cent of the population of all OECD countries is exposed to transport noise, which is regarded by most authorities to be unacceptably high, and the proportions of urban populations exposed to unacceptably high transport noise is often 50 per cent, although overall transport noise levels have tended to stabilise since the late 1970s (Farrington, 1992). Transport noise affects people to varying degrees, ranging from mild annoyance and minor interruptions to everyday activities, to mental and physical damage, as well as an adverse effect on property prices in some cases. As such, noise is a significant aspect of the transport sector's environmental impact, and the clear importance of major transport corridors as sources of noise is shown, in Fig. 15.5, in the urban environment for the Chinese city of Nanjing. On the national level, one estimate of the economic cost of traffic noise, measured by the depreciation of house prices in France, put the cost to the nation at US$0.27–0.45 billion a year (Pearce *et al.*, 1984).

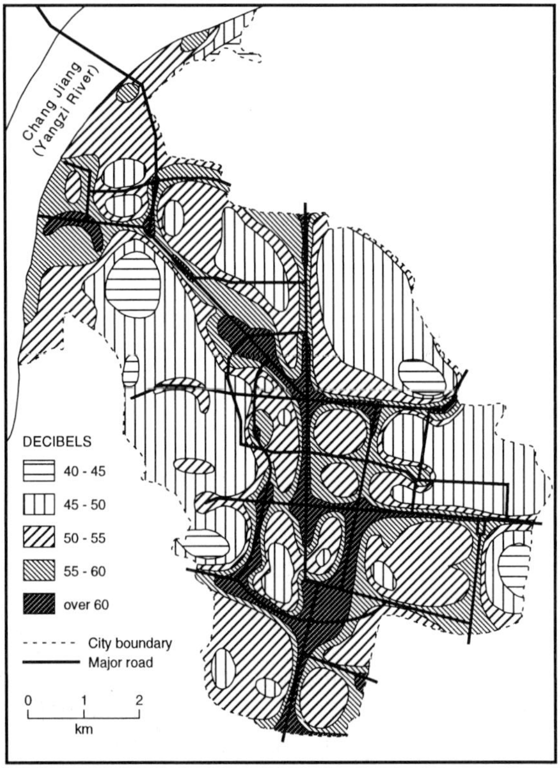

FIGURE 15.5 *Spatial distribution of daytime environmental noise levels in Nanjing, China (reprinted from Noble, A.G. 1980, Noise pollution in selected Chinese and American cities. GeoJournal 4: 573–75. By permission of Kluwer Academic Publishers)*

Although the definition of what level of sound constitutes a problem or a nuisance, and hence when a certain sound becomes unacceptable, is difficult, guidelines are available. Transport noise is usually measured using the logarithmic dB(A) scale, a variation of the decibel scale which is weighted towards the frequencies most affecting the human ear. At 80 dB(A), people standing next to each other would need to shout to be heard, and at 90 dB(A) – a typical level for a heavy lorry passing at 7 m – temporary loss in acuteness of hearing would be felt for a few minutes. The 120 dB(A) noise typically heard at 200 m from a jet aircraft on take-off is approaching

the level at which physical pain and ultimately deafness is experienced. Annoying or unacceptable noise is often taken to be 65 dB(A), although a continuous noise tends to be regarded as less of a nuisance than individual events, and the time of day when a noise is heard is also a factor – night-time noise being less acceptable than that during daylight hours.

Mitigation of noise problems can be approached on four fronts (Nelson, 1987):

- reduction of noise at source – through vehicle design, traffic management, tunnelling and noise abatement procedures for aircraft;
- measures to control noise along its transmission path – mainly by barriers such as fences and embankments, and the use of buildings as noise barriers;
- measures to protect the observer from noise at the point of hearing – such as through building design, ensuring smaller windows on the noisiest façades and incorporating double glazing and acoustic insulation;
- land use planning and zoning – effective on the larger scale by zoning noisy traffic paths away from residential and working areas.

The best results for noise abatement are derived from a combination of all these approaches, and the success of such strategies in reducing the numbers of people affected is well illustrated for US airports in Table 15.6.

TRANSPORT POLICY

The future environmental impacts of transport, as with many other environmental issues, will depend on policy-makers. Passenger travel by public transport and bicycles, all more environmentally sound methods of travel than the car (Table 15.2), depend upon appropriate emphasis being given to such modes. In Britain, the 1990–91 National Transport Survey showed a

TABLE 15.6 *Numbers of people exposed to daytime noise of 62 dB(A) around US airports*

Airport	1970	1980	1985	2000 (estimate)
New York				
JFK	—	—	606 000	571 000
La Guardia	—	—	462 000	283 000
All airports	6.0 million	4.2 million	3.2 million	2.4 million

Source: modified after OECD (1993)

FIGURE 15.6 *More than 5 million bicycles are used in the Chinese capital of Beijing, a city of 6 million people*

22 per cent increase in the average distance people travelled per week since 1985–86, with little change in the number of journeys made. Most of these journeys were made by car. Such a trend reflects policy and planning decisions which feed back on society and change lifestyles: decisions which reinforce dependence upon cars increase car use. A similar attitude

has led to the great increase in road haulage of freight at the expense of rail, for example.

Technological innovations can only go so far in reducing the environmentally damaging aspects of transport, and will need to be complemented with better planning that can reduce the need for travel which is costly in environmental terms. More practical urban and suburban land use patterns can help to reduce car trips, both by shortening them and by reducing the need to travel by car. Public transport, cycling and walking need to be encouraged and provided for as alternatives (Fig. 15.6).

The current obsession with the car which characterises many governments' attitude to transport might change if the true costs of the car and other transport modes is calculated. Lowe (1994: 95) suggests that 'the single most important goal for policy reform may be to ensure that a more accurate tally is used to guide all public and private decisions in the future'. Important, in this respect, is the need to clarify and quantify the true costs of an increasingly mobile society. External costs, which a particular mode of transport imposes on society as a whole, are difficult to quantify. They include traffic congestion, road accidents, pollution, dependence on imported oil, solid waste, loss of cropland and natural habitats, and climatic change.

Many observers believe that most of the environmental issues which make transport unsustainable can only be properly addressed by focusing on ways in which behaviour can be changed, to reduce the demand for the activity which produces the problem in the first place (Whitelegg, 1993). Moderation of the environmental impacts of transport will come about ultimately from government, through regulation, legislation and other ways in which transport modes are encouraged and discouraged. Some suggestions as to the types of approaches which can be taken at different levels of government to ameliorate the problems are suggested in Table 15.7.

TABLE 15.7 *Suggested complementary policy solutions to transport problems for different levels of government in Europe*

Government level	Policies
Local	Improve land use planning and traffic management
Central	Change pricing and tax policies to assist development of alternative fuels, to improve transport efficiency, to encourage shifts to 'greener' modes of travel, and to reduce unnecessary travel
European	Impose 'environmentally optimal' community-wide policies involving fuel and vehicle tax harmonisation, vehicle speed limits and heavy vehicle weights

Source: after Transnet (1990)

FURTHER READING

Carpenter, T.G 1994 *The environmental impact of railways*. Chichester, Wiley. This book details impacts on people, such as noise and pollution, and resources.

Drake, J.A., Mooney, H.A., di Castri, F., Groves, R.H., Kruger, F.J. and Williamson, M. (eds) 1989 *Biological invasions: a global perspective*. SCOPE Report 37, Chichester, Wiley. A comprehensive review of the invasion of natural ecosystems by animals, plants and micro-organisms, the environmental problems caused and attempts to control the process.

Renner, M. 1988 *Rethinking the role of the automobile*. Worldwatch Paper 84. Worldwatch Institute, Washington, DC. A discussion of the environmental impacts of motor vehicles and the need to reduce our reliance upon them.

Whitelegg, J. 1993 *Transport for a sustainable future: the case for Europe.* London, Belhaven. An overview of transport's pollution problems and costs which emphasises the need for fundamental changes in our attitude to personal mobility before transport in Europe becomes sustainable.

16

WASTE MANAGEMENT

Wastes are produced by all living things – excreted by an organism or thrown away by society because they are no longer useful and if kept may be detrimental. But the make-up of waste means that once discarded by one body its constituents may become useful to another. In the natural world, a waste produced by an animal, for example, is just one stage in the continual cycle of matter and energy that characterises the workings of the planet: an animal that urinates is disposing of waste products its body does not need, but the water and nutrients can become resources for other organisms. Similarly, sewage from a town can be used by bacteria which break it down in a river, for example, or an old shirt discarded by one person might be worn by another, often poorer, individual. The term 'waste' is therefore a label determined by ecology, economics and/or culture.

Much of the waste produced by human society consists of natural material and energy, although we have also created products not found in the natural world, such as CFCs and plastics, which can become wastes. The issues surrounding wastes stem from problems of disposal. People's use of resources, and hence production of wastes, has accelerated in the period since the Industrial Revolution, and this fact combined with the growing number of people on the planet has created increasing volumes of waste that need to be disposed of. This disposal requires careful management since too much waste disposed of in a certain place at a particular time can contaminate or pollute the environment, creating a hazard to the health, safety or welfare of living things. Examples of waste products and their impacts on the environment are found throughout this book. They include wastes produced by agriculture (Chapter 5), energy production (Chapter 17) and mining (Chapter 18), and pollution of the atmosphere (Chapters 8, 9 and 15), aquatic environments (Chapters 10, 11 and 12) and urban environments (Chapter 14).

TYPES OF WASTE

Waste can take many different forms: solid, liquid, gas, or energy in the form of heat or noise. The problems associated with its disposal can stem from a range of other properties, including its chemical make-up, whether it is organic or inorganic, the length of time taken for it to be broken down into less-hazardous constituents, and its reactivity with other substances – so-called synergy. The sources of waste are also diverse, stemming from virtually every human activity, including:

- mining and construction
- fuel combustion
- industrial processes
- domestic and institutional activities
- agriculture
- military activities.

Collection of data on waste has had a relatively low priority in many countries, and the methods for their compilation vary widely. Statistics which are available are often given as weights, but the large differences between quantity and quality mean that these figures can only give a general indication of the nature of the waste disposal problem. The waste arisings for the UK, which totalled about 400 million tonnes annually in the early 1990s, are shown by source in Fig. 16.1. The general absence of annually updated data make the recognition of trends difficult.

Since some wastes are inherently more dangerous than others, categories such as 'special', 'controlled' and 'hazardous' are often identified and such wastes dealt with in a more careful manner. In the UK, where the disposal of different types of waste is regulated by different laws, so-called special wastes, for example, are those deemed dangerous to life and are subject to regulations designed to track their movement from production to safe disposal, or from 'cradle to grave'. Although there is no universal agreement over what constitutes hazardous wastes, they include substances that are toxic to humans, plants or animals, are flammable, corrosive, or explosive, or have high chemical reactivity. Such hazardous substances include acids and alkalis, heavy metals, oils, solvents, pesticides, PCBs (polychlorinated biphenyls), and various hospital wastes.

DISPOSAL OF WASTE

All wastes are disposed of into the environment, but some enter the environment in a more controlled manner than others. Some wastes are emitted directly from the source without treatment, others are collected and sometimes treated before disposal. Wastes produced from the combustion of fuel by motor vehicles, for example, are emitted directly into the atmosphere, while domestic sewage wastes are often collected by municipal authorities and disposed into specific locations such as a river or ocean. Three of the compartments into which wastes are emitted – air, rivers and oceans – are usually publicly owned and this common ownership has facilitated unregulated emissions of wastes. In many countries, however, the degradation of these common property resources by wastes has spawned numerous controls on their emission.

This chapter is concerned with wastes that are actively managed and a wide range of management options is available, as Table 16.1 indicates for hazardous wastes. The following sections will look more closely at two of the most commonly used options: landfill and incineration.

Landfill

Dumping of waste is by far the most commonly used method of waste disposal in most

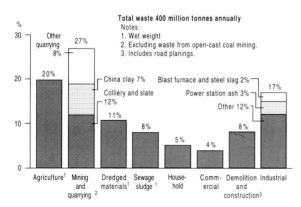

FIGURE 16.1 *Waste arisings by sector in the UK (after DoE, 1992)*

TABLE 16.1 *Treatment and disposal technologies for hazardous wastes*

General approach	Specific technology
Physical/chemical	Neutralisation
	Precipitation/separation
	Detoxification (chemical)
Biological	Aerobic reactor
	Anaerobic reactor
	Soil culture
Incineration	High temperature
	Medium temperature
	Co-incineration
Immobilisation	Chemical fixation
	Encapsulation
	Stabilisation
	Solidification
Dumping	Landfill
	Deep underground
	Marine
Recycling	Gravity separation
	Filtration
	Distillation
	Solvent extraction
	Chemical regeneration

Source: after Tolba and El-Kholy (1992)

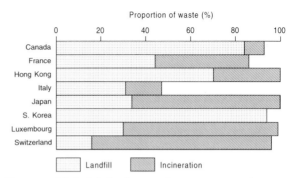

FIGURE 16.2 *Main forms of treatment and disposal of municipal wastes in selected countries (from data in UNEP, 1993)*

If properly designed and managed, landfill sites may do little harm to the environment and can eventually provide a surface for such land uses as playing fields or reforestation, though not usually for heavy structures such as housing. The two main hazards associated with landfills are leakage of toxic leachates which can contaminate surface and groundwater, and leaving refuse open to the air which can allow infestation by rats or fermentation bacteria which can generate methane and cause a fire hazard. Each day's addition must be covered with soil to prevent infestation and fermentation, although in some cases, methane generation is actively encouraged and collected for use as a fuel (Lisk, 1991). Four typical landfill designs are shown in Fig. 16.3. If the site is located on impermeable strata, problems from leaching to groundwater are not faced (Fig. 16.3a), but when sited on an aquifer, various approaches can be taken to prevent leaching, such as an impermeable landfill lining of plastic or rubber (Fig. 16.3b) or by careful control of the local groundwater table (Fig. 16.3c). In these designs the aim of the landfill is to concentrate, isolate and contain wastes to minimise the hazard they represent, but in areas where leachate is not expected to be produced in harmful quantities a dilute and disperse philosophy may be

countries. For example, much of the mining and quarrying wastes arising in the UK, shown in Fig. 16.1, remains as tailings and spoil heaps, while most of the dredged material is dumped at sea. A significant quantity of other wastes is buried in landfill sites, and in many countries it is the most commonly used management method for disposing of municipal wastes (Fig. 16.2) as well as being a frequently used method for hazardous waste disposal. Of the estimated 24 million tonnes of hazardous wastes generated in the OECD member countries of Europe, 70–75 per cent is deposited in landfill (Yakowitz, 1993).

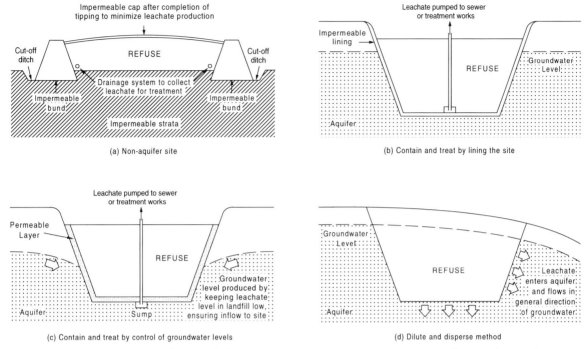

FIGURE 16.3 *Four landfill designs (after Swinnerton, C.J. 1984, Protection of groundwater in relation to waste disposal in Wessex Water Authority.* Quarterly Journal of Engineering Geology *17: 3–8. Reproduced by permission of Blackwell Science)*

adopted, allowing seepage into the aquifer (Fig. 16.3d).

Designs like that shown in Fig. 16.3d have been used all too commonly in the past for disposal of hazardous wastes, and leakage from such old sites is causing increasing concern in many countries. A classic example of the dangers and costs of such sites is the Love Canal, a disposal site in the city of Niagara Falls, northeastern USA, where between 1942 and 1953, a chemical company disposed of more than 20 000 t of chemical wastes. Not long after the landfill was sealed, a school and several buildings were constructed on the site. More than 20 years later, heavy rains in the winter of 1975 and spring 1976 caused land subsidence and created pooling of surface waters which were heavily contaminated with chemicals from the dump. Infiltration of these waters into nearby

residences caused a public outcry over the possible health hazards, and in 1978, residents from 238 houses were relocated. Love Canal was declared a chemical emergency by President Carter, the first executive order which addressed a hazardous waste problem, and subsequent investigation and clean-up of the site has cost about US$100 million (Deegan, 1987).

Subsequent to the Love Canal incident, many of the world's industrialised countries have begun a slow and costly process of identifying and cleaning up potentially dangerous old landfill sites. About 50 000 contaminated sites have been identified in Germany and 4000 in The Netherlands. In the USA, up to 10 000 sites have been identified, of which more than 1000 are the subject of immediate attention. The costs of clean-ups in the USA are shared between the current and previous landowners, those who have dumped waste at the site and those who have transported it there, with some money also provided by the so-called Superfund. Legal arguments over who should pay are proving to be a considerable delaying factor in the process, however.

Although today's landfill sites are subject to much stricter controls in many countries than in the past, the future of burial in landfills is uncertain, given the pressure on available space and opposition from local residents to the creation of new sites. However, the decline of waste disposal at sea, which for many waste forms will become illegal before the end of the decade (see Chapter 10), is likely to increase the pressure for more sites, although some of the additional burden will undoubtedly be absorbed into the incineration option.

Incineration

Incineration is increasingly being used by many countries to reduce the bulk of wastes and to break down hazardous compounds, so rendering them less dangerous. Incineration can also be combined with energy recovery. In Japan, Luxembourg and Switzerland, for example, more than 65 per cent of all municipal waste is incinerated (Fig. 16.2) and the proportion of this waste which also generates energy in these countries is 27 per cent, 100 per cent and 80 per cent, respectively. Incineration is proving to be a good disposal method for tyres, which are otherwise problematic to get rid of. Landfill sites do not welcome them since tyres represent a fire hazard and tend not to stay buried. In practice, they are simply stockpiled, at the rate of 230 million per year in the USA, for example. The high energy content of tyres makes them a good fuel, however: an incineration plant at the largest tyre dump in the USA, near Modesto in California, produces electricity for 15 000 homes and tyres could be more widely used as supplementary fuel in a

number of industries such as cement and paper mills (Barlaz *et al.*, 1993).

Incinerators do, however, release harmful emissions into the atmosphere, such as particulate matter, heavy metals and trace organics. Such emissions can be limited with a range of approaches, such as improved combustion techniques, sorting of wastes prior to incineration, and by fitting pollution control devices. Ash produced in incinerators must still be disposed of, and this material is commonly buried in landfills, although some attempts have been made to use the ash as a component of building materials. Site selection for incinerators, as for landfills, is another controversial issue for the waste disposal industry. While people may recognise that such facilities are necessary, the 'not in my backyard' syndrome is a powerful force in determining where such facilities are located. In practice, many such locally undesirable land uses, or LULUs, are sited in poor and minority communities. While, at first, this pattern may appear to reflect discrimination in siting procedures, research in the USA suggests that the situation may not always be so clear-cut. In some examples, LULUs appear to change community dynamics by driving down property values, resulting in a higher proportion of African Americans and the poor around such sites, so that the dynamics of the housing market are more to blame than any discrimination in the choice of the LULU site (Been, 1994).

Strong public feelings have often been foremost in pushing for more stringent emission standards for incinerators. In many European countries, these standards are now stricter than those applied to other energy sources, and some observers fear that the use of incinerators may be discouraged as a consequence. In practice, the total environmental impact of incinerators should be assessed and compared to similar measures for both alternative energy production and alternative waste treatment facilities (Nilsson, 1991).

International movement of hazardous waste

As controls on the disposal of hazardous wastes have tightened in developed countries, there has been movement of both operations and wastes themselves to areas where legislation is less stringent or poorly enforced. During the 1980s, it became increasingly apparent that such trade in hazardous wastes was on the increase, both between industrialised and less-developed countries, and also between western and eastern Europe (Yakowitz, 1993). International concern over this trend led ultimately to the Basel Convention on the Control of Transboundary Movements of Hazardous Wastes and their Disposal, which came into force in 1992. The convention is based on a series of guiding principles originally adopted by the OECD countries, three of which are particularly important (OECD, 1991a):

- The principle of non-discrimination – OECD members will apply the same controls on transfrontier movements of hazardous wastes involving non-member states as those applied to movements between member states.
- The principle of prior informed consent – movements of waste will not be allowed without the consent of the appropriate authorities in the importing country.
- The principle of adequacy of disposal facilities – movements of waste will only be permitted when wastes are directed to adequate disposal facilities in the importing country.

The Basel Convention also bans exports to countries which have not signed and ratified the treaty, and incorporates requirements to control the generation of hazardous wastes and obligations to manage them within the country of origin unless there is no capacity to do so (Hilz and Radka, 1991).

REUSE, RECOVERY, RECYCLING AND PREVENTION

Although landfill and incineration are currently the most frequently used methods of waste disposal, they are generally considered to be low down on the list of possible techniques devised from an environmental impact perspective, as the US Environmental Protection Agency's hierarchy of treatment techniques for solid wastes indicates (Table 16.2). Reusing a product, as opposed to discarding it, obviously makes environmental sense, and the advantages of waste recovery and recycling have also long been recognised. As for reuse, these advantages include a reduction of resource consumption and a curb to the cost and inadequacies of disposal. In practice, waste products are reused when it is economically viable to do so and viability is assessed for a variety of motives. The history of industrial development is punctuated with examples of waste products being reappraised and turned into valuable resources. In the early nineteenth century, for example, Britain's fledgling chemical industry on Merseyside was using the Leblanc soda process to produce alkalis by treating salt with sulphuric acid, producing highly corrosive hydrogen chloride as a waste product. Realisation that this pollutant was a lost resource, combined with fears of legal action from local landowners over the effects on

TABLE 16.2 *Hierarchy of preferred treatments for solid waste in the USA*

1.	Reuse
2.	Waste reduction
3.	Recycling
4.	Resource recovery
5.	Incineration
6.	Landfill

Source: after USEPA (1989)

surrounding vegetation, spurred industrialists to convert the hydrogen chloride to chlorine which was used to make bleaching powder (Elkington and Burke, 1987). Another example from the nineteenth century concerns a resource which today most people take for granted. In the 1870s, when Nicolaus Otto found that a mix of flammable gas and air could be spark-ignited within the cylinder of a piston machine to generate movement, the flammable gas used was obtained from gasification of a waste product that came from the petroleum refinery process which produced paraffin for lamps. The waste product was called petrol (Svidén, 1993). The perception shift which underlies these and many other examples can be summarised in a formula adopted by the multinational Minnesota Mining and Manufacturing (3M), a company distinguished by its commitment to a constant succession of innovative products:

$$Pollutants\ (waste\ materials)$$
$$+$$
$$knowledge\ (technology)$$
$$=$$
$$potential\ resources$$

Reuse, recovery and recycling

Realisation of the economic value of certain wastes promotes reuse, recovery and recycling. In many developing countries, informal scavenging of municipal wastes from city dumps is widely practised and the materials recovered are put to a variety of uses (Fig. 16.4). It provides employment to individuals who otherwise would have none, particularly important in those countries where social security systems are inadequate or non-existent. Studies of this informal sector have also highlighted the important role of such scavenged resources as inputs to small-scale manufacturing industries. In the Tanzanian capital of Dar es Salaam, these industries provide consumer goods which are

FIGURE 16.4 *A novel reuse for soft drinks cans in Namibia*

much cheaper than if they were made from imported raw materials. Low-cost buckets, charcoal stoves and lamps are all made from scavenged metals from city dumps and sold to people who live in squatter settlements in the city (Yhdego, 1991).

Formal recovery and recycling schemes in more-developed countries are also most advanced for metals and some other materials such as paper and glass, and collection points for such solid wastes have become a familiar sight in many cities in recent years (Fig. 16.5). Some of the benefits of such schemes in environmental terms are shown in Table 16.3.

FIGURE 16.5 *Recycling collection point in a suburb of Berlin. Germany has been a European trendsetter in recycling*

TABLE 16.3 *Environmental benefits of substituting secondary materials for virgin esources*

Environmental benefits	Aluminium	Steel	Paper	Glass
Reduction (%) of				
Energy use	90–97	47–74	23–74	4–32
Air pollution	95	85	74	20
Water pollution	97	76	35	—
Mining wastes	—	97	—	80
Water use	—	40	58	50

Source: Bartone (1990)

In economic terms, the large energy savings gained from metal recycling make the practice worthwhile. In the case of aluminium, the conversion of alumina to aluminium by electrolysis – the Hall–Héroult process – accounts for 80 per cent of the energy used in the entire production process, a stage which is bypassed when aluminium is recycled (Fig. 16.6). These and other benefits have encouraged the increasing use of recycling in those industries that use metals (Table 16.4).

Other critical factors affecting the quantities of material that are recycled are the capacity of the recycling plant and equipment, and the size of the market for recycled materials. In some cases, these factors have been influenced by legislation. Commonly used measures include laws designating a certain level of use of recycled materials and the imposition of a tax on products which do not incorporate a certain amount of recycled material. The Canadian city of Toronto, for example, has a local law which states that daily newspapers must contain at least 50 per cent recycled fibre or the publishers will not be allowed to have vending boxes on the city's streets. Taxes on waste products are in line with the 'polluter pays' principle and can encourage both the reduction of waste and the increase of recycling rates. Taxation has been suggested

TABLE 16.4 *Metal consumption and recycling in the USA, 1990*

Metal	Consumption (thousand tonnes)	Share of consumption provided by recycling (%)
Lead	1297	73
Copper	2168	60
Gold	0.2	47
Aluminium	5263	45
Tin	45	38
Tungsten	8	29
Chromium	423	21
Molybdenum	21	5

Source: Young (1992)

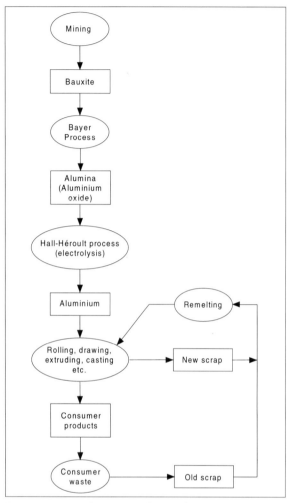

FIGURE 16.6 *Stages in the production of aluminium goods*

as a possible solution to the widespread problem of packaging waste, a priority waste problem in Europe and elsewhere. Imposition of a tax on packaging would effectively incorporate the full social cost of a product into the retail price by including the cost of disposal of its wrapping (Pearce and Turner, 1992).

Manipulation of the retail price of a consumer product has long been used to encourage reuse of certain articles. The charging of a returnable deposit on glass bottles is a good example, providing an economic

incentive to return the bottle for reuse. It is a practice which is still widely used in many developing countries, but one which has declined in some more-developed economies as returnable glass bottles have been largely replaced with throw-away plastic ones. In some countries, such as Germany and Switzerland, recent legislation has been passed to encourage moves back to the refilling of drinks containers, but in some other countries similar measures could significantly increase the price to the consumer, a politically sensitive step to take. In the USA, for example, reusable bottles were common when bottling plants were small and widely distributed throughout the country, but recent decades have seen a centralisation of bottling facilities, in part as a response to the smaller transport costs imposed by lighter plastic bottles. Hence, an increased emphasis on returnable glass bottles would entail considerably higher transport costs. Such economic factors have been partly responsible for the increasing use of the recycling option which has become widespread practice in many developed countries in recent years. For the OECD countries, the average recovery rate for glass rose from 22 per cent in 1980 to 32 per cent in the late 1980s (OECD, 1991a), and in several European countries recycled material makes up more than 50 per cent of all glass used (Fig. 16.7).

Not all waste products are suitable for recycling or reuse in the original process that created them, and the sludge which remains after sewage treatment is an obvious example, although in many places it has some reuse value as a fertiliser due to its high phosphorus and nitrogen content. About half the sewage sludge produced in the UK and one-third of that in Germany is reused in this way, but in these and other countries, further use is limited by the high heavy metal content which often characterises sewage sludge, a function of sewage systems mixing industrial and household wastes. Hence, sewage sludge not used as fertiliser is disposed of by ocean

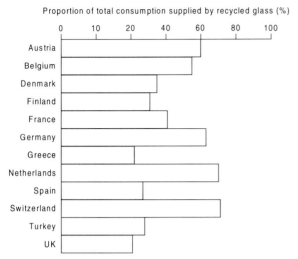

FIGURE 16.7 *Glass recycling in selected European countries, 1991 (from data in OECD, 1993)*

dumping, dumping in landfill and municipal garbage dumps, or by incineration.

As population and industrialisation continues to increase, however, and the treatment of water and sewage becomes more widely used in order to reduce the pollution impacts from untreated sewage outlets, so more sludge will need to be disposed of. An innovative approach to this increasing dilemma of disposal can turn a problem into a benefit by realising the value of the heavy metals themselves. One estimate of the potential yield of some of the high-value metals in sewage sludge suggests that global production of palladium from sewage sludge could be of the same order of magnitude as that from mining production (Table 16.5). The future disposal of sewage sludge should concentrate on the control of pollutants at source and extraction of metals (Lottermoser and Morteani, 1993), an approach which would yield numerous benefits, including:

- more widespread use of sludge as a fertiliser
- revenue from metals extracted
- conservation of geological metal resources
- saving on expensive waste repository space
- prevention of environmental impacts on terrestrial and marine ecosystems.

Waste prevention: cleaner production

There is no doubt that the best way to manage waste is to prevent it at source wherever this is possible. The argument that prevention is better than cure is put by UNEP's Industry and Environment Programme Activity Centre:

> When end-of-pipe pollution controls are added to industrial systems, less immediate damage occurs. But these solutions come at increasing monetary costs to both society and industry and have not always proven to be optimal from an environmental aspect. End-of-pipe controls are also reactive and selective. Cleaner production, on the other hand, is a comprehensive, preventative approach to environmental protection.
>
> (UNEP IE/PAC, 1993: 1)

TABLE 16.5 *Estimated worldwide annual production of noble metals from sewage sludge compared to current production from geological resources*

Metal	Current production (t/year)	Production from sewage sludge (t/year)
Gold	1600	100
Platinum	100	8
Palladium	100	80

Source: after Lottermoser and Morteani (1993)

TABLE 16.6 *Examples of the application of cleaner production*

Country	Industry/process	Cleaner production	Advantages	Payback time of capital costs incurred
France	Metal processing – steel galvanising	Development of a new process for zinc coating of steel	Total suppression of conventional plating waste; lower operating costs and improved quality control	3 years
India	Textile – sulphur black colour dyeing using sodium sulphide to convert sulphur dyes	Substitution of sodium sulphide, with hydrol, a by-product of the maize starch industry	Reduction of sulphide in the effluent; less corrosion in the treatment plant and reduction of odour problems	No capital expenditure
Indonesia	Cement production	Optimisation of kiln operation and improved process control through the installation of expert control systems	Some reduction in NO_x and SO_2 emissions, energy savings and reduction in the quantities of off-specification material	<1 year
Poland	Automobile component manufacture – Cu/Ni/Cr and Zn electroplating of aluminium alloy and steel components	Modification of the rinsing systems and addition of facilities which permit water recycling and raw materials recovery	A decrease in water and raw materials consumption; reductions in contaminant levels in wastewater streams of 80–98%	4 months
Sweden	Metal fabrication – de-greasing of metal sections using trichloroethylene	A switch to an alkaline de-greasing procedure which utilises biodegradable cutting oils	Zero emission of trichloroethylene and reduction in the quantity of hazardous waste sludge produced	11 months
UK	Adhesives	Development of a new range of water-based adhesives in place of solvent-based ones	Compared to solvent-based adhesives, water-based ones are less toxic and require no special handling, yet still offer a wide range of applications	New product

Source: after UNEP IE/PAC (1993)

Cleaner production is achieved by examining all phases of a product's lifecycle, from raw material extraction to its ultimate disposal, and reducing the wastefulness of any particular phase. Hence, cleaner production encompasses such aims as:

- conservation of energy and raw materials,
- reduction in the use of toxic or environmentally harmful substances, and
- reduction of the quantity and toxicity of wastes and pollutant discharges.

A range of industrial examples of such cleaner production is shown in Table 16.6. They range from low-tech changes in production like the example of a textile factory in India to more capitally intensive modifications to large-scale production facilities such as cement and automobile manufacture, and new product development. Nevertheless, for all the examples shown, the payback time for the capital expenditure incurred is 3 years or less, indicating the immediate economic feasibility of such actions.

FURTHER READING

Young, J.E. 1990 *Discarding the throwaway society.* Worldwatch Paper 101. Worldwatch Institute, Washington, DC. An overview of the rise of resource consumption and their wasteful use, focusing on the role of the individual consumer as the ultimate decider of how wasteful a society we live in.

Bradshaw, A.D., Southwood, R. and Warner, F. (eds) 1992 *The treatment and handling of wastes.* London, Chapman & Hall. The papers in this book cover a range of topics on the various types of waste and the methods for disposal.

17

ENERGY PRODUCTION

Energy has long been a basic requirement of human societies. Archaeological evidence from caves in Africa suggests that fire was used by early humans, the hominids, as long ago as 1.5 million years BP, and fire has since fuelled the technologies on which civilised societies have been built. Today, regular supplies of energy drive the motors, appliances, cities, industries and transport on which the lifestyle of most of the Earth's human population depends. Through time, the ability of human cultures to access energy, and the amounts of energy used, are often seen as indicators of society's level of development and resource use. Energy use's attendant environmental problems have long been with us, however. One of the first examples of environmental legislation in England occurred during the reign of Edward I when coal burnt for industrial and domestic purposes caused so great a smoke nuisance that the nobility, strongly backed by London residents, were able to obtain a royal proclamation forbidding coal burning in 1306 (although it proved impossible to enforce). Since then, the generation of energy and the use of fuels in multifarious ways to drive the industries and machinery of modern societies, have facilitated an unprecedented level of both deliberate and inadvertent impacts on the global environment. The production, transportation, conversion and use of energy, particularly that derived from fossil fuels, is responsible for some of the world's most serious environmental problems. These include global climatic change (see Chapter 8), acid rain (see Chapter 9) and pollution from motor vehicles (see Chapters 14 and 15), but human society's harnessing and use of energy lies behind virtually every one of the environmental issues found in this book. This chapter will focus on energy sources, specifically on the possible future of those not based on fossil fuels.

ENERGY SOURCES

Ultimately, most of the energy we use today comes from the sun. This radiant energy is converted by photosynthesis into chemical energy which makes plant life and ultimately all animal life possible. It is also used by human society both as biomass and as fossil fuels which were themselves once living plant matter. Solar radiation is also used directly by human society and indirectly by harnessing the movement of wind and rivers as part of the hydrological cycle. Three other ultimate energy sources are also commonly used: the processes of cosmic evolution preceding the origin of the solar system (nuclear power); the forces of lunar motion (tidal power), and energy from the Earth's core (geothermal power).

These sources of energy are often categorised into renewable and non-renewable resources, based on a timescale of human lifetimes. Hence, although fossil fuels are renewable on a geological timescale, the rate at which we are using them effectively means that they are finite. Although estimates of just how much oil, gas and coal remain for society to use are constantly being revised – and are to an extent a function of technology, price and the rate at which we use them – they are ultimately non-renewable. Current estimates made by the World Energy Council (WEC, 1992) suggest that we have about 40-years-worth of oil at current production rates, 60 years for gas and more than 200 years for coal. While the 1970s were characterised by fears over the imminent exhaustion of fossil energy resources, such fears have dissipated following the re-evaluation of existing reserves and the discovery of new ones. Indeed the WEC stated in 1992 that 'The concepts of exhaustion, or even scarcity, fail to appear anywhere in this survey' (WEC, 1992: 13). Nevertheless, the finite nature of fossil fuels, combined with increasing awareness of the environmental impacts of their use, has spurred interest in the conservation of energy and the development of more sustainable forms of production from renewable resources.

ENERGY EFFICIENCY AND CONSERVATION

Perhaps the most obvious approach to reducing the environmental impacts associated with energy use is to make energy production more efficient and to reduce the amounts of energy that are wasted during its use. Better energy conservation can be promoted in many sectors of society and the adoption of energy-saving technology and methods is usually encouraged by a fairly rapid return in terms of reduced energy bills.

In industry, for example, the conservation of energy is one of the basic aims of campaigns for cleaner production (see Chapter 16), while in the domestic and commercial sectors, substantial savings can be made with relatively simple building design and technologies. In many countries of the developing world, energy lost by burning fuelwood on open fires can be reduced by using closed stoves, which is also safer, while converting wood to charcoal is more energy-efficient still. Heating and hot water often represent the largest proportion of energy use in the home and office of more developed settings, and Table 17.1 illustrates the scale of savings that can be made with readily available technology. On the national level, the potential savings are summed up by Ruckelshaus (1989: 120) who states that: 'Right now more energy passes through the windows of buildings in the US than flows through the Alaska pipeline'. The use of energy for lighting, which consumes about 25 per cent of electricity in the USA, could also be substantially reduced with a more widespread adoption of energy-efficient lighting hardware such as compact fluorescent lamps. Lighting innovations which are commercially available in the USA today could potentially save up to 20 per cent of the country's electricity use (Fickett *et al.*, 1990).

Another area with substantial room for improvement is in the generation of electricity. In many developing countries, power plants which run on fossil fuels use five or six units of fuel to make one unit of electricity. In rich countries, the ratio is about three to one but can reach two to one in combined-cycle gas turbines. Much of the wasted energy is in the form of hot water and use of this water in combined heat and power schemes makes economic and energy-efficient sense for local users. Such combined heat and power units

TABLE 17.1 *Effect of various energy-saving options for heating a three-bedroomed house in Britain*

Heating/ insulation packages	Heat loss (kW)	Average heating bill (£/week)	Area heated
Gas fire in living room	7.22	4.75	Living room only
Gas central heating and loft insulation	5.78	6.60 5.10	Whole house Downstairs only
Gas central heating, loft insulation, cavity wall insulation, draught-stripping and humidistat-controlled extractor fans	4.11	4.50 3.65	Whole house Downstairs only
Gas unit heaters and gas water heaters plus insulation and fans as above	4.11	3.75 3.00	Whole house Downstairs only

Source: after Musannif (1992)

are used to supply factories and small towns and cities, but their use could be more widespread.

RENEWABLE ENERGY

The current use of renewable energy sources is most significant in the developing countries where biomass fuels represent a major energy source. Hydropower provides the main source of electricity for some countries, such as Brazil, Canada and Norway, but is relatively minor on the world scale. Overall, of course, fossil fuels currently make the largest contributions to global energy use (Fig. 17.1).

Although the current cost of renewable energy supplies is not competitive with fossil fuels in many areas, this is largely a function of the lack of finance and effort put into their development and the subsidies given to energy production from oil, coal and natural gas. Given more attention, renewable energy sources could become cost-effective more quickly, and with a number of other advantages in addition to their environmental benefits relative to the burning of fossil fuels.

One such advantage is that most renewable energy equipment is small, allowing much faster construction than conventional technologies. While most large conventional facilities are constructed in the field, most renewable energy equipment is constructed in factories where manufacturing techniques can facilitate cost reduction. The small scale of equipment also makes for a short time between design and operation, so that modifications stemming

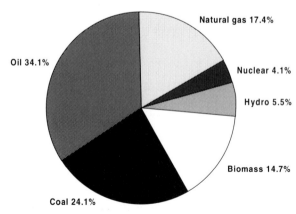

FIGURE 17.1 *Global energy use (Hall* et al., *1993)*

from field tests can quickly be incorporated into new designs. Hence, many generations of technology can be introduced in a short period. There are, however, some difficulties with certain renewable energy sources. The intermittent nature of wind and solar power sources is one such problem, although this can be resolved by appropriate use or storage of energy. It is also important to note that not all renewable sources are appropriate for all types of energy need: wind and wave energy will not fuel a motor vehicle, for example, although electricity so-generated may eventually supply battery-powered vehicles. The following sections will look at the main forms of renewable energy to assess their future potential to contribute to world energy needs.

Hydropower

Hydropower is the only renewable resource used on a large scale today for electricity generation. The main constraints on further development are the social and environmental impacts associated with dams which are outlined in Chapter 13, although many of these problems can be mitigated by improved planning with the inclusion of public participation at an early stage. Although much uncertainty surrounds the estimates of potential hydroelectric power that could be generated, there is agreement that, to date, only a relatively small proportion of this resource has been developed, and the greatest potential lies in the developing world and the countries of the former USSR. The potential for small-scale hydroschemes, designed to serve local needs, is thought to be particularly great (Moreira and Poole, 1993).

Wind energy

The wind's energy has been used by sailing ships and windmills to power human activities for many hundreds of years and wind pumps are still a common feature of rural landscapes today, particularly in dry regions

FIGURE 17.2 *Wind power is widely used in rural areas, particularly to pump groundwater to the surface in regions where surface water is in short supply, as shown here on the Mediterranean island of Malta*

where they are used to pump groundwater to the surface (Fig. 17.2). Modern wind turbines for electricity generation are a comparatively recent phenomenon which some suggest have the potential to supply as much as 20 per cent of global electricity demand (Grubb and Meyer, 1993).

Much of the pioneering work on the large-scale generation of wind-derived electricity has been carried out in California where developments were encouraged by favourable tax policies. California now produces just over 1 per cent of its electricity from wind farms, much of it coming from the 7500 turbines at Altamount Pass. Denmark has also put a great deal of effort into developing its wind power potential, which currently generates 2.5 per cent of its electricity consumption, a portion scheduled to increase to 10 per cent by the year 2005.

There are several environmental impacts associated with wind farms. Noise was a serious concern surrounding earlier generations of turbines, but modern designs make little sound above the rush of the wind. Problems with bird kills persist, however, for some larger species such as eagles, and interference with TV reception and sensitive electronics is also experienced around turbines with steel blades. The land requirements of large wind farms has been another issue of concern, although land between individual machines can still be used for other activities such as farming and ranching. Land values at Altamount Pass, for example, have increased markedly following installation of the wind farms, as royalties have generated extra income for ranchers while ranching has continued. Perhaps the most serious objection to wind power in many areas is on aesthetic grounds, since some people frown upon the idea of landscapes covered in windmills. In the UK, for example, more than fifty well-known literary figures wrote to *The Times Literary Supplement* in 1994 to protest at plans to increase the number of wind turbines on the moors above Haworth because of the associations with the novels of the Brontë sisters.

Solar power

In the very long term – beyond the next century – solar power in its various forms is expected to be able to meet the bulk of the world's energy needs, although in the shorter term the contribution can be expected to be much more modest (WEC, 1993). Solar power can be used passively by designing buildings to optimise the sun's light and heat energy, or actively to heat water and building space and to generate electricity. Photovoltaic (PV) power is one of the most direct methods of converting the sun's rays into electricity; it is produced when individual light particles – photons – absorbed in a semiconductor – create an electric current. PV electricity is thus created with no pollution and no noise. PV systems also need minimal maintenance and no water and have been particularly successful in remote areas such as deserts. Another advantage of PV systems is that they can operate on any scale, from portable modules for remote communications and instrumentation to huge power plants covering millions of square metres, enabling them to be positioned near to the electricity users and reducing the need for transmission systems. They are potentially cost-effective even at high latitudes and in relatively cloudy regions and the land areas required are not great: if the entire electricity needs of the USA were generated by PV systems, for example, the area required would be about 34 000 km^2, less than 0.4 per cent of the US land area (Weinberg and Williams, 1990). The cost to the consumer of electricity from PV arrays has fallen rapidly in the last three decades. They are already cost-effective in remote areas and are likely to rival the cost-effectiveness of conventional energy sources within the next 20 years, as the cost of manufacturing the sheets of semiconductor materials falls.

Another way in which solar radiation is used directly to generate electricity is the solar thermal electric generator, which uses mirrors to track the sun and focus its rays to heat a fluid in a pipe. The heated fluid is then used to create steam that drives a turbine generator, or directly to produce steam and hot water for industrial and domestic purposes. Several designs have been developed, ranging in scale from an individual parabolic mirror which focuses the sun's rays onto its receiver, to a field of sun-tracking mirrors which all

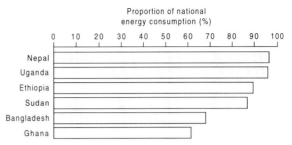

FIGURE 17.4 *Wood fuel as a percentage of national energy consumption in selected developing countries (after UNEP, 1990)*

FIGURE 17.3 *The central-receiver solar thermal power plant at Solar One in California, USA (courtesy of US Department of Energy)*

reflect solar energy to a receiver mounted on a central tower (Fig. 17.3). The nine commercial solar thermal power plants now operating in the Mojave Desert in the USA indicate the viability of the technology, which like other renewable energy sources, has been hampered by a lack of interest and finance from governments.

Biomass

Plant matter, or biomass, is a form of energy which has been used by people for heating, lighting and cooking since the discovery of fire. In many developing countries, fuelwood

is still the main source of energy (Fig. 17.4) and an estimated 2.5 billion people, nearly half the world's population, rely mainly or exclusively on biomass for their daily energy needs (Hall and Rosillo-Calle, 1991). Although in some parts of the developing world fuelwood collection is being carried on unsustainably, with severe environmental impacts (see Chapters 3 and 4), biomass has a large potential to be used as a renewable fuel in many different applications.

Biomass fuels come in a variety of forms, both unprocessed (wood, straw, dung, vegetable matter and agricultural wastes) and processed (e.g. charcoal, methane from biogas plants and landfills, logging waste and sawdust, and alcohol produced by fermentation). These fuels also have a more recent history of use in the industrialised countries of the world. Industrial and agricultural residues have been used for many years to provide electricity in conventional steam-turbine power-generators. In some of the Scandinavian countries, where biomass contributes a fairly high percentage of national energy consumption (Table 17.2), woody waste from the pulp and paper industries is the most significant component. Some countries, notably Brazil, have made considerable progress in substituting fuel derived from sugar-cane for petrol used in motor vehicles (see Chapter 15), while on the local

TABLE 17.2 *Biomass fuels as a percentage of national energy consumption in selected developed countries*

Country	Contribution of biomass to total energy consumption (%)
Austria	4.0
Belgium	0.2
Canada	3.0
Finland	14.0
New Zealand	0.4
Sweden	13.0
Switzerland	1.6
UK	0.3
USA	4.0

Source: after Sipilä (1993)

scale, municipal wastes are burnt in incinerators, and landfills are tapped for emissions of gaseous energy (see Chapter 16).

Sustainably managed, the use of biomass fuels produces no net emissions of carbon dioxide since the amount released into the atmosphere by burning will be taken up by growing plants. Although more efforts can be made to use residues from other activities, and from the harvesting of forests, dedicated energy plantations are probably the largest potential source of biomass. A strong candidate for land to be used for such energy plantations is deforested or otherwise degraded land which could be replanted with trees (Hall *et al.*, 1993). Monocultural plantations come with their own environmental impacts, of course (see Chapter 5), but using degraded lands for plantations would be a considerable improvement on their current ecologically impoverished state.

Although research and development is still needed to identify more accurately suitable plants to grow and where to grow them, biomass has the potential to make a very significant contribution to world energy needs. Some believe that with appropriate encouragement and effort, biomass has the potential to provide as much as 80 per cent of the 1985 world commercial energy use by the year 2050 (Hall *et al.*, 1993).

Tidal power

Energy in the oceans is found in the form of tides, waves, temperature differences, salt gradients and marine biomass, and although the total amounts available are large, only a small fraction is likely to be utilised in the foreseeable future, both because ocean energy is spread diffusely over a wide area and because much of the available energy is distant from centres of consumption. To date, only tidal power has been seriously investigated and technically proven. It is one of the oldest forms of energy exploited by human society – tide mills were operated on the coasts of the UK, France and Spain before 1100 AD – and it has the definite advantage over many other renewable energy forms of being highly predictable. Tidal energy is harnessed in much the same way as that of hydropower, by building a barrage across a suitable estuary. Whereas a hydroelectric dam depends on unidirectional flow, however, a tidal barrage must allow the bay behind it to alternately fill and empty with the tides, the high-tide water being allowed to pass out through turbines which generate electricity.

The amount of energy available from a site depends upon the range of the tides and the area of the enclosed bay. The number of sites suitable for modern commercial-scale energy production are limited, however, and only a few of these have been developed (Table 17.3). The largest is the world's first operational tidal power plant associated with the barrage across the La Rance estuary on the Brittany coastline of northern France, which was built in the early 1960s.

Most of the potential for harnessing tidal energy in Europe lies on the British coast,

TABLE 17.3 *Existing tidal energy plants*

Location	Mean tidal range (m)	Basin area (km²)	Installed capacity (MW)	Date of first use
La Rance (France)	8.0	17	240.0	1966
Kislaya (Russia)	2.4	2	0.4	1968
Jiangxia (China)	7.1	2	3.2	1980
Annapolis (Canada)	6.4	6	17.8	1984

Source: after Cavanagh *et al.* (1993)

where the exploitation of all practical sites could yield up to 20 per cent of the electricity needs of England and Wales. But the environmental effects of such schemes have featured highly in the debate over whether or not to go ahead with barrages across two of the best sites: on the Severn and Mersey estuaries. Despite the environmental advantages of tidally generated electricity, including the lack of pollutants generated and the protection barrages offer to coastlines against storm-surge tides, such schemes inevitably modify the hydrodynamics of their estuaries. Alterations to the tidal range, currents and the intertidal area within the barrage would affect several other environmental parameters, including sediment movement and water quality, which would in turn have an impact on the food chain, ultimately affecting bird and fish populations.

Geothermal energy

Geothermal energy is tapped from the heat which flows outwards from the Earth's interior energy, both from the core as it cools and from the decay of long-lived radioactive materials in rocks. On the global scale, it is most accessible along the boundaries between the Earth's crustal plates, but many geothermal possibilities also exist away from these areas of concentration.

Hot waters from the Earth have been used therapeutically and for their mineral salts since Roman times, when geothermal spring water was also used to heat bath houses. This energy has been exploited commercially, to provide heat, mechanical power and electricity, since the early 1900s. In 1930, for example, large-scale geothermal heating systems for buildings were built in Iceland, and similar operations subsequently constructed in France, Italy, New Zealand and the USA. These and most of the other prevailing geothermal systems are hydrothermal, tapping aquifers of hot water a few kilometres below the surface, while other approaches are largely in the research stage. One such alternative uses energy from hot, dry rocks which is exploited by fracturing them at depth and pumping water down from the surface to be heated and then returned to the surface.

The environmental impacts associated with the use of geothermal energy can be divided into temporary ones due to drilling and exploration, and permanent ones resulting from well maintenance and power plant operation (Palmerini, 1993). Power plants take up land and may attract objections on aesthetic grounds. Air pollution is the greatest concern for most geothermal plants, since gases released during power production are not easily re-injected back into the thermal reservoir, although geothermal plants generally produce fewer gaseous emissions than their fossil fuel equivalents. Other environmental problems include:

- noise
- solid waste and residual water disposal
- microearthquakes
- subsidence

Although geothermal energy is unlikely to contribute significantly to global energy needs in the foreseeable future, it could nonetheless play a key role for some countries where potential is high and energy demands are currently low (e.g. Djibouti, El Salvador, Kenya, Nicaragua, and the Philippines).

The future for renewable energy

It is likely in the foreseeable future that energy demand will continue to rise as economic growth proceeds. Increasing the efficiency with which energy is used can help to offset this rise, but it is unlikely to absorb all the additional needs. Furthermore, the environmental impacts of conventional energy sources, notwithstanding their eventual depletion as non-renewable resources, will continue to play a role in deciding to what degree renewable energy sources contribute. Some believe that, given adequate support, renewables can meet much of the growing demand and gradually replace fossil fuels, at prices lower than those usually forecast for conventional energy sources, to be able to account for three-fifths of the world's electricity market by the middle of the twenty-first century (Johansson *et al.*, 1993). Renewable fuels for motor vehicles are also becoming more feasible technically, with hydrogen being the cleanest option (see Chapter 15). Others are less optimistic about the future for renewables; the World Energy Council (WEC) highlights the degree and extent of change necessary to realise this scenario: 'It is difficult to believe that politicians and policies, energy consumers and behavioural patterns, technology and the capacity to manufacture it and put it into operation on the required scale, will change sufficiently within the required timescale' (WEC, 1993: 93).

These two views highlight the key aspect of renewable energy's future role. Essentially, the future pace of development will be set by political decisions – if we want greater use of renewables then governments need to encourage it. The need for government support, particularly in the research and development phase, but also later in the energy marketplace was well illustrated in 1992 when the world's leading solar thermal electric corporation, Luz International, went bankrupt largely as a result of sudden changes in tax laws and regulations that virtually eliminated financial incentives for solar thermal electric technology in the USA.

NUCLEAR POWER

When energy generated by nuclear fission was first developed for civil uses in the 1950s, having grown out of a small number of national nuclear weapons programmes, it was heralded as cheap, clean and safe. The subsequent image of nuclear power has changed considerably since those times, and today it is one of the most controversial forms of energy from both the economic and environmental perspectives. By the early 1990s, nuclear reactors were in operation or under construction in more than thirty countries and nuclear power stations were supplying about 17 per cent of world electricity production (Fig. 17.5). For some countries the proportion is much higher: France, for example, generates about 70 per cent of her electricity in this way. Ultimately, nuclear power is a non-renewable form of energy production since uranium, the fuel used, is a mineral product. Advances in technology have meant that increasing amounts of uranium can be reprocessed and used again,

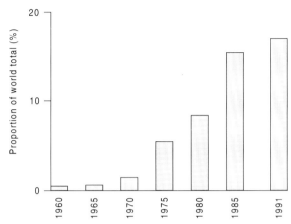

FIGURE 17.5 *Proportion of world electricity generation produced by nuclear power in selected years, 1960–91 (from data in UNEP, 1993)*

perhaps doubling the lifetime of usable uranium from around 40–60 years of supply (WEC, 1993), but in the shorter term the future of nuclear power will be dependent largely upon the associated environmental concerns.

These concerns are based on the fact that nuclear material is radioactive and hence can be highly dangerous to living things, and that for some radioactive elements this radioactivity is very long-lasting. The potential for these materials to be released into the environment can be traced through the nuclear fuel cycle – from mining, conversion, preparation of reactor fuel, and the management of wastes – and can arise again with the decommissioning of old nuclear power plants. Considerable research has been focused on the dangers of increased mortality rates from cancers around operational nuclear plants. While studies of adults suggest that no such effect is associated with nuclear installations (Forman *et al.*, 1987), the situation is less clear-cut for children. The true causes of unusual clusters of childhood leukaemias around the UK reprocessing plants at Sellafield and Dounreay, for example, may well be linked to nuclear emissions, but there may also be an infective cause due to the sudden influx of newcomers

to areas where local communities were previously well protected (Kinlen, 1988). Research into clusters of leukaemia and other cancers around nuclear installations will undoubtedly continue, but greater concerns abound over the disposal of nuclear wastes and the risks of major accidents.

Nuclear waste

Nuclear waste is commonly categorised into three classes (Table 17.4). In most of the countries that produce nuclear wastes, low-level material makes up the large majority of waste by volume, but accounts for a minor proportion of the radioactivity in waste. High-level waste, by contrast, is volumetrically small but accounts for the largest share of radioactivity. Liquid low-level wastes have long been disposed of by adopting a 'dilute and disperse' philosophy. The British nuclear power complex at Sellafield on the Cumbrian coast, for example, has been discharging such low-level liquid waste into the Irish Sea since the early

TABLE 17.4 *Categories of nuclear waste*

LOW-LEVEL WASTE
Liquid and solids lightly contaminated by short-lived radionuclides (e.g. discarded protective clothing, contaminated building materials, uranium mine tailings)

INTERMEDIATE-LEVEL WASTE
Materials contaminated by long-lived radionuclides such as plutonium and transuranic elements, most of which arise from the processes of energy production and reprocessing (e.g. fuel cladding, control rods, liquids used to store spent fuel before reprocessing)

HIGH-LEVEL WASTE
Materials contaminated by highly active radionuclides with long half-lives which may cause significant increases in the temperature of wastes (e.g. non-reprocessed spent fuel, liquid wastes produced during reprocessing of spent fuel)

1950s. The disposal of solid wastes, by contrast, has aimed to contain the material and then dispose of it. Until 1982, some low-level and selected intermediate-level wastes were contained in drums which were dumped at sea, but this practice has been suspended and burial is now the preferred option. For low-level and short-lived, intermediate-level wastes, shallow burial is practised in such countries as the UK and the USA, while in France, surface storage is the preferred disposal option. In Sweden, however, a submarine repository has been constructed for this short-lived material below the bed of the Baltic Sea.

Overall, the volumes of radioactive waste are small when compared to other waste products generated by human activity. Although estimates are difficult to come by, in part because of the political sensitivity which surrounds both the nuclear power and weapons industries, the total volume of high-level waste produced in the UK, for example, will probably have reached about 1200 m³ by the year 2030, enough to fill two small houses. Substantially larger volumes can be expected from decommissioned reactors, but despite the amounts involved, disposal of high-level waste, which will remain dangerous for many thousands of years, is one of the key issues facing the nuclear industry. High-level waste is heat generating and needs to be cooled and then managed safely for a period which is far longer than we believe human civilisation has existed to date. Although the nuclear industry is convinced that burial of high-level and long-lasting, intermediate-level waste material in suitable geological formations, at least several hundred metres deep, is the best method, opponents of the nuclear industry argue that this is not an acceptable solution. They advocate surface storage until a better management method can be devised. The public sensitivity over the issue of nuclear waste disposal is uniquely great, and although the technical aspects of high-level waste disposal have been solved to the satisfaction of most in the nuclear industry, the political aspects remain to be resolved.

Chernobyl

Concern at the potentially huge damaging effects of an accident at nuclear power facilities is the other major environmental issue surrounding nuclear power. Despite the stringent checks and safety measures employed at such facilities, accidents at nuclear reactors do happen, as evidenced by those at Chalk River, Ontario, Canada in 1952, Windscale (now Sellafield), UK in 1957, and Three Mile Island in Pennsylvania, USA in 1979. These events were overshadowed on 26 April 1986 by the accident at the Chernobyl nuclear power station in the Ukraine (then part of the USSR), the most serious accident to have occurred at a nuclear power station to date, and an event which has haunted the world's nuclear industry since.

Radionuclides from the explosion at Chernobyl were dispersed throughout the Northern hemisphere in trace amounts, with particular 'hotspots' in areas where rainfall washed radioactive material from clouds: in parts of the European Soviet republics, Austria, Bulgaria, Finland, Germany, Norway, Romania, Sweden, Switzerland, the UK and Yugoslavia. Most concern has focused on the biomedical dangers presented to humans by the radionuclide deposition. Initial fears over iodine-131 led to the destruction of fruit and vegetables grown in the open air and milk from cows grazing on contaminated grassland, but since iodine-131 has a half-life of only 8 days, attention soon shifted to caesium-134 and -137, the latter with a half-life of 30 years. Caesium accumulates up the food chain from the soil through vegetation to contaminate meat, necessitating special measures to restrict the movement and sale for consumption of livestock as far from the accident as Scandinavia and Britain, for example. In many of the affected areas, such as Cumbria in the UK, these restrictions are still in place. Other

TABLE 17.5 *Main biological effects of the Chernobyl catastrophe*

Level	Short-term response	Long-term response
Biosphere	Global and local changes in radionuclide accumulation and dispersion	Global and local disturbances in genetic and phenetic structure of the biosphere
Ecosystem	Changes in ecosystem diversity, stability and development patterns	Changes in co-evolution process and ecosystem succession
Human population	Changes in birth and mortality rates	Changes in mutation rate, natural selection intensity and adaptive reaction
Human individual	Disturbances in physiology and behaviour	Changes in probability of cancer, hereditary abnormalities and other diseases

Source: after Savchenko (1991)

dangerous radionuclides involved include strontium-90 (half-life, 29 years) and plutonium-239 (half-life, 24 000 years).

The main overall biological effects of contamination from the Chernobyl catastrophe are shown in Table 17.5. Different ecosystems absorb radionuclides in different ways, and forests were some of the most contaminated. Foliage absorbed radionuclides in rainfall along with carbon-14 and tritium from the air by photosynthesis. When foliage fell in the autumn, most of these contaminants were transferred to the ground, concentrating in leaf litter and the thin upper soil layers. Coniferous forests kept radionuclides in their canopy longer, and such trees were consequently felled and disposed of in some heavily affected areas.

In meadows and pastures, radionuclides absorbed in the vegetation were also transferred to the upper soil layers in autumn. In agricultural areas, contaminated upper soil layers were turned over during ploughing and transferred to a depth of 20–40 cm. Natural migration of radionuclides in soil depends upon their type and the speed at which they penetrate the soil, which is of the order of centimetres per year. Rich soils absorb them more easily and keep them for a longer period than poor sandy soils. The solubility of radionuclides is another important factor: strontium-90 has the highest solubility, explaining its immediate uptake by plants.

The impacts of radionuclide contamination of soils depend upon their chemical forms and concentrations, soil type, the presence of normal atoms of radioactive isotopes in soils, their solubility and availability for roots. Caesium-137 moves very slowly through soils, sinking only a few millimetres per year. Strontium-90, by contrast, forms a weaker bond with soil particles and moves in the soil more rapidly. The generally slow movement of radioactive elements through soils has a beneficial side, in that it slows the transfer of contamination to groundwater.

There is generally considered to be a direct relationship between the levels of soil and plant contamination, and a direct relationship between the level of radioactive contamination of agricultural lands and contamination levels in animals, notably in meat and milk products. Concern over these effects has lead to 257 000 ha of agricultural land being taken out of production in Belarus, for example, along with 1.34 million hectares of forest (Marples, 1992).

The human health effects of the Chernobyl disaster are the subject of investigation and monitoring at both national and international levels. The International Atomic Energy Agency (IAEA) has studied radiation doses from the accident in highly contaminated areas where local food restrictions were in force, and found that the external radiation dose exceeded the background values in only 10 per cent of inhabitants (IAEA, 1991). No health disorders that could be directly attributed to radiation were detected by the IAEA. Several assessments of the possible health risks have projected the excess cancer risk for the entire Northern hemisphere to range from zero to 0.02 per cent. Most of the long-term health affects of Chernobyl will be statistically undetectable, however, because they will be spread through a population of hundreds of millions over several decades.

The direct and indirect economic costs of the Chernobyl accident have been estimated to be at least US$15 billion, 90 per cent of which would be in countries of the former USSR (Tolba, 1992). Disruption of agricultural activities, relocation of people, and consequent psychological stress were the main immediate consequences of the accident.

The future for nuclear power

Concern over the environmental effects associated with nuclear power means that the nuclear industry is conducted under some of the strictest controls and regulations of any major industry, and certainly more stringent than for other forms of power generation. While the risks of environmental damage from unsafe waste disposal or accidents are statistically small, these issues continue to rate highly in the public conscience, and the lack of public confidence stemming from these concerns is perhaps the major problem facing the nuclear power industry today. The issues have consequently become subject to a high level of political, media and public relations activity. For some countries, perception of the risks associated with nuclear power have

resulted in national programmes being stopped, scaled down or, in the case of Sweden, subject to a concerted effort to phase it out altogether. Complete abandonment elsewhere is unlikely in the foreseeable future, however, given the very large economic investments nuclear power plants represent and the widespread concern over carbon dioxide and other gaseous emissions from fossil fuel plants – environmental problems to which nuclear power plants do not contribute. As with renewable energy, the future use of nuclear power will be a political decision in which environmental aspects will undoubtedly play a significant role.

FURTHER READING

Johansson, T.B., Kelly, H., Reddy, A.K.N. and Williams, R.H. (eds) 1993 *Renewable energy: sources for fuels and electricity*. New York, Island Press. An encyclopaedic overview of the various forms of renewable energy, including technical, economic and political aspects.

Roberts, L.E.J., Liss, P.S. and Saunders, P.A.H. 1990 *Power generation and the environment*. Oxford, Oxford University Press. This book contains a chapter on the history of power generation and its environmental impact, two detailed chapters on fossil fuels and nuclear fission and a look at future prospects from renewables.

Schipper, L. and Meyers, S. 1992 *Energy efficiency and human activity*. Cambridge, Cambridge University Press. A detailed analysis of trends in energy use since 1970 in manufacturing, transport, the domestic and service sectors with a look to the future, discussing how energy use can be restrained in order to meet goals of environmental and economic development.

Scientific American 1990 Energy for planet Earth. Special issue 263(3). A collection of articles which look at future energy needs for particular sectors, such as buildings and homes, industry, motor vehicles, and particular world regions, as well as the future of fossil fuels, nuclear power, and solar energy.

18

MINING

We only need to look at the names used to describe key periods in the early development of societies – the Stone Age, the Bronze Age and the Iron Age – to realise the long importance of mining as a human activity. People have been using minerals from the Earth's crust since *Homo habilis* first began to fashion stone tools 2.5 million years ago. Today, we are more dependent than ever before upon the extraction of minerals from the Earth. Virtually every material thing in modern society is either a direct mineral product or the result of processing with the aid of mineral derivatives such as steel, energy or fertilisers.

The naturally occurring elements and compounds mined are usually classified into four groups.

- metals (e.g. aluminium, copper, iron)
- industrial minerals (e.g. lime, soda ash)
- construction materials (e.g. sand, gravel)
- energy minerals (e.g. coal, uranium, oil, natural gas).

In terms of volume extracted, construction materials are by far the largest product of the world's mining industry. They are found and extracted in every country. An estimated 11 billion tonnes of stone, and 9 billion tonnes of sand and gravel were taken from the ground in 1991. About 125 000 t of crushed rock, for example, is used to build 1 km of motorway. Metallic minerals are mined in smaller quantities. World production of iron ore in 1992 was 929 million tonnes, phosphate 144 million tonnes, and bauxite 107 million tonnes (Crowson, 1994).

GLOBAL ECONOMIC ASPECTS OF MINERAL PRODUCTION

The location of mineral concentrations in the Earth's crust is a fact of nature, but there are a host of human and natural factors which determine if and when a mineral is recognised as a resource, classified as a reserve, and eventually exploited. Such factors include global and local economic and political influences, as well as the local availability of basic requirements for mining such as energy and water. While the use of minerals dates from before the time of *Homo sapiens*, it was the Industrial Revolution that sparked their large-scale exploitation. Between 1750 and 1900, global mineral use increased ten-fold as the population doubled, and since 1900, use has increased by more than thirteen-fold (Bosson and Varon, 1977).

The rapid increase in society's use of minerals has periodically sparked concern over their imminent depletion, a sentiment expressed strongly in the 1970s by the Club of Rome (Meadows *et al.*, 1972). Although the amount of minerals in the lithosphere is finite, improvements in technology and changing price/cost relationships, as well as shifting

TABLE 18.1 *Estimates of reserves at Palabora Copper mine, South Africa*

	Defined at start-up in 1966	Full potential as seen in 1991
Reserves before mining:		
Tonnage (million tonnes)	259	854
Grade (% copper)	0.69	0.55
Contained copper (million tonnes)	1.79	4.67
Cut-off grade (% copper)	0.3	0.15
Mine life from start-up (years)	24	35

Source: **Crowson** (1992)

perceptions of the political risks of operating in particular countries, mean that estimates of reserves are constantly being re-evaluated. The dynamics of changing resources are illustrated at the level of one mine by the Palabora copper mine in South Africa, which at its start-up in 1966 was estimated to contain an ore deposit with 1.8 million tonnes of copper, giving the mine an operational life of 24 years. By 1991, however, the estimate had changed to 4.7 million tonnes, with a life of 35 years (Table 18.1). Improved technology and economic changes help to explain the increase in reserves, lowering the cut-off grade of exploitable ore. The estimated reserve and lifetime of the mine would be increased still further if an underground mine were developed.

Another important factor is the simple fact that large areas of the world's land surface have not been mapped in detail for their minerals, let alone explored, particularly where underlying as opposed to surface geology is concerned. New ores are regularly discovered and new mines opened, such as the large copper mine at Neves Corvo in southern Portugal which began operations in 1989 and produced nearly 150 000 t of copper concentrate in 1992. These factors combine to balance the fears of those who suggest that the end is nigh for global mineral exploitation. In the words of one mining industry official: 'it is highly improbable that society will run out of minerals over the long term' (Crowson, 1992: iv). Indeed, the recycling of metals already taken from the lithosphere reduces some of the need for exploiting virgin resources, and Crowson suggests that these materials should be regarded as a renewable resource since most metals can be recycled indefinitely.

On the national scale, most countries have adequate reserves of minerals used in the construction industry and these materials (with the exception of cement) are seldom traded internationally. The use of other minerals, particularly metals, is very heavily concentrated in the rich countries, however, and their movement is an important component of international trade. Many developing countries rely heavily upon mineral exports as sources of foreign exchange, with at least sixteen countries earning 40 per cent or more of their export income from non-fuel mineral exports in the early 1990s (Table 18.2). The sources of minerals for major industrial nations vary. Some industrialised countries have considerable domestic reserves of certain minerals: Australia mined 37 per cent of global bauxite production in

TABLE 18.2 *Non-fuel minerals as a percentage of the total value of exports in selected countries in recent years*

Country	Mineral(s)	Share (%)
Botswana	Diamonds, copper, nickel	87
Zambia	Copper	86
Guinea	Bauxite/alumina	82
Sierra Leone	Bauxite/alumina	80
Namibia	Diamonds, uranium, copper	76
Niger	Uranium	75
Zaire	Copper, diamonds	71
Surinam	Bauxite/alumina, aluminium	69
Papua New Guinea	Copper, gold	62
Liberia	Iron ore, diamonds	60
Jamaica	Bauxite/alumina	58
Togo	Phosphates	50
Central African Republic	Diamonds	46
Mauritania	Iron ore	41
Chile	Copper	41
Mongolia	Copper, molybdenum	40

Sources: Young (1992), World Bank (1992b)

1990, for example, while for other minerals less-developed countries are important sources. Generally, industrialised countries import their minerals from proximate areas of the Third World: from Latin America to the USA, Africa to western Europe, and Asia and Oceania to Japan. Most such ores are shipped from their country of origin with minimal processing, since trade barriers often act to prevent developing-country producers from adding value to their raw materials.

In 1990, the European Community, Japan and the USA accounted for 60 per cent of the world consumption of aluminium and refined copper, 58 per cent of refined lead, 55 per cent of tin and 50 per cent of zinc. However, since the Second World War, mineral use in the wealthier nations has not been growing as fast as in developing countries. Between 1950 and 1990, for example, the percentage of global

aluminium used by developing countries rose from 2 per cent to 19 per cent, for zinc the percentage rose from 3 to 25 per cent over the same period, and for steel the rise was from 5 to 25 per cent (Wellner and Kürsten, 1992). This increase is largely accounted for by emerging industrial nations such as Brazil, China, India, Mexico and the 'Asian Tigers' of the Pacific rim.

The slow-down in demand in richer countries is explained by several factors. Industrial economies grew more slowly after the oil crisis of the mid-1970s, and structural change in these economies, with less emphasis on heavy industry and moves towards services and high technology, has meant a slow-down in demand for physical materials. The growth of demand for virgin metals has also declined with more emphasis on recycling and the introduction of new materials such as plastics and ceramics.

ENVIRONMENTAL IMPACTS OF MINING

Concern at the environmental impacts of mining is by no means a recent phenomenon. Georgius Agricola (1556), author of the world's first mining textbook, could have been writing today as he outlined the strongest arguments of mining's detractors in sixteenth century Germany:

> [T]he woods and groves are cut down, for there is need of an endless amount of wood for timbers, machines and the smelting of metals. And when the woods and groves are felled, then are exterminated the beasts and birds, very many of which furnish a pleasant and agreeable food for man. Further, when the ores are washed, the water which has been used poisons the brooks and streams, and either destroys the fish or drives them away.
>
> (Agricola, 1556: 8)

Little appears to have changed in more than 400 years since Agricola. In the late twentieth century, environmentalists often quote this passage as evidence of the long-standing environmental impact of mining. Interestingly, the quote is often used out of context, since Agricola in fact dismissed these anti-mining sentiments on the grounds that the environmental cost was far outweighed by the benefits to society:

> If we remove metals from the service of man, all methods of protecting and sustaining health and more carefully preserving the course of life are done away with. If there were no metals, men would pass a horrible and wretched existence in the midst of wild beasts; they would return to the acorns and fruits and berries of the forest.
>
> (Agricola, 1556: 14)

There can be no doubting that mining involves an environmental impact, and the destruction of habitats suggested in Agricola's textbook is perhaps the most obvious from among a diverse catalogue of ecological changes that the various types of mining operation can cause (Table 18.3). The question as to whether such impacts should be regarded as worth the benefits derived from the minerals extracted is a complex one which will be returned to after consideration of the physical problems involved.

Habitat destruction

Vegetation is stripped and soil and rock moved, both for the extraction process itself and the concomitant building of plant, administration and housing facilities, and the history of mining activity has left considerable areas of disrupted habitat. The scale of damage is greatest at so-called opencast or strip mining sites. For example, opencast phosphate mining on the Pacific Ocean island state of Nauru, one of several small Pacific coral islands exploited for their phosphate, will leave a severely degraded landscape of highly irregular solution-pitted limestone over an estimated 80 per cent of the island's 2100 ha area by the time mining halts around the year 2000 (Manner et al., 1984).

Disposal of waste rock and/or 'tailings' – the impurities left after a mineral has been extracted from its ore – usually destroys still larger areas of natural ecosystems. This is an increasing problem, both because of the growth in demand for minerals and because as rich ores are mined out, so lower-grade deposits are worked, producing more waste per unit of mineral produced. While four centuries ago the average grade of copper ore mined was about 8 per cent (Bosson and Varon, 1977), the average grade in the 1990s was 0.91 per cent, which means about 990 million tonnes of waste generated for the 9 million tonnes of copper produced.

Similar product-to-waste ratios are found in the china clay mines of southwestern England, where the production of 1 t of kaolin

TABLE 18.3 *Environmental problems associated with mining*

Problem	Type of mining operation			
	Open pit and quarrying	Opencast (as in coal)	Underground	Dredging (as in tin or gold)
Habitat destruction	×	×	—	×
Dump failure/erosion	×	×	×	—
Subsidence	—	—	×	—
Water pollution	×	×	×	×
Air pollution*	×	×	×	—
Noise	×	×	—	—
Air/blast/ground vibration	×	×	—	—
Visual intrusion	×	×	×	×
Dereliction	×	—	×	×

×: Problem present
—: Problem unlikely
*: Can be associated with smelting which may not be at the site of ore/mineral extraction
Source: modified after Whitlow (1990)

produces 1 t of mica, 2 t of undecomposed rock ('stent') and 6 t of quartz sand. In Cornwall and Devon, where china clay has been mined since the 1770s, these waste materials are dumped on surrounding land to produce a devastated landscape of heaps, pits and lagoons, with disused land in between, covering an area of 9100 ha (Bradshaw and Chadwick, 1980). On the national scale, such mining wastelands can cover a significant proportion of the national land area. Punning (1993) reports that excavation of Estonia's 4-billion-tonne oil shale reserves, the world's largest commercially exploited deposit, plus sand, gravel and peat extraction, has devastated at least 45 000 ha or 1 per cent of the national land area. Table 18.4 shows the areas disturbed in the countries of the former Soviet Union.

To date, however, the largest areas of strip mining in the world are for coal (Fig. 18.1). In

the USA, which produces about 60 per cent of its coal by this method, it has affected vast areas of the Appalachian Mountains and large areas of the states of Ohio, West Virginia and Kentucky. On average, surface mining for coal consumes 10–30 ha of land for every million tonnes of coal produced in the USA, but in other parts of the world, local conditions and available technology can result in much higher land-consumption-to-production ratios. In southern Russia's Kuznetsk Basin, for example, the figure is about 166 ha per million tonnes (Bond and Piepenburg, 1990).

In areas where smelting takes place near the extraction site, further damage can be caused by cutting trees to fuel smelters. In Copper Basin, Tennessee, USA an area of 14 500 ha was deforested in the nineteenth century to smelt copper ore in vast open-air pits, and the charcoal requirements for iron ore smelting in the Grande Carajás Project in Pará state, Brazil

TABLE 18.4 *National areas of land disturbed by mining in the countries of the former Soviet Union*

Country	Area of disturbed land, beginning 1989 (thousand hectares)	Years required to reclaim
Russia	1180	11
Ukraine	198	8
Belarus	120	13
Uzbekistan	52	32
Kazakhstan	167	16
Georgia	3	7
Azerbaijan	21	19
Lithuania	32	27
Moldova	3	6
Latvia	44	44
Kyrgyzstan	7	16
Tajikistan	6	15
Armenia	8	82
Turkmenistan	5	7
Estonia	47	78

Source: after Bond and Piepenburg (1990)

will spur the cutting of an estimated 50 000 ha of tropical forest each year during the mine's expected 250-year life (Fearnside, 1989). Although fuel for smelting was originally proposed to be grown on plantations, in practice, much of the charcoal has been made from natural forest.

The environmental impacts of mining are not just limited to terrestrial ecosystems. The mining of coral and sand for road and building materials is a widespread threat to coastlines in many parts of the world (Fig. 18.2) and particularly to island reefs in the Pacific Ocean. Apart from the physical destruction of reefs to provide limestone, and the introduction of toxic substances to the marine environment during the mining process, the dredging of sand causes beach erosion and alters water circulation, leading to sedimentation on reefs (IUCN, 1988). When exploitation of polymetallic (manganese) nodules from the deep-sea bed becomes a commercial proposition, although this is likely to be a long time in the future, it too will bring disturbances to the world's oceans. Impacts will occur at different levels in the water column: at the sediment surface due to collector impact, immediately above the sea bed due to the plumes of sediment raised, and at other levels as abraded nodule materials and sediments are discharged (Thiel and Schriever, 1990). The most commercially significant concentrations of nodules are found in the Pacific and Indian oceans.

Geomorphological impacts

Mining involves the movement of quantities of soil and rock, and thus by definition it is a human-induced geomorphological process. The scale of the process has reached the level whereby estimates suggest that as much as a

FIGURE 18.1 *Open cast mining at the Cordero coal mine; southern Powder River Basin, Wyoming, USA (courtesy of The RTZ Corporation plc)*

hundred times more material is stripped from the Earth by mining than by all the natural erosion carried out by rivers (Goudie, 1993b). The impacts of mining, however, are much more localised than the denudation caused by rivers. The opencast copper mine at Bingham Canyon in Utah, USA, reputed to be the largest human excavation in the world, has involved the removal of more than 3400 million tonnes of material over an area of 7.2 km², to a depth of 770 m. Nevertheless, such features can attain significant proportions on the national scale: surficial scars on the landscape of Kuwait, created by the oil-generated construction boom, cover large parts of the national land area, with some individual quarries measuring up to 40 km in length (Fig. 18.3). Reconstruction of the country's urban fabric after the Gulf War will continue to enlarge these surface features.

New landforms are also created by the deposition of wastes, and these new landforms can create new geomorphological hazards, such as landslides and subsidence. The death of 150 people at Aberfan, South

FIGURE 18.2 *Sand mining from beaches in Oman, in areas where building sand is scarce, is threatening breeding sites of endangered green and hawksbill turtles. Coastal erosion studies indicate that many of these beaches are ancient and slow to generate, hence removal of sand threatens their existence. Some beaches are earmarked for tourist development, another reason for their protection*

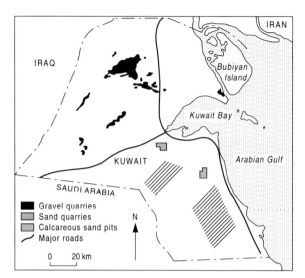

FIGURE 18.3 *Distribution of sand and gravel quarries in Kuwait (after Khalaf, F.I. 1989, Desertification and aeolian processes in Kuwait.* Journal of Arid Environments *12: 125–45. Reproduced by permission of Academic Press Ltd)*

Wales in 1966, when waste piles from the Merthyr Vale coal mine suddenly collapsed, graphically illustrated the dangers of unregulated waste tipping. Surface subsidence is caused by underground workings, and is most commonly associated with coal mining where whole seams of material, frequently several metres thick, are removed, resulting in overlying strata collapsing downwards. Damage to mining equipment, buildings, communications, and agriculture through the disruption of drainage systems are the most common results. In the USA, for example, subsidence due to underground coal mining affects 800 000 ha in thirty states (Gray and Bruhn, 1984). Collapse also occurs when water is pumped out of mine workings, lowering local water tables which results in the drying out and shrinkage of materials such as clays, and through the reduction in underground fluid pressure which results from oil and natural-gas extraction. Blunden

(1985) notes that marine inundation is a danger caused by subsidence from oil and gas extraction in areas near sea level and quotes examples from oil fields on the shore of Venezuela's Lake Maracaibo, coastal locations in Japan and the Po delta in Italy. A particularly dramatic example occurred in the Wilmington oilfield in southern California where abstraction between 1928 and 1971 caused 9.3 m of subsidence at Long Beach, Los Angeles (Table 18.5).

Pollution

Pollution from mineral extraction, transportation and processing can also present serious environmental problems affecting soil, water and air quality. Examples of pollution emanating from the transport of oil are given elsewhere (see Chapters 10 and 11), while here the emphasis is on mine workings and smelting.

Down and Stocks (1977) note four major water pollution problems associated with mining, all of which can cause serious damage to aquatic life and, on occasion, to human health:

- heavy metal pollution
- acid mine drainage
- eutrophication
- deoxygenation.

Many tailings contain metal ores and other contaminants, formerly locked up in solid rock, which can be leached into soils and waterways and blown into the surrounding atmosphere. Some of the highest heavy metal concentrations are derived from uranium tailings and pumped water from iron mines, but one of the most notorious examples of deleterious effects on human health came from a lead–zinc mine in Toyama prefecture, Japan. Cadmium-polluted water from the mine, which was used in rice paddy fields, led to an outbreak of 'itai-itai' disease, a

TABLE 18.5 *Examples of ground subsidence due to oil and gas abstraction*

Location	Area affected (km²)	Maximum subsidence (m)	Period	Rate (mm/year)
Inglewood, USA	—	2.9	1917–63	63
Maracaibo, Venezuela	450	5.03	1929–90	82
Wilmington, USA	78	9.3	1928–71	216

Source: after Cooke and Doornkamp (1990); Goudie (1993b)

cadmium-induced bone disorder. About a hundred deaths were reported from the disease among long-term residents of the area between 1947 and 1967 (Nriagu, 1981).

Acid mine drainage was originally thought to be associated only with coal mining, but any deposit containing sulphide, and particularly pyrite, can be a source, and it is produced from the working of many other minerals. Reactions with air and water produce sulphuric acid, and pH values of polluted waterways may be as low as 2 or 3. Acid mine drainage is not just a problem of operational workings, but may continue as a pollution source for some years after mine closure (Robb, 1994). Surface tailings can also produce acid effluents; Hägerstrand and Lohm (1990) report that acid produced from sulphides in tailings around the Falu copper mine in central Sweden, which has been worked for at least 800 years, have seriously damaged several lakes in the region and are still producing sulphuric acid.

When tailings dry out, dust blow can also cause localised pollution. Fine material blown from the quartzite dumps that surround Johannesburg, a city that grew up on the gold mining activity in South Africa's Witwatersrand, used to stop work at food processing factories and cause traffic chaos due to decreased visibility on windy days before the dumps were stabilised with vegetation (Bradshaw and Chadwick, 1980).

The processing of minerals can also release dangerous compounds into surrounding environments. In Brazil, where an estimated 650 000 mostly small-scale prospectors are involved in gold mining, the use of mercury as an amalgamate to separate fine gold particles from other minerals in river bed sediments contaminates waterways in 'gold rush' zones. Miners released an estimated 100 t of mercury – an extremely toxic metal that accumulates in the food chain and can cause birth defects and neurological problems – into the basin of the Madeira River, a tributary of the Amazon, each year. Fish eaten by the local population were found to contain mercury at concentrations up to five times the Brazilian safety limit (Malm *et al.*, 1990).

Sulphur dioxide (SO_2) released from open-air smelting of nickel ore at Sudbury, Ontario, Canada in the early years of this century, destroyed vegetation over a 10 000 ha area. Construction of a closed plant with a 387.5 m stack in 1972 turned a local problem into a regional one, affecting lakes up to 60 km downwind, some with a pH of 5.5 or less. Nickel emissions from the stack have also been high. Nickel concentrations are naturally low in lake waters, with a global mean of 10 g/l. Some lakes in the Sudbury area have been found with nickel concentrations of 300 g/l. The combination of acid rain and heavy metal pollution has rendered many of the lakes in the area biologically dead (Blunden, 1985).

Miscellaneous impacts

The attraction of mining operations as sources of employment may have deleterious effects on labour-intensive areas of the economy such as agriculture. Lewis and Berry (1988) cite Zambia's Copper Belt as an example, and Barrow (1991) suggests that Brazil's Pelada gold mine has had a similar effect. Itinerant miners venturing into little-explored terrain also carry with them new diseases and problems associated with alcohol, bringing often disastrous consequences for local populations. Gold prospectors in Amazonia, so-called *garimpeiros*, have spread virulent strains of malaria into the reserves occupied by the Yanomani indians in northern Amazonas. Local tribes with little resistance to the new strains of parasites have suffered severe outbreaks of malaria (Smith *et al.*, 1991), and met with previously unknown diseases such as measles.

There are also cases where mining operations have had serious political ramifications. A classic example in this respect can be cited from the Panguna copper mine on the island of Bougainville in Papua New Guinea, where discontent over the distribution of mining revenues was a factor in precipitating a revolt out of a long-simmering dispute between the islanders and the PNG government. Discontent over the unequal distribution of the income derived from mining was not helped by the fact that 130 000 t of metal-contaminated tailings were dumped into the Kawerong and Jaba rivers every day, severely depleting biological activity in the waters. The resulting conflict led to guerillas taking over the island and the subsequent closure of the mine in 1989.

REHABILITATION AND REDUCTION OF MINING DAMAGE

Many techniques are available for the amelioration, curtailment or restoration of damage caused by mining and its associated activities (e.g. Down and Stocks, 1977; Bradshaw and Chadwick, 1980; Blunden, 1985), and there is little doubt that for a variety of reasons the mining industry in general has recently adopted a more benevolent approach to its environmental impact than in times past. Site surveys prior to mining, retention and replacement of topsoil after excavation, and careful reseeding with original species can be employed to return environments to something close to their original states (Fig. 18.4).

Waste dumps and tailings can be stabilised with vegetation both to ameliorate their visual impact and to reduce the risks of slope failure (Fig. 18.5), or flattened and properly drained as in the case of the waste tips at Aberfan. Mining wastes in many countries are used for a variety of purposes, particularly in road construction and for landfill (Table 18.6), but transportation costs to areas of high demand often limit their use. Of the 27 million tonnes of china clay waste produced each year in Devon and Cornwall, about 1.5 million tonnes are used by local industries, but only a small proportion of the 12 million tonnes of china clay sands which have potential for use as aggregates are used, because local demand is low (DoE, 1992).

Many of the efforts made to reduce pollution originating in mining wastes have stemmed from the realisation that the pollutants are themselves economically valuable. A 40 per cent reduction in SO_2 emissions from the Sudbury nickel smelter, for example, was achieved in the late 1960s when the gas was converted into sulphuric acid. Similarly, the use of new technologies for recovering wasted resources, such as microbes to 'bioleach' metals from tailings, have also been developed in response to the lower grades of accessible ores and lower operating costs. The recycling of minerals, particularly metals, also helps to reduce environmental impacts, of course (see Chapter 16).

FIGURE 18.4 *Rehabilitated land at the Cordero coal mine, southern Powder River Basin, Wyoming, USA. Compare with Figure 18.1 (courtesy of The RTZ Corporation plc)*

FIGURE 18.5 *An area of rehabilitated waste rock dumps at Bingham Canyon copper mine, Utah, USA (courtesy of The RTZ Corporation plc)*

In recent decades, a growing public concern in the industrialised nations at the environmental cost of mining has led to increasing legislative controls on the activities of mining companies, including the requirement to conduct Environmental Impact Assessments (Ellis, 1989). The generally higher public profile of environmental issues and the pressures brought to bear by environmentalists, particularly in developed countries, has encouraged multinational mining companies to give environmental concerns a higher profile. The formation in 1991 of the International Council on Metals and the Environment, a group of more than twenty-five major metal mining companies, is clear indication of this trend. The council aims to promote sound environmental and related health policies and practices to ensure the safe production, use, recycling and disposal of metals.

Increasingly, responsible mine operations are conducted with considerable attention to the possible pollution effects, involving regular monitoring of groundwater, air quality and noise, to maintain acceptable standards. In some cases, such as at the Neves Corvo copper mine, Portugal, tailings are delivered to special ponds to reduce dust blow and to prevent oxidation of the contained sulphur which could give rise to acidic water. The dam is operated on a no discharge basis, with any overflow during storms caught and pumped back. Environmentally sound management and clean-ups cost money, however, and legal obligations are often necessary to force appropriate practices, and while many such legal obligations are now well established in most developed countries, they are often inadequate in developing countries where the importance of mining to the national economy (Table 18.2) makes the imposition of costly rehabilitation measures unattractive. The use of legally enforced pollution control equipment on copper smelters in the USA means that for each tonne of copper produced, twelve times less SO_2 is emitted than from smelters in Chile (UNECLA, 1991).

Legal instruments also cover international pollution problems. Massive salinisation of

TABLE 18.6 *Production and use of mine and quarry wastes (waste rock and tailings) in selected countries*

Country/ mine type	Production/year (thousand tonnes)	Use
AUSTRALIA		
Coal	60	Small amounts for road construction and fill
BELGIUM		
Coal	4	Lightweight aggregate manufacture
FINLAND		
Various	8	Road construction; aggregate for concrete
INDIA		
Gypsum	2	Building plaster
Mica	>1	Mica insulation bricks
Coal	10	Mine backfilling (large percentage); road construction and fill
SOUTH AFRICA		
Gold ore	45	Road construction and aggregate for concrete (50%); silicate bricks
UK		
China clay	22	Aggregate for concrete, silicate bricks, road construction and fill (<5%)
Tin ore	0.5	Aggregate for concrete (minor percentage)
Coal	50	Road construction, bricks, cement, lightweight aggregate manufacture (10%)
USA		
Copper	860	Road construction; bitumen filler
Iron ore	55	Road construction
Miscellaneous quarrying	68	Amenity banks

Source: after Blunden (1985)

the River Rhine by chloride pollution (Fig. 18.6), 30 per cent of which is discharged by potash mines in the Alsace region of France, has become the subject of an international convention (Kiss, 1985). The Convention on the Protection of the Rhine against Chlorides was signed in 1976 by Switzerland, the former West Germany, France and The Netherlands but was not ratified by France until 1985 because of protests over proposals for the alternative disposal of the Alsatian salt. Under the convention, the French Government was required to install a subsoil injection system to dispose of the salt, paid for by the affected states in proportion to the degree to which they were polluted, in contradiction to the 'polluter pays' principle, but local farmers resisted the salt injection plan. The French now propose to store the salts above ground.

Care must be taken, when establishing clean-up targets, to set appropriate standards. Runnels *et al.* (1992) point out that levels of

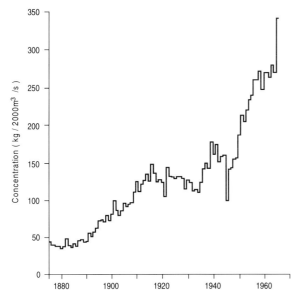

FIGURE 18.6 *Increase in chloride concentrations in the River Rhine, 1875–1963 (after CIPRCP, 1984)*

minerals and pH in rivers can be naturally high in areas of undisturbed mineral deposits due to weathering and leaching of metalliferous ores. Ironically, in the course of time, some areas previously damaged by mineral extraction have come to be regarded as sites of beauty and environmental importance with minimal active rehabilitation. Some former quarries and gravel pits in the UK are now designated nature reserves, such as the Felmersham Gravel Pits in Bedfordshire, a Site of Special Scientific Interest noted for its large number and wide variety of dragonflies, while the Norfolk Broads are entirely the work of medieval peat cutters (Lambert *et al.*, 1970).

In other situations, mining may even be advocated as an option for revitalising ecosystems damaged by other forms of development. Björk and Digerfeldt (1991) describe how peat extraction in two former wetland areas of Jamaica has allowed their re-establishment. The Black River and Negril wetlands were drained and canalised for the cultivation of rice and cannabis, respectively, but successful revitalisation of the wetland habitat after peat cutting on the Negril has led to similar proposals for the Black River.

THE TRUE COSTS OF MINING

The value to society of the products of mining are undeniable, providing as they do the basic building blocks of everyday life for virtually the entire human population. Mining inevitably entails disruption to the natural environment, although given minerals' importance to society and the relatively localised nature of the impacts, the disruption is on a far smaller scale than that of agriculture, for example.

Many positive steps are being made to limit the environmental impacts of mining. The recycling of metals, for example, means less new damage and often makes economic sense even when the economics only take extraction, processing and distribution into account. Other new technologies for recovering wasted resources, such as the use of microbes to 'bioleach' metals from tailings, for example, have also been developed in response to lower-grade ores. Operational mines, too, are increasingly incorporating environmental rehabilitation into their plans. Although such action often takes a higher priority in countries where legislation demands it, multinationals operating in countries with less-stringent requirements also yield to pressures for more environmentally sustainable practices called for by shareholders and environmental pressure groups (Banks, 1993). For multinational companies whose largest stock in trade is their reputation, public feeling over the importance of the environment has undoubtedly influenced practice both directly and indirectly through legislation.

Ultimately, it is for society to decide whether the environmental cost is worth the societal benefits of mining output. Bingham Canyon, which has operated since 1904 and has another 30 years of life, has long provided about 15 per cent of US copper and made a significant contribution to the economy of the state of Utah. It is for American society to decide whether the environmental cost, in terms of a large hole in the ground and associated disruption to the local environment, is worth this contribution to society.

In this respect, as for numerous other environmentally damaging societal practices, there have been calls for the true costs, including those of clean-up and rehabilitation, to be incorporated into the price of the minerals themselves (e.g. von Below, 1993). In practice, the concept of 'full cost pricing' is difficult to quantify, particularly for the mining industry where costs vary widely among producers, reflecting widely differing mineral endowments, and where prices set in international auction markets bear only a tenuous relationship to costs. Irrespective of the difficulties in putting economic values on certain aspects of the environment (see Chapter 21), the cost–benefit equation which balances environmental and use value of minerals will vary from country to country and society to society. In practice, it is impossible to set objective 'correct' prices (Humphreys, 1993), although environmentalists are right to point out that some countries have a freer hand to assess this cost–benefit than others, given the heavy reliance of some developing countries on income from their mineral exports.

Problems may occur, however, when those who benefit are not fully coincident with those who pay the costs. Banks (1993) points out that generally the greatest costs in environmental and social terms fall on local populations around mine sites, while the benefits, usually measured in economic and political terms, tend to be concentrated at the national and international scales. When the cost–benefit equation is unbalanced, difficulties may arise, as the experience of Bougainville illustrates. Redressing the imbalances can also act retrospectively: a global precedent was set in 1993 when Australia paid off Nauru to settle the island's claim against the British Phosphate Commission for environmental degradation of two-thirds of its territory.

FURTHER READING

Blunden, J. 1985 *Mineral resources and their management*. London, Longman. A comprehensive review of mineral resource management, including techniques for rehabilitation and reduction of mining's environmental impact.

Bosson, R. and Varon, B. 1977 *The mining industry and the developing countries*. New York, Oxford University Press. This book focuses on the role of mineral extraction in Third World economies.

Down, C.G. and Stocks, J. 1977 *Environmental impact of mining*. London, Applied Science. A comprehensive review of the full range of environmental problems associated with mining and the techniques developed to combat them.

Young, J.E. 1992 *Mining the earth*. Worldwatch Paper 109. Worldwatch Institute, Washington, DC. This pamphlet documents the environmental impacts of mining and argues for more efficiency in mineral use, less government subsidy for mining and the need for developing countries to find less-mineral-dependent modes of development.

19

WAR

Aggression appears to be a fundamental characteristic of human nature, and violence has been used to resolve disputes since prehistoric times. Warfare is no less characteristic of the world today than in the past: since the end of the Second World War, more than 130 wars and violent internal conflicts have raged in more than eighty countries, most of these in the developing world. With the course of history, however, technological developments have greatly enhanced humankind's capacity for destruction and while the devastation to human life and civilisation is well appreciated, the ecological impacts of war are less well documented. They are nonetheless wide-ranging and not always obvious. In the broad perspective, war and the preparations for war are the antithesis to development, squandering, as they do, scarce resources and eroding the international confidence necessary to promote development. While competition for the control of resources has long been a reason for conflict, in recent years, resource degradation is increasingly being seen as a cause of war, and one that looks set to become more important with an increasing world population.

COST OF BEING PREPARED

Keeping armies operational and their weaponry updated represents a huge drain on economic, human and natural resources. Over the past two decades, the world spent about US$17 trillion on military activity at 1988 prices, averaging US$850 billion per year (SIPRI, 1990) and although military spending has fallen slightly since the mid-1980s in industrialised countries, it has continued to increase in most developing countries. The military employs between 60 and 80 million people worldwide and arms imports by developing countries, half of which are financed by export credits, account for 30 per cent of the debt burden of developing countries (UN, 1989b). The scale of military spending, some of its contradictions and trade-offs with social and environmental priorities are quoted by Tolba (1992):

- The UN Environment Programme spent US$450 million in the decade 1980–90, less than five hours-worth of global military spending.
- One Apache helicopter costs US$12 million, a sum that could pay for the installation of 80 000 hand pumps to give Third World villages access to safe water.
- One day spent on the 1991 war over Kuwait – US$1.5 billion – could have funded a 5-year global child immunisation programme against six deadly diseases, thereby preventing the death of 1 million children per year.

Although data on the military's consumption of resources are, like much information on the

TABLE 19.1 *Estimated military consumption of selected non-fuel minerals as a proportion of total global consumption in the early 1980s*

Mineral	Share (%)
Copper	11.1
Lead	8.1
Aluminium	6.3
Nickel	6.3
Silver	6.0
Zinc	6.0
Fluorspar	6.0
Platinum group	5.7
Tin	5.1
Iron ore	5.1

Source: after Kim (1984)

military, difficult to obtain, an indication of the scale of impact is given in the estimates collated by Renner (1991). The US military consumes 2–3 per cent of total US energy demand, while worldwide nearly 25 per cent of all jet fuel is used for military purposes. An estimate of the military consumption of non-fuel mineral resources is shown in Table 19.1. These estimates suggest that the quantities of aluminium, copper, nickel and platinum used by the military worldwide exceeds the entire Third World's demand for these minerals. Between 0.5 and 1 per cent of the world's land surface is thought to be used by the military during peacetime. Land used for training and weapons testing can be severely degraded, as well as sites used to store hazardous wastes from military activities. In the USA, the military is thought to be the country's largest generator of hazardous wastes, and more than 1500 military bases with consequent environmental problems have been identified. The largest of such contaminated sites, the Rocky Mountain Arsenal in Colorado, has been used to dump 125 different dangerous chemicals over a 25-year period.

Such problems are not confined to the national territories of the polluting armies. Soviet and US forces, particularly, maintained numerous overseas bases during the Cold War years, many of which have been vacated in the 1990s. Twenty-three years of Soviet troop presence in the former Czechoslovakia, for example, has left a legacy of widespread groundwater contamination by waste oil, and similarly at sites in Moravia contaminated by poisons used in chemical warfare (Carter, 1993a). The question as to who should pay for clean-ups on these bases will remain a sensitive political issue for some time to come.

Some of the worst environmental degradation has been associated with the development of nuclear weapons. Testing of nuclear weapons in the atmosphere has caused the dispersion of radioactive material across the globe, adding a small increment to the background exposure of the world's population. A partial ban on atmospheric testing was signed in 1963, but it was not until 1980 that the last atmospheric test was made in northwest China. Underground testing of nuclear weapons, which often causes small to moderate earthquakes, still continues. The former Soviet nuclear test ground near Semipalatinsk in eastern Kazakhstan, for example, covers about 2000 km² of contaminated land.

Serious contamination from weapons-producing plants has also come to light in recent times. By far the most serious accident involving radioactive wastes in the former Soviet Union, prior to Chernobyl, was an incident which occurred in 1957 at a secret nuclear weapons processing plant in the southern Urals known as Production Association Mayak, located between the towns of Kasli and Kyshtym, 80 km northwest of Chelyabinsk. The fact that the accident was not officially admitted to until 1989, when the then Soviet Government stated that 2 million curies of radioactive elements had been released from an exploded waste tank and deposited over an area of 90 000 ha, gives some indication of the secrecy surrounding such military operations. More than 10 000 people were evacuated from the area and parts of the region were still restricted more than 30 years later (Pryde, 1991).

It has also been revealed that Production Association Mayak routinely dumped medium-level and high-level radioactive wastes into the River Techa over the period 1949–56, totalling much greater doses of radioactivity for local populations, particularly those using the river for drinking water. Another serious incident occurred in 1967 when radionuclides from the dry banks of Lake Karachay, an open depot of radioactive waste from the plant, were blown into the surrounding region, severely contaminating 2700 km². The medical and ecological consequences of these incidents are the subject of continuing investigations (Akleyev and Lyubchansky, 1994).

DIRECT WARTIME IMPACTS

The relationship between people and their environment can be changed significantly during wartime. Priorities are altered, and certain resources are used more rapidly than during peacetime in order to fuel the war effort. In the time of Henry VIII, for example, many of England's oak trees were cut down to build warships. In addition to the destruction of agricultural land and woodland during the prolonged trench warfare on the coastal plains of France and Belgium during the First World War, wide areas of forest were felled beyond the battle zones to supply the war effort. In total, Belgium lost most of its remaining forests and France lost about 10 per cent of its forested area at this time (Graves, 1918).

Deliberate destruction and manipulation of the environment has also been used by armies to gain military advantage for centuries. So-called 'scorched earth' policies, in which

vegetation and crops are deliberately destroyed to prevent their use by the enemy, is an age-old military tactic: in Biblical times, for example, Samson tied burning torches to the tails of 300 foxes and drove them into the olive groves and grain fields of the Philistines. In modern times, the scope for environmental modification for military purposes is indicated in Table 19.2, a list submitted during the negotiations for the 1977 Environmental Modification Convention (see below), although it should be noted that this list is by no means exhaustive in that it does not include the effects of numerous biological and chemical weapons which have been developed.

A number of these modifications have been employed in conflicts. During the Viet Nam War, for example, the US army diverted water from the Mekong delta to drain the Plain of Reeds where an enemy base was located, and a classified rain-making programme was conducted from 1967–72, seeding clouds from the air using silver and lead iodide to increase normal monsoon rainfall over the Ho Chi Minh trail, a vital supply line for North Vietnamese forces (Jasani, 1975). The Viet Nam War was also the scene of the most extensive systematic destruction of vegetation ever undertaken in warfare. About 72 million litres of herbicides were sprayed over thousands of square kilometres of the country between 1961 and 1971 in an effort to deprive North Vietnamese forces of the cover provided by dense jungle. Most spraying was targeted on forest but cropland was also sprayed, and although missions were only conducted during light winds to avoid non-target contamination, drifting sprays actually caused damage over much wider areas (Orians and Pfeiffer, 1970). An example of repeated wind-drift spray damage to rubber trees on the 9 km² Plantation de Dautieng is shown in Fig. 19.1.

About half of Viet Nam's mangrove forests were completely destroyed by aerial spraying,

TABLE 19.2 *Military techniques for modifying the environment*

Technique	Military application	Feasibility
ATMOSPHERE		
Fog/cloud dispersion	Make target areas more visible	Relatively easy for supersaturated fog, difficult for warm fog
Fog/cloud generation	Protect target areas from attack and nuclear flash	Depends on availability of equipment and materials
Hailstone generation	Damage thin-skinned equipment, power/communication wires and antennae	Restricted to hail cloud conditions
Material to alter electrical properties	Interrupt communications and remote sensing	Unknown
Introduction of electric fields	Interrupt communications and remote sensing	Very high energies needed
Generating and directing destructive storms	Damage battlefields, ports, airfields	High energies needed but some success in hurricane dispersal
Rain and snow making	Inhibit mobility, block routes, hinder communications	Very dependent on cloud systems, highly localised, short duration
Lightning control	Start fires, destroy communications antennae	Conceptually may be possible
Climate modification	Strategic effect on food production and ecology	Very large energy input
Change of high atmosphere or ionosphere	Strategic effect on food production and possibly on human survival	Uncertain
OCEANS		
Change of physical, chemical and electrical parameters	Affect acoustic paths, possible effect on potential food supplies	Uncertain but large energies may be necessary
Addition of radioactive material	Long-term effect on humans who may have ingested contaminated food	Uncertain and difficult
Tsunami generation	Destroy low-lying areas, surface fleets	Uncertain and difficult
LAND		
Earthquake/tsunami generation	Damage battlefields major military bases and strategic facilities	Specific to certain locations and mechanism not fully understood
Burning of vegetation	Generally destructive	Easily started but dependent on flammability of vegetation and weather conditions
Generation of avalanches and landslides	Disrupt communications	Only in mountainous areas or in regions of unstable soils

TABLE 19.2 *(continued)*

Technique	Military application	Feasibility
Modification in permafrost areas	Destroy roads, rail foundations, alteration of stream sources	Relatively easy by removal of insulating cover but effect limited to melting season
River diversions	Flooding, river navigation hazards	Large engineering effort necessary
Stimulation of volcanoes	Ash and gases may act as precipitation nuclei affecting certain communications and remote sensing	Uncertain but results difficult to predict

Source: after Goldblat (1975)

an area which has shown few signs of subsequent natural recovery. The effects on wildlife are not well documented but a few studies and much anecdotal evidence suggest decreased abundance of many animal species, and herbicide-related illness in domestic livestock (Westing, 1984). In total, an estimated 22 000 km² of farmland and forest was destroyed, mainly in the south, by tactical spraying, intensive bombing and mechanical clearance of forest. A further 1170 km² of forest was destroyed by cratering from 13 million tonnes of bombs and another 40 000 km² by bombardment (Collins *et al.*, 1991). The conflict left an estimated 26 million bomb and shell craters across 170 000 ha of Indochina, most in South Viet Nam, representing displacement of 2.6 billion m³ of earth (Westing and Pfeiffer, 1972).

The military experience of defoliant chemical use in Viet Nam was subsequently repeated on a smaller scale in El Salvador during the 1980s, where government forces, with US backing, laid waste areas held by or sympathetic to anti-government guerillas. The term 'ecocide' has been coined for such widespread destruction (Weinberg, 1991).

Another military method involving large-scale devastation to the natural environment, sometimes with that very aim in mind, is the destruction of various forms of infrastructure. Dams have been common targets in this respect. Allied bombing of the Möhne, Eder and Sorpe dams in the Ruhr valley of Germany during the Second World War, for example, aimed to cripple the German economic heartland. Water released from the damaged Möhne and Eder dams killed 1300 people, left 120 000 homeless and ruined 3000 ha of arable land as well as numerous factories and power plants. But the most devastating example of deliberate dam breaching occurred during the Second Sino-Japanese War of 1937–45 when the Chinese dynamited the Huayuankow dike on the Huang Ho River near Chengchow to stop the advance of Japanese forces. The mission was successful in this respect and several thousand Japanese troops were drowned, but the flow of water also ravaged major areas downstream: several million hectares of farmland were inundated, as well as eleven cities and more than 4000 villages. At least several hundred thousand Chinese drowned and several million were left homeless (Bergström, 1990).

Oilfields have also been the target of deliberate damage. Retreating Austrian forces systematically destroyed facilities on Romanian oilfields in 1916–17, as did Iraqi forces before being driven out of Kuwait in

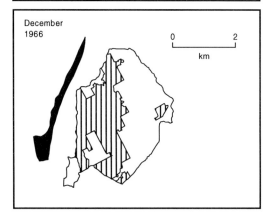

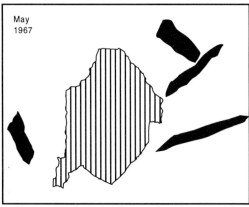

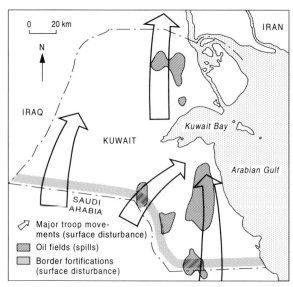

FIGURE 19.2 *Military action in Kuwait 1990–91 and its impacts on surface terrain (after Middleton, N.J.,1991,* Desertification. *Reproduced by permission of Oxford University Press)*

FIGURE 19.1 *Damage to rubber trees on the Plantation de Dautieng by wind drift of aerially sprayed defoliants during the Viet Nam War (reprinted with permission from Orians, G.H. and Pfeiffer, E.W. 1970, Ecological effects of the war in Vietnam.* Science *168: 544–54. Copyright 1970 American Association for the Advancement of Science)*

1991. In the latter example, an estimated 7–8 million barrels of oil were discharged, more than twice the size of the world's previously largest spillage caused by a blow-out in the Ixtoc-1 well in the Bay of Campeche off the Mexican coast in 1979. The resulting oil slick contaminated large areas of Kuwait and a 460-km stretch of the northern coast of Saudi Arabia. An estimated 30 000 seabirds died in the immediate aftermath of the spills, and the long-term effects on the Gulf's ecology are still being monitored (Readman *et al.*, 1992). Fires started in 613 Kuwaiti oil wells caused the burning of 4–8 million barrels a day and resulted in massive clouds of smoke and gaseous emissions, although fears of effects on the climate were unfounded. Other disturbances to the desert surfaces of Kuwait were caused by the large-scale movements of military vehicles and the building of an earth embankment along the border with Saudi Arabia (Fig. 19.2). Experience of similar surface destabilisation during the desert campaigns of North Africa in 1941 and 1942 saw wind erosion dramatically increased during the height of the fighting, but deflation subsided back to prewar levels within a few years of the end of the campaign (Oliver, 1945).

There are many other environmentally damaging consequences of warfare. In Croatia, for example, Richardson (1993) documented the numerous dangers caused by damage to industrial and municipal plants in the war that followed the break-up of Yugoslavia in the early 1990s. One of the most dangerous and possibly long-lasting hazards was caused by the release of PCBs from damaged transformers in many Croatian towns, including fifty in Dubrovnik alone. Contamination of soils, rivers and groundwater was expected. PCBs bioaccumulate up the food chain, they are highly toxic to fish, and there is some evidence linking PCBs to the occurrence of certain cancers in humans.

Numerous impacts on wildlife have also been documented. Some mammal species have been brought close to and beyond extinction due to the direct and indirect effects of war. The last semi-wild Père David's deer was killed by foreign troops during the Boxer Rebellion in China in 1898–1900, although the species still survives in captivity. Similarly, the European bison was pushed close to extinction during the First World War by hunting to supply troops with food; it recovered in numbers during the interwar years but its numbers were again decimated in the Second World War. Many Pacific island endemic bird taxa were also lost or pushed towards extinction by warfare and its associated disturbances during the Second World War. On the twin islands of Midway, for example, extinctions of the Laysan finch and the Laysan rail occurred (Fisher and Baldwin, 1946). In Central Africa, war has several times in recent years threatened the survival of the 650 mountain gorillas in the Virunga volcanoes which straddle the borders of Rwanda, Zaire and Uganda.

FIGURE 19.3 *An area near Port Howard in the Falklands Islands still closed off to human use due to landmines which remain uncleared since they were laid during the 1982 conflict in the South Atlantic (courtesy of Mark Carwardine)*

Conversely, however, some species of wild animals can benefit during wartime because of decreased exploitation. Westing (1980) notes the cases of various North Atlantic fisheries during the Second World War and a number of fur-bearing mammals in northern and northeastern Europe, such as the polar bear, red fox and wolf. During the Viet Nam War, tigers increased in abundance in some areas thanks to the ready availability of food provided by corpses.

Disruption and destruction does not necessarily cease with the end of hostilities. On average, 10 per cent of all munitions used in any war fail to explode and remain to threaten humans, livestock and wildlife, and impede postwar reconstruction and rehabilitation efforts in agriculture, forestry, fishing, mining and other related activities (Fig. 19.3). Landmines are a particular problem which are present on a huge scale in some countries (Table 19.3) and can take many years to clear. In Poland, more than 84 million items of explosive ordnance from the Second World War (including more than 14 million mines) had been disposed of by the early 1980s, with

TABLE 19.3 *Estimated number of landmines in the worst affected countries and territories, early 1990s*

Country	Number of land mines (million)
Afghanistan	9–10
Angola	9
Iraq	5–10
Kuwait	5
Cambodia	4–7
Western Sahara	1–2
Mozambique	1–2
Somalia	1
Bosnia and Herzegovina	1
Croatia	1

Source: Ruel (1993)

350 000 being located and destroyed every year (Westing, 1984).

Nuclear war

There is little direct information on the possible environmental effects of a nuclear war, but some insight can be gained from the detonation of the only two nuclear devices used in warfare to date, in Japan at the end of the Second World War (Table 19.4), although these were relatively small devices by today's standards. The most important cause of death and physical destruction was the combined effects of blast and thermal energy. The fireball created by the blasts was intense enough to vaporise humans at the epicentre and burn human skin up to 4 km away. Ionising radiation comprised about 15 per cent of the explosive yield of the bombs, and among its effects was radiation sickness among many survivors of the blast.

There is a limited amount of information on the ecological effects of nuclear explosions which has been gleaned from observations made after above-ground tests. Severe damage to vegetation caused by detonations decreases more or less geometrically with distance from the blast epicentre. After explosions, plants reinvade damaged areas at a rate of succession that would be expected after a severe disturbance, as Shields and Wells (1962) documented for a Nevada Desert test site after numerous above-ground detonations.

Considerable research has been undertaken into the possible global effects of a major

TABLE 19.4 *Destruction caused by the first atom bombs dropped in wartime*

	Hiroshima	Nagasaki
Date of detonation	6 August 1945	9 August 1945
Type	Uranium 235	Plutonium
Height of explosion (m)	580	503
Yield (kiloton TNT)	12.5	22
Total area demolished (km²)	13	6.7
Proportion of buildings completely destroyed (%)	67.9	25.3
Proportion of buildings partially destroyed (%)	24.0	10.8
Number of people killed (by 31 December 1945)	90 000–120 000	70 000

Source: Ohkita (1984)

nuclear exchange, much of it focusing on the potential climatic effects. Although results vary with different scenarios, simulations using general circulation models (GCMs) have raised the spectre of a 'nuclear winter' in which the aftermath of a nuclear war would be characterised by darkened skies over large parts of the Earth due to smoke and dust injected into the atmosphere. Such effects could continue for weeks or months, blocking sunlight, reducing temperatures and affecting the global distribution of rainfall. Such climatic effects could have wide-ranging impacts on agriculture and major biomes, with serious implications for food production and distribution far beyond the areas of immediate devastation (SCOPE, 1989).

INDIRECT WARTIME IMPACTS

The dislocation of economies and societies caused by war can result in many perturbations to the way people interact with their environment. Forced mass migrations, disruption to transport systems and many other aspects of the economy can, for example, leave people less able to cope with hazards of the natural environment, increasing their vulnerability to such disasters as famine. On occasion, deliberate destruction of the environment and infrastructure can also lead to famine. A recent example is the mass starvation that swept southern Somalia in 1992–93 which followed a 6-month period in which grain stores were plundered, and water pumps and other agricultural infrastructure and machinery were destroyed in the Bay region by forces loyal to former president Siad Barre (Drysdale, 1994).

Numerous indirect environmental impacts also occur due to the mass movements of refugees away from areas of conflict. During medieval Europe's Hundred Years' War, for example, whole regions were depopulated, leaving abandoned farmland to be colonised

by brush and woodland. Refugees tend to concentrate in safer zones, in cities, towns or refugee camps in rural areas, exerting intense pressures on local resources. Deforestation for fuel and to build shelters, resultant soil erosion and local pollution, particularly of water sources, are impacts commonly associated with large concentrations of refugees. In Central America during the 1980s, for example, about 20 000 refugees fleeing the conflict in Nicaragua cleared extensive areas of rain forest for camps on the Miskito coast of Honduras, some of them decimating part of the Rio Platano forest reserve protected under the United Nations' Man and the Biosphere Programme (Weinberg, 1991).

Many such examples can also be quoted from the Horn of Africa, scene of numerous conflicts over the last 20 years. In lowland Ethiopia, clearance for shelter for 100 000 refugees is estimated to deforest about 330 ha, plus the destruction caused by an annual fuelwood need of 85 000 t. Water demands for the 400 000 long-term Somali refugees in the arid Hararghe region of the country has necessitated the building of numerous earth dams and a daily truck delivery of 800 000 litres of water a day to the 210 000 refugees in the two camps at Hartisheikh, with unknown implications for local ecology (Tsehai, 1991).

In northwestern Tanzania, the camp of 300 000 refugees from Burundi at Benaco absorbed 410 000 Rwandans fleeing civil war in a month in mid-1994. Suggestions from aid workers to ease the pressure on local forests included the trucking of refugees outside the 4–5 km radius round the camp which was rapidly denuded in the quest for fuelwood, the marking of trees that would take a long time to regenerate, and the distribution of types of food aid that needed less cooking, such as maize powder instead of maize grain (UNHCR, 1994). Although refugee assistance agencies tend to be aware of the potential environmental degradation around camps,

FIGURE 19.4 *Collecting fuelwood near the village of Maúa in northern Mozambique. Civil war that lasted from the mid-1970s to 1992 meant a great concentration of dislocated people around relatively safe settlements, dramatically increasing pressure on local resources*

such effective action to combat the problem is rarely forthcoming in practice (Black, 1994).

The scale of the refugee problem imposed on relatively safe neighbouring countries can reach high levels. Malawi in southern Africa, already one of the region's most densely populated countries, accommodated a million Mozambican refugees during the late 1980s and early 1990s, in addition to its own population of 10 million. Within Mozambique, where conflict was more-or-less continuous between 1964 and 1992, no less than 4.5 million people were estimated to have been displaced within the country by the time a peace accord was signed between the government and the opposition forces of Renamo. Natural resources around safer villages, towns and cities suffered for years from the increased population pressures (Fig. 19.4). Displaced people living in the capital city of Maputo had to travel up to 70 km into the dangerous

surrounding countryside in the search for fuelwood, and little forest was left in the early 1990s in the Beira corridor area which was protected by Zimbabwean troops during the war. On the coast, mangrove forests were severely exploited, presenting a serious threat to prawn and shrimp fisheries, one of Mozambique's major exports, which reproduce in the mangroves.

Cultivated land was not allowed adequate fallow periods, which traditionally last for up to 7 years, due to high population densities. Additionally, the plots allocated to displaced people were often too small for a family, leading to further pressure and overutilisation (Dejene and Olivares, 1991). The war's impact on wildlife in Mozambique is also thought to have been severe. At independence in 1975, there were more than 65 000 elephants in the country; initial postwar estimates suggest that 15 000 remain. All the evidence points to

TABLE 19.5 *Major multilateral arms control agreements with a bearing on the environmental effects of warfare, 1970–90*

Agreement	Signed	Entry into force
Treaty prohibiting emplacement of nuclear weapons and other weapons of mass destruction on the sea bed	1971	1972
Bacterial and Toxin Weapons Convention prohibiting development, production and stockpiling of biological and toxin weapons, and/or their destruction	1972	1975
Environmental Modification Convention prohibiting military or other hostile use of environmental modification techniques	1977	1978
Inhumane Weapons Convention prohibiting or restricting use of certain conventional weapons deemed to be excessively injurious or to have indiscriminate effects	1981	1983
Treaty of Rarotonga (South Pacific nuclear-free treaty)	1985	1986

Source: after Goldblat (1990)

Renamo having been heavily involved in the ivory trade, selling tusks to South Africa in return for arms.

Similar extreme degradation has been noted in restricted coastal zones and larger urban centres in Angola, another former Portuguese colony wracked by civil war in its post-independence years. Conversely, however, much of the countryside depopulated during the conflict has suffered little damage, with some exceptions such as the giant sable, only found in a small area of central Angola and now highly endangered since its habitat was a central area of conflict. Indeed, in the view of one international conservation agency, the likelihood of widespread environmental degradation will be much greater with ensuing peace than during the past two decades of war (IUCN, 1992). The economic paralysis caused by the war has functioned as a brake on the wholesale exploitation of certain natural resources, and greater danger lies in postwar rapid and unchecked economic exploitation. Nevertheless, there is no great competition between wildlife and people for

land use in Angola, the key problem throughout much of the rest of Africa, and the clear valuation put on wildlife which is hunted by rural people for meat is a strong positive conservation aspect offering great potential.

LIMITING THE EFFECTS

Numerous treaties, conventions and agreements have been adopted to prevent utter human and environmental devastation in times of war, although they do not specifically cover nuclear weapons, and the effectiveness of such agreements as a deterrent and during the course of war are difficult to evaluate and enforce. Some of the most important relevant conventions are shown in Table 19.5 (Goldblat, 1990). Other parts of multilateral agreements also have a bearing on the environmental impacts of war, such as the two additional 1977 protocols to the Geneva Conventions of 1949, which bind belligerents to respect and protect the natural environment 'against widespread, long-term and severe damage' and prohibit the destruction of works and installations which harbour dangerous forces, including dams, dikes and nuclear generating stations. Such treaties are only binding on those states which are a party to them, however, and the protocol which bans widespread, long-term and severe environmental damage has not been ratified by major powers such as the USA, France or the UK since these nations are concerned that it could be interpreted as outlawing nuclear weapons. Some treaties meant to exclude any military activities from certain areas have also been adopted. These include the Spitzbergen Treaty (1920), the Åland Convention (1921) and the Antarctic Treaty (1959).

The widespread ecological effects of the Gulf War prompted renewed calls from conservation agencies for a new Geneva Convention to limit the environmental effects of military conflicts, and the idea of a Green Cross – a new, impartial international organisation, similar to the Red Cross and the Red Crescent, to assess environmental damage resulting from war – has also been suggested.

Although idealists might suggest that effective protection for the environment can only be achieved by global demilitarisation, this prospect seems unlikely. Two sovereign nations (Costa Rica and Japan) are non-militarised in a formal sense by virtue of their national constitutions, although in fact both support more-or-less potent armed forces. Indeed, the role played by environmental resources in causing conflict can be interpreted as a reason for increasing military might in a world of rising population.

ENVIRONMENTAL CAUSES OF CONFLICT

The environment has often been a cause of political tension and military conflict. Countries have fought for control over raw materials, energy supplies, land, river basins, sea passages and other environmental resources. In the Second World War, for example, Japan sought control over oil, minerals and other resources in China and South East Asia; rich phosphate deposits have been a root cause of the extended conflict in the Western Sahara; Israel's long occupation of a large portion of the west bank of the River Jordan has been both for strategic military purposes and to secure access to large groundwater reserves, and if the Gulf region had not been such a key supplier of oil supplies to the West it seems unlikely that Western troops would have intervened to such an extent over the Iraqi invasion of Kuwait in 1990–91.

Some observers suggest that environmental change, particularly in the form of degradation and dwindling renewable resources, is set to become an increasingly potent cause of conflict as world population is likely to exceed 9 billion within the next 50 years and global economic output may increase five-fold (Homer-Dixon et al., 1993). The degradation and loss of productive land, fisheries, forests and species diversity, the depletion and scarcity of water resources, and pressures brought about by stratospheric ozone loss and potentially significant climatic change may precipitate civil or international strife.

Such root causes of conflict have been identified from past eras. Pennell (1994), for example, has suggested that environmental degradation and economic isolation, and consequent impoverishment, made piracy an attractive option for inhabitants of the Guelaya Peninsula in northwestern Morocco in the mid-nineteenth century. A key aspect of such scenarios is the role played by certain groups who are faced with dwindling finite resources. Such groups may be within a certain country or they may be identifiable on the global scale as countries with unequal access to resources. A proposed chain of factors in this equation which leads to violence is suggested in Fig. 19.5.

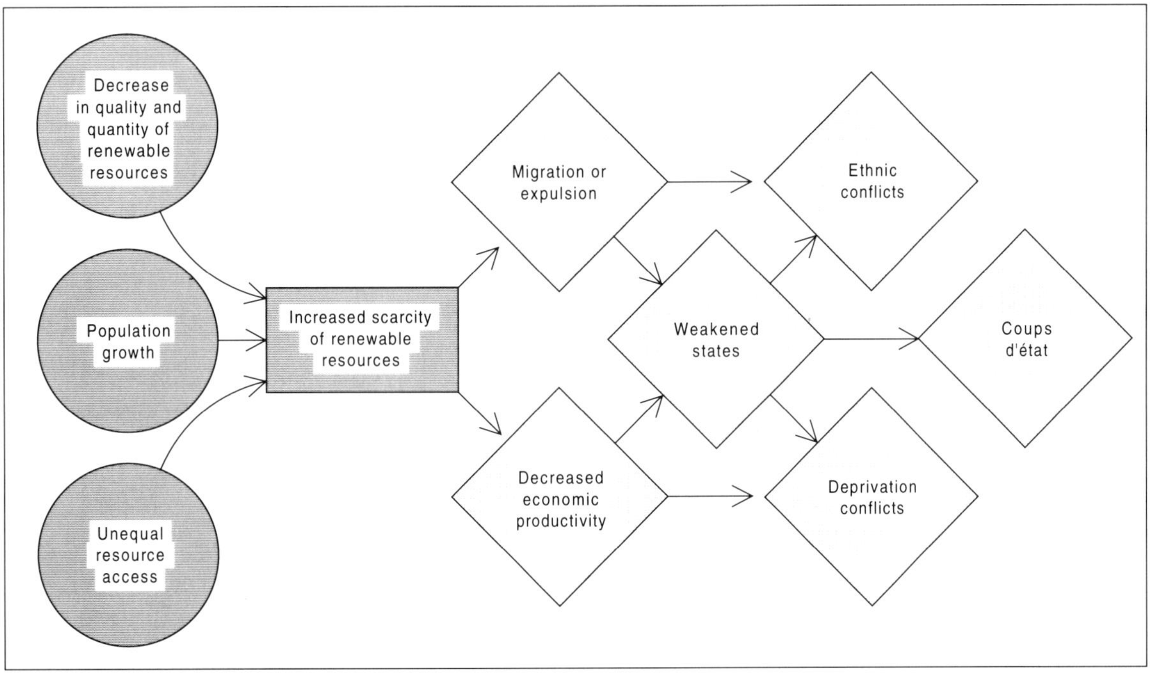

FIGURE 19.5 *Some sources and consequences of renewable resource scarcity (from* Environmental change and violent conflict *by Homer-Dixon, T.F., Boutwell, J.J. and Rathjens, G.W. Copyright © 1993 by Scientific American Inc. All rights reserved)*

Some researchers have highlighted examples of such events in more recent history. Bennett (1991) presents a number of studies of African conflicts in this vein, while in Central America, Durham (1979) considers that the root causes of the 1969 'Soccer War' between El Salvador and Honduras ran much deeper than the national rivalry over the outcome of a football match which ignited the conflict. He traces the origins of a war in which several thousand people were killed in a few days to changes in agriculture and land distribution in El Salvador that began in the mid-nineteenth century, forcing poor farmers to concentrate in the uplands. Despite their efforts to conserve the land's resources, growing human pressure in a country with annual population growth rates of 3.5 per cent reduced land availability and resulted in deforestation and soil erosion on the steep hillsides. Many farmers consequently moved to Honduras and it was their eventual expulsion that precipitated the war. In the years following the Soccer War, this competition for land was not addressed and *campesino* support for leftist guerrillas provided a powerful contribution to the country's subsequent 10-year civil war.

FURTHER READING

Myers, N. 1993 *Ultimate security: the environmental basis of political stability*. New York, Norton. An analysis of the importance of environmental issues in world politics.

Renner, M. 1991 Assessing the military's war on the environment. In *State of the world 1991*. New York, W. W. Norton. This article presents a catalogue of the military's environmental effects, mainly during peacetime.

SCOPE (Scientific Committee on Problems of the Environment) 1989 *Environmental consequences of nuclear war. Volume II: ecological and agricultural effects*, 2nd edn. SCOPE Report 28, Chichester, Wiley. A detailed compilation of the possible effects of a global nuclear exchage on the natural environment and our ability to produce food.

Westing, A.H. (ed.) 1990 *Environmental hazards of war: releasing dangerous forces in an industrialised world*. London, Sage. A collection of articles on the dangers of military conflict, particularly in industrialised countries.

20

NATURAL HAZARDS

Natural hazards such as earthquakes, floods, tropical cyclones and disease epidemics are normal functions of the natural environment. They do, therefore, affect all living organisms but they are usually only referred to as disasters or catastrophes when they impact human society to cause social disruption, material damage and loss of life. As such, natural hazards should be defined and studied both in terms of the physical processes involved and the human factors affecting the vulnerability of certain groups of people to disasters. Although some places are more hazardous than others, all locations are at risk – there is always the chance of a disaster – from some natural hazard or other. All places also have some natural advantages, however, and the presence or absence of human activities in any location is the result of weighing up the risks relative to the advantages. In many locations, the physical phenomenon responsible for a hazard also offers some of the advantages: a river, for example, will flood, which may be hazardous, but it is also a source of water and its floodplain is a location with fertile soils. Hence, the hazard should be seen as an occasionally disadvantageous aspect of a phenomenon which is beneficial to human activity over a different timescale. To complicate matters, however, not all people who choose to locate in a particular area enjoy the same degree of choice – a flexibility which may be dictated by the economic means at their disposal, among other things. To complicate matters further, there is a wide range of ways in which people respond to hazards and again the rich often have more options in this respect than poorer members of society. The study of natural hazards is very much a geographical subject, therefore, since it includes analysis of both the physical and human environment. It is also a topical issue, since the 1990s has been declared the United Nations International Decade for Natural Disaster Reduction.

HAZARD CLASSIFICATIONS

Natural hazards can be classified in many different ways. Physical geographical approaches may divide them according to their geological, hydrological, atmospheric and biological origins as shown in Fig. 20.1. They can also be classified spatially, as certain hazards only occur in certain regions: avalanches occur in snowy, mountainous areas, for example, and most volcanic eruptions and earthquakes occur at tectonic plate margins. Another approach divides hazards into rapid-onset, intensive events (short, sharp shocks such as tornadoes) and

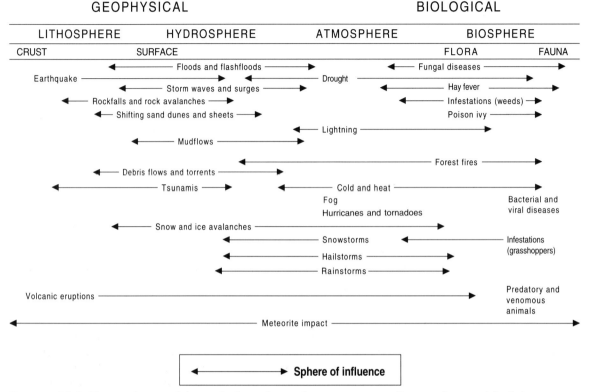

FIGURE 20.1 *Types of natural hazards, by sphere of occurrence, medium and materials (after Gardner, 1993. Reproduced with modifications by permission of McGill-Queen's University Press)*

slow-onset pervasive events which often affect larger areas over longer periods of time (such as droughts). Table 20.1 is an attempt to classify hazards in this way and gives some indication of the frequency and predictability of events in time.

Although these classifications provide a useful summary of hazard types, many disasters involve composite hazards. Hence, an earthquake may cause a tsunami wave at sea, landslides or avalanches on slopes, building damage and fires in urban areas, and flooding due to the failure of dams, as well as ground shaking and displacement along faults. Similarly, many natural disasters cause disruption to public hygiene and consequently result in heightened risks of disease transmission.

Other approaches to defining and classifying disasters focus on such factors as the areal extent or the scale of damage: Burton *et al.* (1978), for example, postulate that a 'disaster' must cause more than US$1 million in damage or the death or injury of more than a hundred people. Such an absolute approach has disadvantages, however, since a certain level of damage will have a different impact on society depending upon the strength of that society to cope with the disaster, a recognition used to assess disaster-proneness at the national level (see below). In practice, very poor societies commonly suffer the highest numbers of casualties while very rich societies suffer the highest property damage, and there is a case for defining a disaster at any scale, from the personal to the regional, national and global.

TABLE 20.1 *Natural disasters classified according to frequency of occurrence, duration of impact and length of forewarning*

Disaster	Frequency or type of occurrence	Duration of impact	Length of forewarning (if any)
Lightning	Random	Instant	Seconds–hours
Avalanche	Seasonal/diurnal; random	Seconds–minutes	Seconds–hours
Earthquake	Log-normal	Seconds–minutes	Minutes–years
Tornado	Seasonal	Seconds–hours	Minutes
Landslide	Seasonal/irregular	Seconds–decades	Seconds–years
Intense rainstorm	Seasonal/diurnal	Minutes	Seconds–hours
Hail	Seasonal/diurnal	Minutes	Minutes–hours
Tsunami	Log-normal	Minutes–hours	Minutes–hours
Flood	Seasonal; log-normal	Minutes–days	Minutes–days
Subsidence	Sudden or progressive	Minutes–decades	Seconds–years
Windstorm	Seasonal/exponential	Hours	Hours
Frost or ice storm	Seasonal/diurnal	Hours	Hours
Hurricane	Seasonal/irregular	Hours	Hours
Snowstorm	Seasonal	Hours	Hours
Environmental fire	Seasonal; random	Hours–days	Seconds–days
Insect infestation	Seasonal; random	Hours–days	Seconds–days
Fog	Seasonal/diurnal	Hours–days	Minutes–hours
Volcanic eruption	Irregular	Hours–years	Minutes–weeks
Coastal erosion	Seasonal/irregular; exponential	Hours–years	Hours–decades
Soil erosion	Progressive (threshold may be crossed)	Hours–millenia	Hours–decades
Drought	Seasonal/irregular	Days–years	Days–weeks
Crop blight	Seasonal/irregular	Weeks–months	Days–months
Expansive soil	Seasonal/irregular	Months–years	Months–years

Source: after Alexander (1993)

Another difficulty arises in distinguishing between purely 'natural' events and human-induced events. This chapter is not concerned with such obviously human-induced hazards and catastrophes as industrial accidents or pesticide poisonings. However, in one sense, all 'natural' disasters can be thought of as human-induced since it is the presence of people which defines whether an event creates a disaster or not. Many of the natural physical processes which cause disasters can also be triggered or made worse by human action: while a volcanic eruption is a purely natural process, for example, the failure of a slope, producing a landslide, can be induced by many human activities, such as road-cutting, construction or deforestation. The composite nature of many disasters also blurs the distinction: the major cause of death due to an earthquake is usually crushing beneath buildings, so is the disaster natural or human-induced? Some researchers believe that the difficulty of making this distinction has made the division pointless, and prefer to talk of 'environmental' hazards which refer to a spectrum with purely natural events at one end and distinctly human-induced events at the other (e.g. Smith, 1992).

RESPONSE TO HAZARDS

Responses can be made by individuals, particular groups or the whole society, and they are related to the information available on past events and the probability of recurrence, the perception of that information, and the awareness of particular opportunities. One form of response might be to change the land use in a particular area or to move an activity to another location. Other responses can be categorised into those that aim to reduce losses and those that accept losses. The range of possible adjustments in these categories that can be made in response to the hazard from volcanic lava flows is indicated in Table 20.2. In practice, the most effective responses often involve a combination of measures.

TABLE 20.2 *Examples of possible adjustments to the hazard of lava flows from volcanoes*

Modify the event	Modify the risk	Modify the loss potential	Spread the losses	Plan for losses	Bear the losses
Not possible	Protect high value installations	Introduce warning systems (requires monitoring)	Public relief from national and local government (available in most countries)	Individual family or company insurance (possible in more developed countries but still limited)	Individual family, company or community loss-sharing (traditional and still widely practised)
	Alter lava flow direction (e.g. by bombing or barriers)	Prepare for disaster through civil defence measures (emergency evacuation plans)	Government sponsored and supported insurance schemes (available in more developed countries)		
	Arrest forward motion (e.g. by watering flow margin)	Introduce land use planning (based on hazard-zone maps)	International relief (greatest benefit in developing countries)		

Source: after Chester (1993)

Some approaches take action to reduce the risk of impact by preventing or modifying events. Since there is no known way of altering the eruptive mechanism, modifying the event itself is not an option for lava flows, but other hazard events can be prevented or modified. Floods, for example, can be prevented over a long period by impounding water behind a dam, the impact of an individual hurricane can be modified by cloud-seeding, and frosts can be prevented by installing a heating system in an orchard. Other approaches aim to modify the loss potential through prediction, either in time, so issuing hazard warnings to enable evacuation, or in space, by producing hazard zone maps which can be used to regulate land use in hazard-prone areas. These examples illustrate the ways in which responses can operate over a variety of timescales, from the long-term view of building a dam or regulating land use, to the immediate efforts involved in evacuating a threatened area. Most of these responses are more generally characteristic of the more-developed countries which have sufficient economic resources, infrastructure and trained personnel to organise and carry out such actions.

Other responses effectively accept the losses imposed by a hazard and cope with these either by planning for them in the form of economic insurance, or by relying on help from a wider community when they happen, whether from fellow members of a family or from local, national or foreign governments and aid agencies. Economic insurance is again an option usually available only in the more prosperous economies, and in some cases it can actually encourage individuals and societies to persist with activities in hazardous areas.

The adoption of a particular response is determined by assessment of the risk. Assessment may be made scientifically, by calculating the probability of a particular hazard causing a disaster, based on information of past event frequency and area of impact plus the current vulnerability of people and property. An important measure of frequency is the 'return period' for a particular magnitude of event. Engineers calculate the statistical probability of a certain size of event occurring, and hence speak of the 50-year flood, for example, and design structures to withstand the magnitude of hazard likely to affect the structure within its lifetime. It is not a fail-safe measure, of course, because land use changes in a catchment can alter its flood response characteristics, and because it is still possible that a bridge built to withstand a flood which happens on average only every 50 years is, in fact, hit by a 100-year flood. Conversely, assessments are often made on the basis of how a particular hazard is perceived. For many reasons, not least the lack of long-term records for many areas, perception is involved in most hazard risk assessments.

HAZARDOUS AREAS

While people have always been at risk from natural hazards, the perception that the world is becoming a more hazardous place is widespread. Proving this hypothesis is not possible since there is still no comprehensive worldwide database of hazards or disasters. Rising trends worldwide in the numbers of natural catastrophes, and the resulting cost of damages compiled by the insurance industry (Fig. 20.2), is at least partly due to improved reporting of events (sigma, 1994), although there are well-founded fears, and some evidence, that certain types of hazard such as tropical cyclones are increasing in frequency with global warming (see Chapter 8). With an increase in world population, however, it is also logical to suppose that the numbers of people living in high-risk areas is also rising, although the effects of efforts to reduce the risks must also be considered.

Some studies of particular locations have concluded that hazard potential is increasing.

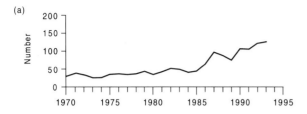

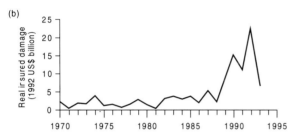

FIGURE 20.2 *Global natural catastrophes, 1970–93: (a) frequency; (b) real insured damage at 1992 prices (after sigma, 1994)*

In a comprehensive account of the geophysical hazards facing Los Angeles county in California, where mid-winter storms are a main disaster catalyst, Cooke (1984) concluded that increasing risk was being driven mainly by rapid population growth and the spread of settlement into hills and mountains. In many individual cases, this spread is driven by aesthetic desires for a less-crowded environment with better views. Conversely, several examples of poorer sectors of society living in more hazardous urban zones, because they have no other option, are documented in Chapter 14.

On the national scale, some countries are perceived to be more prone to disasters than others. The impact of a particular event depends on where the event occurs and how prepared the area is to cope with the disaster. In the case of an earthquake in an urban area, for example, the impact will vary according to such factors as the quality of the buildings, the ability of the society to respond to any destruction, and the timing of the event. In many ways, the overall impact of the disaster depends upon the economic strength of the

country in which it occurs, because an event which affects a certain number of people and causes a certain amount of damage will have a greater overall impact in a poorer country than in a richer one.

This logic has been followed in an attempt to quantify the risk associated with the effects of disasters on the national scale by the UN Disaster Relief Organisation (UNDRO). Using data on disasters over a 20-year period (1970–89), the study assessed the economic impact of disasters on the economies of 195 countries by expressing the damage costs of each disaster as a proportion of the annual Gross National Product (GNP) of the country concerned in a disaster proneness index:

$$\text{Index} = \frac{\text{Damage (US\$)}}{\text{Annual GNP (US\$)}} \times 100$$

Only 'significant' disasters were included in the study: that is those causing damage in financial terms amounting to 1 per cent or more of a country's total annual GNP. The Total Index was then calculated as the sum of the individual indices for each significant disaster, providing a measure of the total effect of disasters over 20 years expressed as a percentage of annual GNP. An Average Index was also derived by calculating the mean of the individual indices for each significant disaster during the 20-year period, providing a measure of the impact of the average significant disaster which struck the country during the period, again as a percentage of the country's GNP (Kerpelman, 1990).

The most disaster-prone countries ranked according to the Total Index are shown in Table 20.3. Several small-island developing countries (the Cook Islands, Montserrat and Vanuatu) appear high in the ranking because they are particularly prone to tropical cyclones, and in two of these cases their high indices are based on a single disaster during the 20-year study period. Countries which are prone to frequent disastrous events, such as

TABLE 20.3 *The most disaster-prone countries according to the UNDRO Disaster-Proneness Index*

Country	Number of disasters (1970–89)	Average Index (% 1980 GNP)	Total Index (% 1980 GNP)
Montserrat	1	1200	1200
Vanuatu	4	58	228
Nicaragua	8	26	207
Burkina Faso	4	48	191
Dominica	3	47	141
Cook Islands	1	119	119
Chad	5	18	92
Bolivia	10	8	84
St Lucia	2	41	81
Yemen	1	67	67

Source: Kerpelman (1990)

Nicaragua and Bolivia, are highly disaster-prone according to their Total Index ranking, but the average impact of a significant disaster is not so great. Other countries prone to frequent disasters, such as India and the Philippines, do not appear in the top seventy countries according to their Total Index, indicating that their stronger economies are better able to absorb the effects of the disasters. Richer countries also have more to lose in economic terms, of course, but are generally better able to absorb these losses. Japan is a case in point here, experiencing the greatest absolute economic loss of any country during the study period (US$82 billion in two significant disasters) but this cost represented just 8 per cent of their 1980 GNP.

EXAMPLES OF NATURAL DISASTERS

The following sections take a closer look at a number of major natural causes of disaster. The selection is based on an analysis of the causes of deaths from all types of disasters over the period 1900–90 compiled by the US Office of Foreign Disaster Assistance (Blaikie *et al.*, 1994). Civil strife and famine were by far the largest causes of death, accounting for nearly 88 per cent of all casualties. Although natural events such as drought and flooding are often implicated as triggers for the onset of famine (see Chapter 4), the most significant direct natural causes of death were earthquakes, volcanoes, cyclones, epidemics and floods.

Earthquakes

Each year, around 1 million earth tremors are recorded over the Earth's surface, but the vast majority of these are so small that they are not felt by people. The handful of major earthquakes which occur each year can cause widespread damage, however, causing some of the world's most devastating natural disasters. China has suffered some of the worst fatalities in major earthquakes: the 1556 event in Shensi caused more than 800 000 deaths, and the Tangshan earthquake of 1976 left up to 750 000 fatalities.

Most earthquakes are caused by abrupt tectonic stress release around the margins of the Earth's lithospheric plates, although they also occur at weak spots near the centre of plates. Whilst the locations of plate boundaries and most associated faults are well known, the timing of earthquakes is less easy to predict. A combination of seismic monitoring and observations of natural phenomena, including animal behaviour, were used in the only successful prediction of a large-scale earthquake which devastated the Chinese city of Haicheng in Liaoning Province in February 1976. An evacuation of Haicheng and two other cities was ordered 48 hours before the main earthquake, almost certainly saving many thousands of lives (Adams, 1975).

The main environmental hazard created by seismic earth movements is ground shaking, which is related to the magnitude of the shock and normally assessed on the Richter scale, a complex logarithmic scale which measures the vibrational energy of the shock. Local site conditions have an important effect on ground motion, with greater structural damage usually found in areas underlain by unconsolidated material as opposed to rock. Particularly severe damage was caused during the earthquake of 1985 in the central parts of Mexico City which is built on a dried lake bed. Similarly, a relatively modest magnitude earthquake, registering 5.4 on the Richter scale, in San Salvador in 1986, caused unusually large destruction since most of the city is built on volcanic ash up to 25 m thick.

Another serious earthquake hazard associated with soft, water-saturated sediments is soil liquefaction, in which intense shaking causes sediments to temporarily lose strength and behave as a fluid. Loss of bearing strength can cause buildings to subside and soils can flow on slopes greater than 3°. Mudflows, landslides, and rock and snow avalanches triggered by ground vibrations often play a major role in earthquake disasters, particularly in mountainous areas. In a study of major earthquakes in Japan, landslides were found to cause more than half of all earth-

quake-related deaths (Kobayashi, 1981). Tectonic displacement of the sea bed is the main cause of large sea waves, or tsunamis, which can travel thousands of kilometres at velocities greater than 900 km/h and cause great damage when they hit coastlines. Tsunamis are most common in the Pacific where one of the largest and most damaging tsunamis, measuring 24 m in height, drowned 26 000 people in Sanriku, Japan in 1896.

Some attempts have been made to regulate earthquake intensity by injecting pressurised water along faults to stimulate many small tremors and thus reduce the probability of major movements, but the technique still runs the risk of itself inducing a major earthquake. Hence, responses to the hazards associated with earthquakes focus on reducing risk and coping with the losses. Since the failure of structures is a major cause of injury during seismic shaking, much effort has gone into building location and design. Earthquake-hazard zoning maps have been produced for many countries and some of the approaches used for building in particularly hazardous sites are shown in Fig. 20.3. Building codes are only effective if enforced, however, and several recent examples of major damage and loss of life in urban areas have occurred in cities where such codes have not been followed (see Chapter 14).

Volcanic eruptions

Volcanic eruptions have killed around 250 000 people in the last 400 years due to a range of associated hazards which include falls of rock and ash, the force of lateral blasts, emission of poisonous gases, debris avalanches, lava flows, mudflows known as 'lahars', and 'pyroclastic flows' made up of suspended rock froth and gases (Chester, 1993). Only about 500 volcanoes are thought to have erupted in historical time, although the traditional categories of 'active', 'dormant' and 'extinct' are not infrequently found wanting when

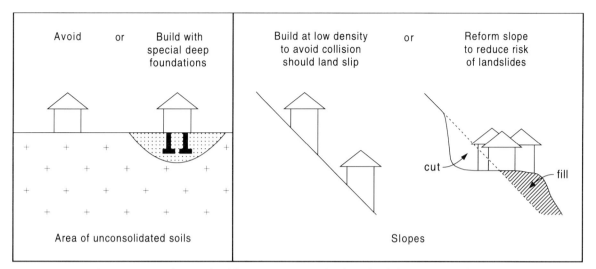

FIGURE 20.3 *Some approaches to building on sites at high risk of disturbance during earthquakes (after Smith, K. 1992* Environmental hazards: assessing risk and reducing disaster. *Reproduced by permission of Routledge)*

supposedly inactive volcanoes erupt, sometimes causing widespread destruction (e.g. Mount Pinatubo, the Philippines, 1991).

Lava flows, one of the most characteristic volcanic hazards, are rarely fast enough to cause death but can completely destroy anything in their path. The distinction has been made above between responses to the lava flow hazard typically employed in more-developed countries and those more common in less-developed countries (Table 20.2). Developing countries tend to rely upon hazard warning and evacuation, whilst some wealthier nations have attempted to modify the hazard by bombing to divert flows, erecting barriers to protect inhabited areas, and controlling the flow using water (e.g. in Hawaii, Iceland, Italy and Japan), although these countries also employ warning and evacuation measures. Such a distinction can also be made with regard to the whole suite of volcanic hazards. Chester suggests that 'many deaths are now avoidable often through the application of fairly simple methods of prediction' (1993: 186) since the activity of most volcanoes builds up to a

major eruption and the build-up can be detected, given adequate monitoring.

However, a stark contrast can be drawn between the successful prediction and evacuation of the area around Mount Pinatubo on Luzon Island in the Philippines in 1991, and the failure to evacuate in time for the 1985 eruption of Nevado del Rúiz in the Colombian Andes. Seismic and other anomalies in activity were reported from Nevado del Rúiz for almost a year before the eruption, and numerous geologists and seismologists were called in to assess the situation, but poor communication and bureaucratic inefficiency prevented action being taken in time. A hazard map was prepared (Fig. 20.4), the final version of which was due to be presented to the authorities the day after the eruption started on 13 November. The eruption melted part of the summit's snow and ice cap and a torrent of meltwater and pyroclastic debris flowed from the summit, coalescing in lower channels to entrain vegetation, water and mud to form rapidly flowing lahars. Nearly 70 per cent of the population of Armero (about 20 000 people) was killed in the mudflow and

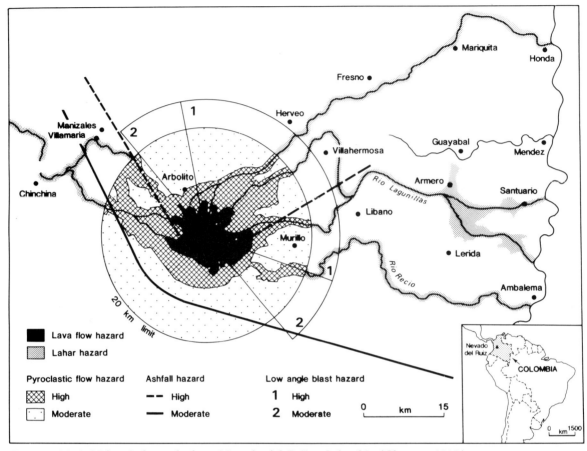

FIGURE 20.4 *Volcanic hazards from Nevado del Rúiz, Colombia (Chester, 1993)*

another 1800 died on the volcano's western flanks. Fifty schools, two hospitals, fifty-eight industrial plants, roads, bridges and power lines were destroyed or damaged, and 60 per cent of the region's livestock perished. The total cost to the economy of Colombia was estimated at US$7.7 billion, or 20 per cent of the country's GNP in that year. As Voight (1990: 151) concludes: 'the catastrophe was not caused by technological ineffectiveness or defectiveness, not by an overwhelming eruption, or by an improbable run of bad luck, but rather by cumulative human error – by misjudgement, indecision and bureaucratic short-sightedness'. At the time, the Colombian Government had other problems on its mind.

On 6 November, the president ordered troops to storm Bogota's Palace of Justice which had been occupied by guerillas. One hundred people were killed, including ten senior judges.

Tropical cyclones

Tropical cyclones, which are also known as hurricanes in the Atlantic and typhoons in the western Pacific, are among the most destructive of all natural hazards. They are generated at sea over surface water with a temperature of more than 26°C. An indication of the cost of tropical cyclone damage for some of the worst recent events is given in Table 20.4

TABLE 20.4 *Some of the most damaging tropical cyclones, 1970–93*

Start date	Name	Country	Insured damage (1992 US$ million)
4 Aug 1970	Celia	USA	1185
18 Sept 1974	Fifi	Honduras	1439
23 Sept 1975	Eloise	USA	403
3 Sept 1979	Frederic	USA	1469
17 Aug 1983	Alicia	USA	960
30 Aug 1985	Elena	USA	716
15 Sept 1989	Hugo	North America	4935
27 Sept 1991	Mireille	Japan	5341
23 Aug 1992	Andrew	USA	15 500

Source: after sigma (1994)

which is compiled from a list of the insurance industry's single most costly catastrophe in each year from 1970 to 1993: tropical cyclones were the most-damaging events in 9 of the 24 years.

The destruction caused by tropical cyclones occurs in a combination of very strong winds (the threshold wind speed is 33 m/s but winds can exceed 80 m/s), torrential rainfall (which can cause flooding, landslides and mudflows), high waves, and particularly by storm surge which results from the shearing effect of wind on water to cause local rises in sea level and inundation of coasts where cyclones make landfall.

Meteorological satellites provide some of the most important instruments for detecting and monitoring tropical cyclones, and early-warning systems are the most widely used response. Improvements in forecasting, warnings and evacuation plans have reduced the human death toll in many countries in recent decades, although property damage has risen over the period. In the USA, 70 million people are estimated to be at risk from hurricanes, nearly 10 per cent of them resident in Florida. Several observers have noted the marked difference in impact between developed and developing countries, the former tending to be better organised and able to cope with communicating warnings and facilitating evacuation.

The devastating impact of tropical cyclones on the economies of some small-island developing countries has been shown above (Table 20.3), but one of the countries most frequently and severely affected is Bangladesh, which was hit by sixty-one severe cyclones, thirty-two of which were accompanied by storm surges, between 1797 and 1993 (Khalil, 1992; sigma, 1994). Some of these events have been characterised by huge fatalities. The cyclone of October 1876, which produced a storm surge 12 m high, resulted in the loss of 400 000 lives; 500 000 people were killed in a 10 m storm surge in November 1970, and 150 000 people lost their lives in April 1991. Bangladesh's high population density has meant that survivors quickly reoccupy the coastal lowlands after cyclone disasters strike, and the Cyclone Protection Project, part of Bangladesh's Flood Control Action Plan (see below), aims to rehabilitate coastal embankments built in the 1960s and 1970s to provide some protection for these zones. However, eradication of the threat from cyclones and their associated storm surges is impossible, and much more effort has gone into communicating warnings to the most rural populations so that the additional 3400 multipurpose community shelters that are to be constructed will be used. More effective protection and replanting of threatened areas of mangrove forest in the Sundarbans (see Chapter 11) and elsewhere along the coast has also been called for, to provide a quasi-natural defence against storm surges (Khalil, 1992). The long-term effectiveness of these measures is questioned, however, by some of the predicted changes in the region due to climate and sea-level change (see Chapter 8).

Epidemics

Infectious diseases, many of which are preventable and curable, are the greatest cause of morbidity and mortality on the global scale. They account for nearly half of all deaths in developing countries, where children under five are the most susceptible, whereas in developed countries diseases of old age and abundance, such as heart disease and cancers, are the most common killers. This clear difference is linked to the fact that far greater numbers of people in developing countries suffer from malnutrition, inadequate water supply and sanitation, poor hygiene practices and overcrowded living conditions.

An infectious disease is caused by a host being invaded by a parasite or other pathogen, which might be a bacterium, virus or worm. The pathogen carries out part of its lifecycle inside the host and the disease is often a by-product of these activities. Many pathogens also need a vector, an 'accomplice', to help them spread from host to host, a role often played by an insect. Many of the major Third World diseases are associated with water, an association in which three typical situations can be distinguished (Table 20.5), although some diseases may fall under more than one category. Although any disease represents a natural hazard to individuals and groups of people, the emphasis here will be on epidemics, relatively sudden outbreaks of a disease in a certain area. Two diseases associated with water will be examined: cholera and malaria.

Cholera is caused by a bacterium known as *Vibrio* which can survive outside the human gut in such environments as water, moist foodstuffs, human faeces and soiled hands, all of which can act as media for its transmission. The disease originated in South Asia, notably in the Ganges–Brahmaputra delta and appears to spread and recede geographically in cycles. It first reached Europe in the

TABLE 20.5 *Classification of diseases associated with water*

Role of water	Comments and examples
Water-borne	Diseases arise from presence in water of human or animal faeces or urine infected with disease pathogens which are transmitted when water is used for drinking or food preparation (e.g. diarrhoea, dysentery, cholera, typhoid, guinea worm)
Water-hygiene	Diseases result from inadequate use of water to maintain personal cleanliness (e.g. water-borne diseases above as well as infestation with lice or mites)
Water-habitat	Diseases for which water provides the habitat for vectors. The pathogen may enter the human body from the water directly (e.g. bilharzia is transmitted when a fluke leaves a snail vector and penetrates human skin as people wash or swim), or indirectly via a fly (e.g. trypanosomiasis – sleeping sickness) or mosquito (e.g. malaria)

Source: after GEMS (1988)

nineteenth century, causing six major pandemics, and realisation of the key transmission role played by dirty water was a central force behind the public health movement in many European countries in the nineteenth century. These measures have helped to all but eradicate cholera from most developed countries but a seventh pandemic spread from Indonesia in the early 1960s, reaching Africa and southern Europe within 10 years. Its rapid spread through Africa caused many deaths after a period of about 75 years when the disease was a relatively minor ailment (Stock, 1976). The pandemic reached South America in January 1991 where it is thought to have been brought from Asia in a ship's ballast water. From its landfall in Peru, where over 1700 cases per day were reported during the first 4 months, the disease spread throughout much of South America within a year (WHO, 1992b). The outbreak of a recent epidemic in Bangladesh at the end of 1992, involving a new, hardier strain of the cholera *Vibrio* first identified in a coastal algal bloom, has raised fears of an eighth pandemic (Levins *et al.*, 1994). Although it may not be possible to prevent totally the emergence of new cholera strains and the cyclic spread of the disease, there is no doubt that the introduction of basic sanitary measures, the provision of safe drinking water and the promotion of less-crowded living conditions in developing countries would do a great deal towards limiting the spread of this often fatal disease.

Unlike cholera, the transmission of malaria requires a vector. The disease is caused by a single-celled pathogen which is transmitted through a number of species of blood-sucking mosquito. Until just a few decades ago malaria was a virtually global disease, common in parts of Europe, for example, such as the Rhine delta and low-lying parts of Mediterranean countries until after the Second World War. In England, the disease was known as 'marsh ague' or 'marsh fever', and it was a common cause of death among many inhabitants of wetlands in the nineteenth century (Dobson, 1980). The widespread drainage of wetlands, an ideal habitat for mosquitoes, along with concerted campaigns using DDT in the 1950s, has helped to eradicate malaria from most temperate and many subtropical areas, and the disease has been all but wiped out in Europe and North America (although in more recent times, increasing international travel to the tropics has resulted in cases among some temperate zone residents).

Malaria, however, is still endemic in most parts of the developing world and despite numerous attempts at global eradication since the 1950s, it has reached epidemic proportions in a number of countries in association with high population densities, poor sanitation, environmental degradation and civil unrest. Worldwide, more than 2 billion people are at risk from the disease and an estimated 170 million people are infected (WHO, 1992a). More than 1 million people die from malaria each year in Africa south of the Sahara alone. Resistance of the pathogen to drugs and resistance of mosquitoes to insecticides, as well as the rudimentary level of health services in many areas, have combined to increase the malaria problem. Much research into antimalarial drugs was conducted in the USA during the Viet Nam War, but since then, less effort has been focused on a disease which is just one of many which continue to plague the inhabitants of the poorer parts of the world.

Floods

Floods are one of the most common natural hazards and are experienced in every country. They occur when land not normally covered by water becomes inundated, and they can be classified into coastal floods, most of which are associated with storm surges (see above and Chapter 11), and river channel floods. Atmospheric phenomena are usually the primary cause of river flooding. Heavy

TABLE 20.6 *Flood hazards and their abatement*

Critical flood characteristics	Adjustments to floods	Flood hazard abatement	
		Structural	**Non-structural**
Depth	Accept the loss	Levées, dikes, embankments, flood walls	State laws
Duration ordinances	Public relief		Zoning
Area inundated	Abatement and control	Channel capacity alteration (width, depth, slope and roughness)	Subdivision regulations
Flow velocity			Building codes
Frequency and recurrence	Land evaluation and other structural change	Removing channel obstructions	Urban renewal
			Permanent evacuation
Lag time	Emergency action and rescheduling	Small headwater dams	Government acquisition of land and property (creation of floodway and public open space)
Seasonality		Large mainstream dam	
Peak flow	Land use regulation	Gully control, bank stabilisation and terracing	
Shape of rising and recession limbs	Flood insurance		Fiscal methods: building financing and tax assessment
Sediment load	Controlled flooding of low-value land		Warning signs and notices
			Flood insurance

Source: after Alexander (1993)

rainfall is most commonly responsible, but melting snow and the temporary damming of rivers by floating ice can be important seasonally. The likelihood of these factors leading to a flood is affected by many different characteristics of the drainage basin, such as topography, drainage density and vegetation cover, and human influence often plays a role – especially by modifying land use (including the effects of urbanisation) and through conscious attempts to modify the flood hazard.

Techniques for predicting floods are well developed. Most floods have a seasonal element in their occurrence and they are often forecasted using meteorological observations, with the 'lag time' to peak flow of a particular river in response to a rainfall event being calculated using a flood hydrograph. Return periods for particular flood magnitudes can be calculated for engineering purposes, although long periods of historical data are required which are by no means available for all river basins, and developments can alter a river's flood characteristics. Flood-hazard maps are commonly used for land use zoning.

Many of the human responses to the flood hazard shown in Table 20.6 have been put into practice on the Mississippi River in the USA, one of the world's largest and most intensively managed rivers. Channel straightening, for example, was initiated in the 1930s to increase the gradient and velocity of stretches of the Mississippi so that river water would erode and deepen the channel, thus increasing its flood capacity. Engineering structures have

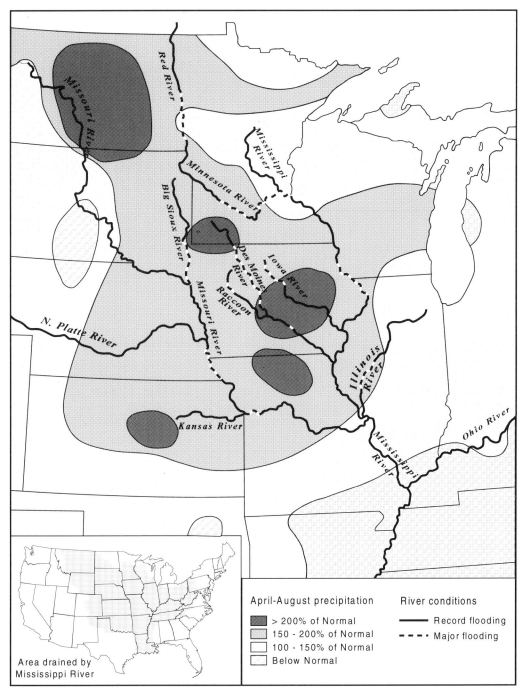

FIGURE 20.5 *Precipitation anomalies and flooding on the upper Mississippi and tributaries in 1993 (after Williams, J. 1994, The great flood. Weatherwise 47: 18–22. Reprinted with permission of the Helen Dwight Reid Educational Foundation. Published by Heldref Publications, 1319 Eighteenth Street, NW, Washington DC 20036-1802. Copyright © 1994)*

also been widely employed; more than 4500 km of the Mississippi are lined with embankments or levées designed to confine floodwaters.

Despite the numerous attempts at hazard mitigation, the Mississippi still overflows its banks with devastating consequences. The 1993 flood on the river's upper reaches was the most recent example. High soil-moisture content from heavy rains the previous year enhanced runoff during the period between April and August 1993, when rainfall totals were well above average in north-central states. As levées failed in more than 1000 places, record flooding occurred along many stretches of the Mississippi, Missouri and several tributaries (Fig. 20.5). Floodwaters washed over an area of about 4 million hectares, destroying or seriously damaging more than 40 000 buildings (Williams, 1994), with the period of inundation prolonged in many places by the levées which prevented the return of water to the channel once the peak had passed. The cost of damage to insurance companies totalled US$755 million, while the total damage is estimated to be US$12 billion. Forty-five people were killed (sigma, 1994).

The inability of structures to prevent river flooding has been highlighted in discussions over a controversial proposal to build large embankments along stretches of the major

FIGURE 20.6 *The River Buriganga in flood outside Dhaka in Bangladesh. Not all floods are considered as hazards by rural Bangladeshis, since some enhance food production (courtesy of Charles Toomer)*

rivers of the Bangladesh delta. About 80 per cent of the national territory is floodplain, prone both to cyclonic flooding and floods from the major rivers (Ganges, Brahmaputra and Meghna) which flow through the country. Recent major river floods occurred in 1987 and 1988, inundating 40 per cent and 57 per cent of the country, respectively (Rasid and Pramanik, 1990). These events had return periods of 100 years or more, and their effects inspired the proposal as part of a comprehensive Flood Control Action Plan coordinated by the World Bank.

Not least among the arguments put forward against the embankments, an expensive high-tech, top-down solution, is the fact that they are designed to prevent all floods. Yet Bangladeshi villagers distinguish between beneficial rainy season floods, known as 'barsha' in Bengali, and harmful floods of abnormal depth and timing which are termed 'bonna' (Fig. 20.6). 'Barsha' floods water the paddy fields and enhance soil fertility both through sediment inputs and nitrogen-fixing algae which thrive in the floodwater. Fish caught in flooded fields are also the main source of animal protein for rural inhabitants (Boyce, 1990).

While there is widespread agreement that Bangladesh needs to improve resource management, opponents of the embankment proposal suggest that less-technological approaches, which aim to improve flood management and actively involve local inhabitants, would be more appropriate. New ponds could be constructed for fishing and water storage, and low embankments would check flooding only during the early growth stages of the rice crop. Better preparation for unusually high and rapid flooding would complement these approaches. In addition to these measures, others advocate a more deep-rooted strategy which seeks to alleviate the factors that cause people to be vulnerable to flooding (Blaikie *et al.*, 1994; see also Fig. 2.7). Better access to land, replacements for land lost to erosion, and compensation for animals and other assets lost to natural disasters would all protect livelihoods and reduce vulnerability, as would better health care and facilities.

FURTHER READING

Alexander, D. 1993 *Natural disasters.* London, University College London Press. After documenting the range of geophysical hazards, the author looks at the impacts on human society and the options for disaster management.

Blaikie, P., Cannon, T., Davis, I. and Wisner, B. 1994 *At risk: natural hazards, people's vulnerability and disasters.* London, Routledge. This book emphasises the importance of the human factors behind hazards that become disasters, looking at the social, political and economic causes of people's vulnerability to natural events.

McCall, G.J.H., Laming, D.J.C. and Scott, S.C. (eds) 1992 *Geohazards: natural and man-made hazards.* London, Chapman & Hall. A collection of papers focusing on geophysical hazards, including those associated with earthquakes, volcanoes and soils.

Smith, K. 1992 *Environmental hazards: assessing risk and reducing disaster.* London, Routledge. All major rapid-onset events of both the natural and human environments – including seismic, mass movement, atmospheric, hydrological and technological hazards – are covered in this book, which emphasises the physical aspects of hazards and their management.

21

CONSERVATION AND SUSTAINABLE DEVELOPMENT

All the environmental issues covered in this book are, by definition, made up of physical and human elements. The approach throughout has been to analyse both the physical problems and the human responses to understand why the issues are issues, and how solutions can be found. Although particular issues are apparent at particular spatial scales, all have some manifestation on the global scale. In the case of climatic change brought about by atmospheric pollution, the issue affects the entire planet because climate is a global phenomenon. In other cases, more localised issues have become so commonplace that cumulatively they occur on a worldwide scale. Environmental problems associated with agriculture, urban areas, industrial pollution, waste management, warfare, deforestation and soil erosion are just a few examples of such cumulative global issues. Since so many environmental issues now have a global nature, many believe that the time has come for a complete rethink of the way we view these issues, for a change in the philosophy that lies behind the ways in which people interact with the environment, and for a change in the methodology with which the interactions occur. The approach most widely advocated by this rethink is sustainable development, and this chapter looks in more detail at the need for sustainable development and some of the ways in which it might work.

THE NEED FOR CHANGE

The socio-economic system is just one part of the natural ecosystem in which materials are transformed and energy converted to heat (Fig. 21.1). The operation of the socio-economic system is dependent upon the ecosystem as a provider of energy and natural resources and as a sink for wastes. The ecosystem also provides numerous 'environmental services' by virtue of its processes. These include the operation of climate and the hydrological cycle, recycling of nutrients, the generation of soils, pollination of crops, and

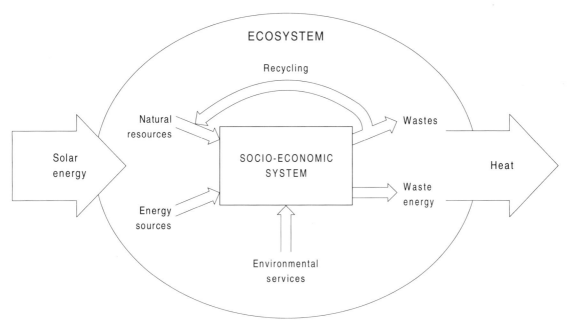

FIGURE 21.1 *The socio-economic system as part of the global ecosystem (after Folke and Jansson, 1992; Daly, 1993)*

so on. This ecological, economic perspective emphasises the fact that resource use and waste disposal take place in the same environment and that both activities affect the life-support functions of the environment. Hence, the socio-economic system cannot expand indefinitely since it is limited by the finite global biosphere. Until recently, for example, the number of fish that could be sold at market was limited by the number of boats at sea, now it is limited by the number of fish in the sea.

Through much of the history of human occupation of the planet, the socio-economic system has been small, relative to the biosphere, so that resources were plentiful, the environment's capacity for assimilating wastes was large and biospheric functions were relatively little affected by human impact. Today the situation is different. As Goodland *et al.* (1993a: 298) put it: 'now the world is no longer "empty" of people and their artefacts, the economic subsystem

having become large relative to the biosphere'. As the socio-economic system has grown larger, fuelled by increased 'throughput' of energy and resources, its capacity for disturbing the environment has increased. Since this disruption ultimately feeds back on the operation of society itself, society's behaviour must conform more closely with that of the total ecosystem because otherwise it may destroy itself (Folke and Jansson, 1992).

The potential for socio-economic degeneration and eventual collapse to occur when the human system operates too far out of harmony with the environment can be illustrated from both contemporary and historical examples, since although human activity has until recently been within the capacity of the environment on the global scale, breakdown has occurred on more local scales. Folke and Jansson (1992) note the recent rapid self-generated collapse of coastal shrimp industries in Taiwan and Thailand, within a decade of their beginnings, due to clearance of

FIGURE 21.2 *The ruins at Palenque in southern Mexico, a reminder of the collapse of the Mayan civilisation. No one is sure of the true reasons for the downfall of the Mayas, but one theory implicates human-induced environmental degradation as a key factor*

mangroves which act as nursery and feeding areas for shrimps, and the deterioration of water quality through eutrophication and disease. A lack of harmony between the socio-economic system and the natural ecosystem has also been suggested as a cause of collapse for several ancient cultures. The decline of the Mayan civilisation in Central America that began around the year 900 may have been due to excessive use of soils and an over-reliance on maize, which failed due to a virus (Fig. 21.2). Similarly, the civilisation that flourished in the tenth–twelfth centuries around Angkor Wat in present-day Cambodia was based on a sophisticated irrigation system, but forest clearance for cropland resulted in high silt loads which clogged the canals in rainy season floods. Irrigation channels were abandoned to become stagnant swamps where mosquitoes bred prolifically, and malaria epidemics swept through the city. This weakened Angkor Wat's capacity to adapt to change and the city was abandoned (McNeely, 1994).

A schematic representation of three ways in which human society interacts with the environment is shown in Fig. 21.3. The model can be applied on various scales, from the global to the local, and across differing timescales. In cycle A, which is typical of the global economy in historical times, wealth is accumulated largely by degrading the environment. This wealth has brought numerous advances to most parts of the world,

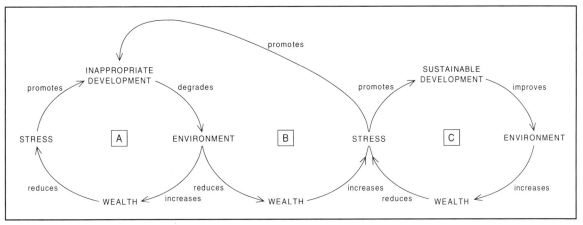

FIGURE 21.3 *Three cycles showing the relationship between modes of development and the environment*

which reduce 'stress', a term which is used here in a wide sense to reflect the general well-being of society. Such advances include sanitation and other facilities, improved health and higher living standards. These improvements have promoted further inappropriate development to continue on cycle A.

At some point in time, however, cycle A crosses an environmental threshold and the society may enter cycle B. The degraded environment begins to feed back on socio-economic wealth, and stress is increased. Increased stress in cycle B promotes further inappropriate development, particularly when the society in question has limited options as examples of poor, vulnerable groups quoted in this book have shown (e.g. accelerated soil erosion caused by farmers in Haiti and Ethiopia is discussed in Chapter 6), and on the national scale where rapid defor-estation in many tropical countries has been pursued in response to the burden of debt and declining commodity prices (see Chapter 3). Continuing on cycle B can ultimately lead to socio-economic collapse, as in the examples of modern shrimp industries and ancient civili-sations cited above.

Many people believe that globally we have now entered cycle B due to the sheer numbers of people on the Earth and the resulting scale of human activity. Currently, as much as 20–40 per cent of the potential global net primary productivity in terrestrial ecosystems is estimated to be diverted to human activities (Wright, 1990), and there are numerous examples throughout this book which indicate that the productivity of many the world's renewable resource-producing systems has reached its peak and in some cases is in decline. Increased stress is manifested by, amongst other things, increasing levels of world poverty, fears over the effects of human-induced global warming and the depletion of the ozone layer.

The only appropriate way to exit cycle B is to enter cycle C. To use a biological metaphor, the socio-economic system needs to be adjusted from its present dominantly parasitic relationship with the environment to a more symbiotic one. The global community does have this option; by adopting sustainable development we can live within the confines of the biosphere, enhance environmental quality and achieve greater wealth by chang-ing the pattern of resource consumption.

Adopting cycle C must be a permanent long-term strategy, however, because there is still a danger that reduced stress encourages a reversion to the old ways of cycles A and B.

Examples of societies that have managed to attain a sustainable use of resources can be quoted from both historical and recent times, although the view that all traditionalists were conservationists is somewhat romantic, as evidence of pre-Hispanic accelerated soil erosion in Mexico (see Chapter 6) and the widespread species extinctions that occurred in Hawaii prior to the arrival of European settlers (see Chapter 7) indicate. Contemporary examples of commercial resource exploitation, in which local communities have had to develop social methods to sustain the resources on which they depend, can also be cited (e.g. Acheson, 1975; Berkes, 1989). The experience of these societies can be put to good use by other groups seeking to adopt sustainable development.

SUSTAINABLE DEVELOPMENT

Despite the fact that the term sustainable development has become common currency among many groups, it is a confused and sometimes contradictory idea and there is no widespread agreement as to how it should work in practice (Dovers and Handmer, 1993). The concept has developed in the international forum from a document called the World Conservation Strategy (IUCN/UNEP/WWF, 1980), which argued that three priorities should be incorporated into all development programmes:

- maintenance of ecological processes
- sustainable use of resources
- maintenance of genetic diversity.

Sustainable development gained further credence thanks to the World Commission on Environment and Development (also known as the Brundtland Commission after its chair Gro Harlem Brundtland of Norway) which was formed by the UN in 1983 and reported in 1987 (WCED, 1987). The Commission emphasised that the integration of economic and ecological systems is all-important if sustainable development is to be achieved, and coined a broad definition for sustainable development which is often quoted: 'development which meets the needs of the present without compromising the ability of future generations to meet their own needs' (WCED, 1987: 43). Subsequently, similar calls for sustainable development have been made by various international groupings (e.g. IUCN/UNEP/WWF, 1991), notably at the UN Conference on Environment and Development (UNCED), otherwise known as the Earth Summit, in Rio de Janeiro in 1992.

Valuing environmental resources

Much research and thinking about sustainable development has focused on modifying economics to better integrate its operation with the workings and capacity of the environment, to use natural resources more efficiently and to reduce flows of waste and pollution. The full cost of a product, from raw material extraction to eventual disposal as waste, should be reflected in its market price, although in practice such a 'cradle to grave' approach may prove troublesome for materials such as minerals (see Chapter 18). Many economists contend that a root cause of environmental degradation is the simple fact that many aspects of the environment are not properly valued in economic terms. Commonly owned resources, such as the air, oceans and fisheries, are particularly vulnerable to overexploitation for this reason, but when such things have proper price tags, it is argued that decisions can be made using cost–benefit analysis. Putting a price on environmental assets and services is one of the central aims of the discipline of environmental economics. This can be done by finding out how much people are willing to pay for an aspect of the environment or how much people would accept in compensation for the loss of an environmental asset. One of the justifications of environmental pricing is the fact that money is the language of government treasuries and big business, and thus it is appropriate to address environmental issues in terms that such influential bodies understand.

There are problems with the approach, however. People's willingness to pay is heavily dependent on their awareness and knowledge of the resource and of the consequences of losing it. Information, when available, is open to manipulation by the media and other interest groups. In instances where the resource is unique in world terms – such as an endangered species, or a feature such as the Grand Canyon – who should be asked about willingness to pay? Should it be local people, national groups or an international audience? Our ignorance of how the environment works and the nature of the consequences of environmental change and degradation also present difficulties. In the case of climatic change due to human-induced atmospheric pollution, for example, all we know for sure is that the atmospheric concentrations of greenhouse gases have been rising and that human activity is most likely to be responsible. However, we do not know exactly how the climate will change nor what effects any changes will have upon human society. We can only guess at the consequences, so we can only guess at the costs. Economists are undeterred by these types of problem: 'Valuation may be imperfect but, invariably, some valuation is better than none' (Pearce, 1993: 5).

McNeely (1988) has classified such economic values into two groups:

■ *direct values*, which can be subdivided into values for consumptive use and values for productive use; and

■ *indirect values*, which are made up of non-consumptive value use, option value, and existence value.

Table 21.1 shows some examples of values which, where they have been calculated, appear to provide compelling arguments for the conservation and proper management of natural systems. For example, the value to the national economy of wilderness areas in Kenya, when conserved as National Parks, is estimated to be at least an order of magnitude greater per hectare than the land would yield if put to pastoral use, for example. However, differing values potentially derived from particular areas can be of benefit to different sectors of society. In the Kenyan example, management of the Masai Mara Reserve as a game park principally benefits the operators of tourist lodges and tours, while the pastoral Masai derive only 1 per cent or less of the accruing revenue. For the Masai, a more financially worthwhile use of the reserve would be to open it up to pastoralism and/or for poaching ivory and other wildlife products (Holdgate, 1991).

Other difficulties stem from the different ways groups perceive environmental value. Some groups may consider that certain parts of the environment lie outside the economists' realm. Adams (1993) gives the example of a mining company which wanted to exploit a site in Australia's Kakadu National Park, an area which is sacred in Aboriginal culture. In this case, the application of cost–benefit analysis amounts to immorality: 'The aborigines say "It is sacred." The economist replies "How much?"' (Adams, 1993: 258). It is for these sorts of reasons that some people reject the economic approach to valuing and hence preserving the environment and biological diversity. They argue that economic criteria of value can change and are opportunistic in their practical application. The greed which underlies such systems will ultimately destroy them (e.g. Ehrenfeld, 1988).

TABLE 21.1 *Illustrative values of biological resources*

Resource	Location	Value
DIRECT VALUES		
Consumptive use		
Firewood, dung	Nepal, Tanzania, Malawi	90% of total national primary energy
Wild animals as food	Ghana Nigeria	Main protein source for 75% of population 20% of animal protein in rural areas
Wild pigs	Sarawak	US$100 million/year
Productive use		
Wild plants and animals	USA	4.5% of GDP (US$87 billion/year 1967–80)
Timber (exports)	Asia, Africa, South America	US$8.1 billion/year (1981–83)
Plant-based drugs	OECD countries	US$43 billion (1985) but rising to US$200 billion to 1.8 trillion if benefits from better health, etc., costed in
INDIRECT VALUES		
Non-consumptive use		
Each lion	Kenya, Amboseli National Park	US$27 000/year
Each elephant herd	Kenya, Amboseli National Park	US$610 000/year
National parks (via tourism)	Kenya	US$40/ha/year
Wetlands area as flood protection	Boston, USA	US$17 million
Option		
Wild relatives of crop species with future potential (e.g. teosinté maize)	Mexico	Incalculable by definition until used, but e.g. US$6.8 billion/year contribution to perennial maize hybrid
Existence		
Aesthetic value of existence of species	Worldwide	US$100 million/year taken by WWF

Source: McNeely (1988)

Growth and development

A key issue in the sustainable development debate is the relative roles of economic growth – the quantitative expansion of economies, and economic development – the qualitative improvement of society. In its first report, the World Commission on Environment and Development suggested that sustainability could only be achieved with a five-fold to ten-fold increase in world economic activity in 50 years. This growth would be necessary to meet the basic needs and aspirations of a larger future global population (WCED, 1987). Subsequently, however, the WCED has played down the importance of growth (WCED, 1992). This change of thinking has been reflected in the two types of reaction to calls for sustainability that have been made to date. One the one hand, to concentrate on growth as usual, though at a slower rate, and on the other hand, to define sustainable development as 'development without growth in through-put beyond environmental capacity' (Goodland *et al.*, 1993a: 297). The idea of controlling throughput, the flow of environ-mental matter and energy through the socio-economic system (see Fig. 21.1), can be referred back to the ways in which human society interacts with the environment, as shown in Fig. 21.3: cycles A and B are based on increased throughput, while in cycle C throughput is controlled to a level within the environment's capacity to support it. Sustainable development – cycle C – means that the level of throughput must not exceed the ability of the environment to replace resources and withstand the impacts of wastes (Table 21.2). This does not necessarily mean that further economic growth is impossible, but it does mean that growth should be achieved by better use of resources and improved environmental management (Daly, 1987) rather than by the traditional method of increased throughput.

One indication of the degree of change necessary to make this possible is in the ways we measure progress and living standards at the national level. Measurements such as Gross National Product (GNP) and Gross Domestic Product (GDP) are the principal means by which economic progress is judged, and thus form key elements in government policies. But GNP is essentially a measure of throughput and it has severe limitations with respect to considerations of environmental and natural resources. The calculation of GNP does not take into account any depletion of

TABLE 21.2 *Key rules for the operation of environmental sustainability*

OUTPUT RULE

Waste emissions should be within the assimilative capacity of the environment to absorb, without unacceptable degradation of its future waste-absorptive capacity or other important services

INPUT RULE

Renewable resources: Harvest rates of renewables used as inputs should be within the capacity of the environment to regenerate equal replacements

Non-renewable resources: Depletion rates of non-renewables used as inputs should be equal to the rate at which sustained income or renewable substitutes are developed by human invention and investment. Part of the proceeds from using non-renewable resources should be allocated to research in pursuit of sustainable substitutes

Source: after Goodland et al. (1993b)

natural resources or adverse effects of economic activity on the environment, which have feedback costs on such things as health and welfare. The need to introduce environ-mental parameters into national accounting systems is now widely recognised, and adjusted measures of 'green GNP' now being worked upon could provide a good measure of national sustainability (Pearce and Mäler, 1991).

Further investigation of the limiting throughput approach to sustainable develop-ment can be conducted by representing human society's impact on the environment by the equation

$$I = P \times A \times T$$

where I is impact, P is population, A is afflu-ence (measured as the consumption of environmental resources per person) and T is technology. These three factors can provide the keys to keeping throughput within environmental capacity. This can be done by one of three means (Goodland *et al.* 1993a):

- limit population
- limit affluence
- improve technology.

In practice, combinations of these goals will be appropriate and some of these will now be considered at a regional level.

Regional perspectives

A distinction can be made between three global regions which should have different priorities in their contributions to global environmental sustainability (Goodland *et al.*, 1993a). The North must concentrate on reducing its long history of environmental damage due to overconsumption and affluence; the main priority in the South should be to stabilise population growth; and in the countries of the East, the former communist bloc, modernisation of wasteful and polluting technology should be the central priority. To promote these strategies, it is in the self-interest of the North to accelerate the transfer of technology to the East and South.

Measures of population density and relative affluence (which also has a major bearing on technological capacity) have been used to classify some major world regions, as shown in Fig. 21.4. More detailed design and implementation of sustainable development strategies for these regions will vary according to their conditions and characteristics (Clark, 1989). Low-income, low-density areas such as Amazonia and Malaya–Borneo include many of the world's remaining settlement frontiers where large-scale clearance for agriculture,

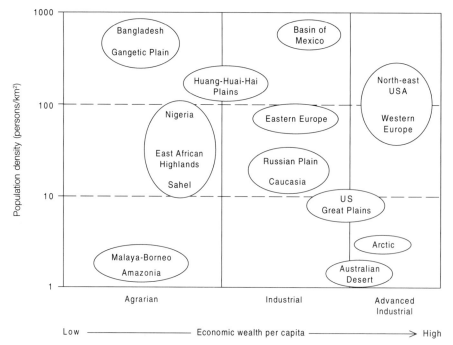

FIGURE 21.4 *Characteristics of regional environmental transformation (after Clark, 1989; Kates* et al. *1990)*

grazing and timber has begun only recently, after a longer history of resource use in shifting cultivation, small-scale plantation and mining sites. The widespread poverty of farmers engaged in land clearance, and the poorly developed state of local institutions which might guide sustainable development, means that appropriate management in such areas will be particularly difficult to attain.

Regions of low income and high population density, such as on the Ganges–Brahmaputra floodplains of the Indian subcontinent and the Huang–Huai–Hai plains of China, by contrast have long histories of human agricultural endeavour. This agricultural character has been augmented in recent decades by rapid industrial development and urbanisation, bringing the types of pollution problems that have been faced by industrial European countries within the last 100 years. The introduction of employment opportunities which can relieve pressure on agricultural land, but which do not simultaneously exacerbate urbanisation and industrial pollution problems, is the critical management challenge.

Other developing world areas such as the Basin of Mexico, with an agricultural history over hundreds of years, have experienced very rapid population growth, speedy industrialisation and a burgeoning consumer culture in recent decades. The grave problems of pollution and congestion are typical, if extreme, for primate cities of the developing world. Development in such cases can only be made sustainable, and problems eased, by a concerted national effort at decentralisation.

Areas which bear the more direct mark of environmental impact from the relatively wealthy countries include such frontier regions as the Arctic and the Australian desert, environments perceived as 'harsh' and 'sensitive' by the residents of more temperate climes. Exploitation, primarily for mineral resources, in these areas of low population density involve sophisticated technology and large economic investments. Although our knowledge of such environments remains poor, and hence the consequences of inappropriate actions can be great, the involvement of relatively few powerful corporate and governmental institutions means that the potential for introducing sustainable development practices is relatively good.

The greatest potential and the largest responsibilities for appropriate change lie in the densely populated wealthy industrialised regions of the world. Such regions have perpetrated a disproportionately large environmental impact, both locally and at the continental and global levels, through such pathways as emissions of acid rain precursors and greenhouse gases and through the export of environmentally damaging practices to other parts of the world. In recent times these areas have also achieved some successes in improving environmental quality locally. Some of the leads these countries have taken are considered in more detail in the following section.

National efforts

While some of the technologies and efforts to curb environmental problems have been introduced in the preceding chapters, numerous instances of environmental improvements being hampered by economic and political forces have also been noted. Hence, it is appropriate to look at some of the ways in which governments and other institutions have worked to promote sustainable practices. The rise in interest in environmental issues and sustainable development has inspired numerous efforts to gather information on national environmental situations and to target development efforts more appropriately. These include country status quo reports on environment and development issues, submitted to UNCED from both developed and developing nations, and several schemes aimed primarily at developing countries, such as National Environmental Action Plans prepared with World Bank assistance, National Conservation Strategies prepared with assistance from IUCN, an international NGO, and a number of similar initiatives sponsored by bilateral aid agencies, such as the US Agency for International Development's Country Environmental Profiles.

Data collection has provided the basis for a set of environmental indicators which are being developed on the national level (e.g. Environment Canada, 1991) and by member countries of organisations such as the Organisation for Economic Co-operation and Development (OECD, 1991b). Such indicators are being used to monitor the sustainability of resource use. In the forestry sector, for example, an index of intensity of forest use is calculated as the ratio of the total annual harvest to the annual growth of national forest stock. Hence, a ratio of one or less reflects a sustainable use of forest resources.

These and other types of procedures have been used to produce the first national strategy for achieving sustainable development in The Netherlands. The Dutch National Environmental Policy Plan (NEPP) and the National Environmental Policy Plan Plus provide the framework for a systematic revision of national policies to drive the country towards sustainability. The NEPP has three broad, long-term objectives for achieving sustainable development:

- consumption of no more energy than can be recovered from the sun,
- treatment of all wastes as raw materials,
- promotion of 'high-quality' products that last, can be repaired and are suitable for recycling.

Realistically, such long-term goals can only be achieved by incremental improvements, but specific targets for such improvements have been set (Adriaanse, 1993): CFCs are to be phased out by 1998, for example, and pesticide use is to be cut to half 1985 levels by the year 2000. Sustainability levels have been calculated for a number of environmental pollutants in a similar manner to that described above for forestry management. In the case of eutrophication, an index has been developed to reflect emissions of phosphorus and nitrogen from synthetic fertilisers, manure and waste dumping. The national index was around 300 during the 1980s but has begun to drop, with the aim of further reduction by more than two-thirds by the year 2000 (Fig. 21.5a), primarily through improvements to the treatment of manure and sewage. Agricultural improvements will also contribute towards the goal of reducing acidification, along with lower emission ceilings for sulphur dioxide. The acidification index reflects the acid deposition from the three main sources affecting The Netherlands (sulphur dioxide, nitrogen oxides and ammonia), and stepped targets for its reduction have been set (Fig. 21.5b).

In the immediate term, the NEPP has adopted the 'polluter pays' principle to implement the targets, but in the longer term the aim is to move away from the command and control approach of environmental laws used in most countries, to a structure of economic incentives, social institutions and self-regulation to achieve sustainable development. The likelihood of success in this approach has been enhanced by the widespread participation of business, government and local communities in the development of the plan.

Nonetheless, The Netherlands, like all countries, does not operate in a vacuum. The country is affected by pollutants which arrive from outside its borders, such as water pollution in the River Rhine and acid rain precursors from many parts of Europe. The low-lying country is also particularly vulnerable to any rise in sea level that may be consequent upon human-induced global warming. Dutch imports of resources from

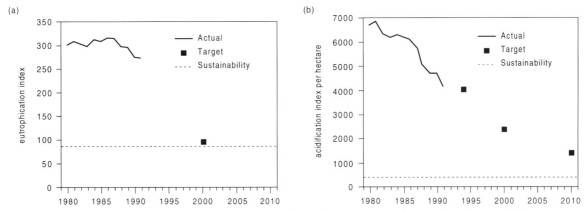

FIGURE 21.5 *Recent performance and future goals under the NEPP in The Netherlands (after Adriaanse, 1993) for: (a) eutrophication; (b) acidification. See text for explanation of index in each case*

other countries effectively contribute to environmentally damaging practices elsewhere, and Dutch exports produced under a sustainable regime may be less competitively priced than equivalent products derived from unsustainable methods. While the NEPP is undoubtedly a very significant step in the right direction at the national level, its measures can only work effectively if paralleled by similar initiatives elsewhere. Hence, it is appropriate to look again at international efforts to promote sustainable development.

Global dimensions

Numerous international attempts to promote conservation and sustainable development have been mentioned already in this and other chapters. These include the establishment of the UN Environment Programme, the recommendations of organisations such as the World Commission on Environment and Development, and many international agreements such as the conventions on biological diversity and climate change signed at UNCED. These and other initiatives have all made a positive contribution, but they have done little to address one of the major underlying difficulties in achieving global sustainability: poverty and the unequal distribution of resources.

Inequalities exist at all levels (see Chapter 2), but on the global scale an obvious imbalance is evident between North and South. Some minor, if significant, attempts have been made to address both debt problems and environmental problems in developing countries by converting part of the external debt of a country into a domestic obligation to support a specific programme. Some examples of existing 'debt-for-nature swaps' are detailed in Table 21.3. An international conservation group, or in some cases a national government, purchases part of a debtor country's foreign debt on the secondary market, at a fraction of the theoretical face value of the debt. The low cost of the original debt on the secondary market is a reflection of the fact that creditors have a low expectation of the debt itself being repaid. Once the debt has been acquired, the 'investing' conservation organisation agrees a rate of exchange with the debtor country so that the debt is converted into local currency which is then used to finance the intended aims.

TABLE 21.3 *Examples of established debt-for-nature agreements*

Debtor	Date	Debt face value (US$ million)	Cost (US$ million)	Investor	Terms and aims
Bolivia	7/87	0.65	0.1	Conservation International	Full legal protection of Beni Biosphere Reserve, set up of buffer zones and local management fund
Costa Rica	4/89	24.5	3.5	Government of Sweden	Expansion and protection of Guanacaste National Park
Madagascar	1/91	0.12	0.06	Conservation International	Interest from endowment fund for ecosystem management programmes, species inventories and environmental education
Poland	1/90	0.05	0.01	WWF-International and WWF-Sweden	Clean up of River Vistula and development of Biebrza National Park
Sudan	12/88	0.8		Bank donation	Midland Bank donated US$0.8 million in loans to UNICEF to provide 10 village water wells with hand pumps

Source: modified after WCMC (1992: Table 32.11)

Supporters of debt-for-nature swaps point out that they are agreements in which all parties involved stand to gain. Conservation organisations increase the spending power of their money, through the difference between the secondary market cost and the face value of the debt purchased, and are able to exert influence on conservation policies in developing countries. The debtor gains by reducing its foreign debt, reducing the government's need to raise foreign currency to service its debts, and also wins finance to support conservation programmes which the government has some control over.

Not all developing nations support such agreements, however. Brazil, for example, has argued that debt-for-nature is an infringement upon national sovereignty, passing some control and influence over national resources to foreign interests. Despite debt-for-nature swaps, part of the packages of development aid which flow from North to South, the problems of indebtedness and unequal trading relations remain essentially unchanged. Redistribution from rich to poor countries on any significant scale still appears to be politically impossible, yet it is the resolution or otherwise of this issue, perhaps more than any other, which will dictate the pace of global change towards sustainability.

THE SUSTAINABLE FUTURE

It is important to realise that sustainable development does not mean no human impact on the environment. Such a situation is impossible to achieve so long as there are people on the planet. The ideal scenario to strive for is one in which all environmental impacts can be undertaken consciously, in the full knowledge of the costs and consequences, even though this situation is a long way off, not least because we still have much to learn about the operation of nature. And even when and if such a status quo is attained, accidents will still happen. Seventeen years before the supertanker *Exxon Valdez* released 36 000 t of oil into Prince William Sound in Alaska in March 1989, causing widespread environmental damage, the author of a report on human activities in the region declared that: 'the volume of tanker traffic in this area (Prince William Sound at Valdez) makes occasional massive oil spills a distinct likelihood' (Price, 1972: 77). Despite stringent safety procedures which can reduce the chances of catastrophe, the best-laid human plans can malfunction for unexpected reasons, among them human error. Hence, there will always be a need for continual vigilance and the existence of contingency plans for when the unexpected does occur.

However, one of the central themes of this book has been the need to change the ways in which socio-economic systems work in order to reduce environmental impacts which feed back on the operation of society. The emphasis on economic growth, which relies on increasing the amount of resources channelled through society, must be replaced by an emphasis on sustainable development – the qualitative improvement in human welfare. There is no doubt that great modifications to

society are necessary to achieve a globally sustainable future. Some suggest that the scale of alteration is comparable to only two other changes in the history of humankind: the agricultural revolution of the late Neolithic period and the Industrial Revolution of the past two centuries (Ruckelshaus, 1989). However, while these previous revolutions were gradual and largely unconscious, the sustainable revolution will have to be a conscious one. It needs to be constantly flexible, guided by scientific and economic information and interpretation of events, and steered by governmental and private policy.

A basic prerequisite for this revolution is a fundamental revision of the way in which we view human society and the operation of nature. The need to evaluate natural systems properly, in economic and other terms, has been mentioned, as has the need to realise that maintaining a sustainable environment is a global challenge which must be based on a long-term perspective, and is required for the entire world's population, not just a select and privileged few. If a large portion of the members of our species is poor, we cannot hope to live in a peaceful and sustainable world, and if the transfer of technology and wealth from rich to poor is not forthcoming on a much greater scale than at present, then efforts by less-developed nations to improve their conditions along the same lines as the rich nations have followed will result in increasing worldwide ecological damage. At the same time, however, many lessons can be learnt by the richer countries, which are often over-reliant upon polluting and unnecessary technology, from more appropriate innovations in developing countries (Fig. 21.6).

Another fundamental change that many have to make is the need to return to the realisation that humans are part of nature and not separate from it. Industrial and urban societies have benefited greatly from their histories of economic development, but one of the casualties has been this basic truism.

FIGURE 21.6
Mowing the lawn in front of the Indian parliament building in New Delhi: an image which symbolises the fact that the world's developing countries have much to offer richer nations in terms of appropriate use of nature and technology

FIGURE 21.7 *Natural materials on sale for medicinal, magical and religious uses in Agadez, Niger. The contrast between this scene and its equivalent in urban-industrialised societies illustrates the real and perceived gap which has grown between many people and nature. Sustainable development requires a return to the realisation that we are a part of, and not apart from, nature*

Traditional societies have maintained their environments through spiritual connections with plants and animals and other aspects of the natural world, which although not lost by all (Fig. 21.7), has been widely replaced by a notion of human domination of nature. This notion must change. In simple and not overly dramatic terms, the continuance of the human race depends upon our ability to abstain from destroying the natural systems which regenerate our planet.

These changes in values will also have to be encouraged, cajoled and in some cases enforced. Existing powerful institutional forces thus need to change their direction too, and be supplemented with new ones. Money, trade and national defence are the concerns of today's most influential institutions, but the priorities of governments change and the existence of organisations such as the World Bank, transnational corporations and NATO, has a relatively short history. As realisation of the importance of environmental concerns continues to grow, historical experience suggests that such power bases can adjust accordingly. The alternative, which also has historical precedents, is institutional collapse.

Perhaps, above all, it is important to realise that a sustainable future lies in our hands. The need to alter values, beliefs and behaviour should by now be clear. Anyone who has read this book as far as this point should be well aware of that.

FURTHER READING

Odum, H.T. 1989 *Ecology and our endangered life-support systems.* Sunderland, MA, Sinauer Associates. An ecological view of sustainable development.

WCED (World Commission on Environment and Development) 1987 *Our common future.* Oxford, Oxford University Press. This is a key report on the environmental and development problems faced by the world's human population with proposals for their solution.

WCED 1992 *Our common future reconvened.* London, WCED. This is an update on WCED's 1987 report in which some of the proposals have been modified in the light of more recent thinking on the subject.

World Bank 1992 *World development report 1992: development and the environment.* New York, Oxford University Press. This book takes a realistic look at the environmental problems associated with economic development, assesses the current situation, and suggests appropriate policies to combat them.

abiotic: non-biological (see also **biotic**)

albedo: the reflectivity of a surface to sunlight

anadromous: fish which migrate from the sea into fresh water for breeding (e.g. salmon)

anaerobic: the term applied to organisms that live, or processes that occur, in the absence of gaseous or dissolved oxygen. Anaerobic bacteria, for example, obtain oxygen by breaking down such things as vegetable matter

atmosphere: the gaseous layer surrounding a planet. The Earth's atmosphere is divided into four parts distinguished according to the rate of change of temperature with height: the **troposphere**, **stratosphere**, mesosphere and thermosphere

benthic: applied to organisms living close to the bottom of a lake or sea

bioaccumulation: the increasing concentration of a substance up a food chain

biochemical oxygen demand (BOD): the amount of oxygen used in biological/chemical processes which decompose organic material in waste effluent. BOD is often used as an indicator of the polluting capacity of wastes

biodegradation: the decomposition or breakdown of organic matter by micro-organisms, particularly by oxygen-using or 'aerobic' bacteria

biodiversity: a term that refers to the number, variety and variability of living organisms. It is commonly defined in terms of genes, species and ecosystems, corresponding to three fundamental levels of biological organisation

biological classification: the arrangement of organisms into a hierarchy of groups that reflect evolutionary relationships. Usually the smallest group is the **species**, although species may be subdivided into subspecies and varieties. Similar species are grouped into **genera**, genera into **families**, families into **orders**, orders into classes, classes into phyla (singular: phylum) for animals – the equivalent for plants is divisions – and these into kingdoms. The formal Latin name given to an organism has two parts: the European crested newt, for example, is known as

Triturus cristatus, indicating that it is the species *cristatus* in the genus *Triturus*

biomass: the total mass of living organisms usually expressed in weight per unit area

biome: ecosystem of a large geographical area with characteristic plants and typical climate (e.g. tundra, savanna)

biosphere: the part of the Earth and its atmosphere in which organisms live

biotic: living or biological in origin, as opposed to **abiotic**

carcinogen: a substance which has the potential to cause cancer

cations: atoms that have lost one or more electrons and are thus left with a net positive charge

cetacean: a whale, dolphin or porpoise. The cetaceans are an **order** of aquatic mammals

CFCs (chlorofluorocarbons): organic compounds which include atoms of carbon, chlorine and fluorine. CFCs have been created for use in industrial processes and manufacturing and have become a cause for environmental concern due to their inadvertent release into the atmosphere. Breakdown of CFCs in the stratosphere releases chlorine which is thought to contribute to ozone depletion

coliform organisms: a group of bacteria abundant in the intestines of humans and other warm-blooded animals. Their concentration in water is used as an indicator of water quality

commodity: any useful material or product which might be traded

curie: the former unit of measurement for radioactivity, now replaced by the 'becquerel' (1 curie = 3.7×10^{10} becquerel)

degradation: a reduction in quality or decline in usefulness of an environmental resource. Human causes of degradation occur when a resource is used in an unsustainable manner

denitrification: the breakdown of **nitrates** by soil bacteria resulting in the release of nitrogen

DNA (deoxyribonucleic acid): the principal material of inheritance, found in chromosomes

ecosystem: a community of organisms and the **abiotic** environment which they inhabit and interact with. Ecosystems are recognised at many different geographical scales from a pond, for example, up to the scale of a **biome**

endemic: an endemic species is one found only in a particular geographical region, due to such factors as isolation, particular soils or climate

epidemic: an unusual outbreak of a disease affecting a number of people in a relatively short time

epiphyte: a plant that grows on the outside of another plant, using it for support and not as a source of nutrients (e.g. lichen on trees)

erodibility: the susceptibility of soil to erosion

erosivity: the power of an agent of erosion (water, wind, ice, gravity) to move soil material

eutrophication: the addition of mineral nutrients to an ecosystem, so raising primary productivity. The process occurs naturally, particularly in saltwater and freshwater bodies over long periods, but can also be human-induced, in which case it is sometimes referred to as 'cultural eutrophication'. Cultural eutrophication occurs by the often inadvertent addition of nutrients in sewage, detergents and from run-off contaminated with fertilisers

evapotranspiration: combined term for water lost as a vapour from soil or open water (evaporation) and water lost from the surface of a plant, mainly through stomata (transpiration)

family: a group of similar **genera** (see **biological classification**)

genera (singular: genus): groups of similar species (see **biological classification**)

GNP (Gross National Product): a measure of the market value of an economy's production over a given period (usually a year). GNP is adjusted to give GDP (Gross Domestic Product) by removing the value of profits from overseas investments and those profits from the economy which go to foreign investors

greenhouse effect: the process by which atmospheric gases trap long-wave radiation

emitted from the surface of a planet to warm its atmosphere

heavy metals: a broad group of metals of atomic weight higher than that of sodium and having a specific gravity in excess of 5.0. Those recognised as being of environmental concern, when present at certain concentrations which are harmful to living organisms, include copper, cadmium, mercury, tin, lead, antimony, vanadium, chromium, molybdenum, manganese, cobalt and nickel

humus: decomposed organic matter in soils

hydrosphere: the water present at or near the Earth's surface as a liquid, solid or gas

***K*-strategy:** a species life-style in which the reproduction rate is relatively low and individuals relatively long-living, which allows a species to persist in a particular place for a long period (cf. ***r*-strategy**)

keystone species: a species whose removal from the **ecosystem** of which it forms a part leads to a series of extinctions in that ecosystem

leachate: solution formed when water percolates through a permeable medium

leaching: removal of soil materials in solution

lithosphere: the Earth's crust and the upper part of the mantle

macrobenthic: large **benthic** organisms

Mesolithic (Middle Stone Age): A transitional period in the development of human societies between the **Palaeolithic** and **Neolithic** periods. The Mesolithic occupies different time spans in different places

methylation: the addition of a methyl group (CH_3-) to an element or compound, usually through the activity of **anaerobic** bacteria. The process of methylation has important implications when it involves **heavy metals** since it releases them from materials, so enabling them to enter the food chain

monoculture: the cultivation of a single-species crop

morbidity rate: the number of people in a population with physical injury and disease in a given time interval

mortality rate: the number of deaths in a population in a given time interval

Neolithic (New or Late Stone Age): a level of human development characterised by the use of polished stone tools and the beginnings of settled agriculture. The Neolithic occupies different time spans in different places

nitrate: a chemical compound containing nitrogen, an essential **nutrient** for plants

non-point source: a term applied to a source of pollution meaning a dispersed source, such as a field, as opposed to a 'point source' such as an industrial waste pipe

nutrient: an element or compound needed by a living organism. A distinction is often made between 'macronutrients' which are needed in relatively large quantities (e.g. carbon, hydrogen, oxygen, nitrogen, potassium and phosphorus), and 'micronutrients' needed in smaller amounts (e.g. iron, manganese, zinc, copper)

onchocerciasis: a tropical disease caused by the worm *Onchocerca volvulus* and transmitted by a biting fly which breeds in fast-flowing streams. The worm's larvae can enter the eye causing blindness, hence the disease's common name 'river blindness'

order: see **biological classification**

organochlorines: organic compounds containing chlorine which tend to **bioaccumulate** and are often long-lasting in the environment. They include several types of pesticides (e.g. aldrin, dieldrin, DDT) and **PCBs**

Palaeolithic (Early Stone Age): the stage in the development of a human society when people obtain food by hunting and fishing and gathering wild plants, as opposed to engaging in settled agriculture. The Palaeolithic occupies different time spans in different places

palaeontology: the study of fossils

pandemic: a disease prevalent over a whole country of the world. Commonly used to describe global **epidemics**

PCBs (polychlorinated biphenyls): a group of chlorinated hydrocarbons (see **organochlorines**) used mainly in high-voltage transformers. There is evidence that these compounds are highly persistent and toxic when released into the environment

pelagic: organisms which inhabit the open water at sea or in a lake

permafrost: permanently frozen ground where temperatures below 0°C have persisted for at least two consecutive winters and the intervening summer. Permafrost covers about 26 per cent of the Earth's land surface

pheromones: chemicals used by an animal as a social cue

phytoplankton: see **plankton**

phytotoxic: poisonous to green plants

plankton: animals and plants, many of them microscopically small, that float or swim in fresh or salt water. The animals (zooplankton) are mainly protozoa, small crustacea and larval stages of molluscs. The plants (phytoplankton) are almost all algae

podsolic soils (podsols): a soil type formed at an advanced stage of **leaching** which has left an acid **humus** layer

primary/secondary forest: primary forest is that which has not been disturbed by human action while secondary forest is a forest ecosystem which has regrown following human disturbance. In practice, it can be difficult to distinguish in the field between the two types of some forests, particularly where secondary forest is more than, say, 60 years old

***r*-strategy:** a species life-style in which the reproduction rate is relatively high and individuals relatively short-living, which means that a species is a good coloniser but tends to persist in a particular place for a short period (cf. ***K*-strategy)

radionuclide: an unstable nucleus of an atom that undergoes spontaneous radioactive decay, thereby emitting radiation and eventually changing from one element into another

raptor: a bird of prey that hunts by day (e.g. eagle, goshawk, kite, osprey, peregrine falcon, vulture)

schistosomiasis: any one of a group of diseases caused by infestation by blood flukes of humans and some other mammals. The diseases are common in low latitudes and are also known as bilharzia

species: groups of living individuals which look alike and can interbreed, but cannot interbreed with other species (see **biological classification**)

Stone Age: the period in the development of human societies during which tools and implements were made of stone, bone or wood, and no metals were used. The dates of the period vary widely from place to place, and the term is sometimes used to describe present-day pre-agricultural peoples, although they may use imported metal implements. See also **Palaeolithic**, **Mesolithic** and **Neolithic**

stratosphere: the layer of the **atmosphere** above the **troposphere** which extends on average between 10 km and 50 km above the Earth's surface

succession: the progressive natural development of vegetation towards a so-called 'climax' in which vegetation is supposedly in equilibrium with the existing environmental conditions. A distinction is made between 'primary' succession which takes place on land with no previous vegetation, and 'secondary' succession which occurs on land where vegetation has been partially or wholly destroyed

symbiotic: situation in which two species live together and both of them benefit from the relationship

technology: a term used to describe the total system of means by which people interact with their environment and each other. It includes tools, information and knowledge, and the organisation of resources for productive activity

tectonic movements: deformation within the Earth's crust

toxic: poisonous

transnational corporations (TNCs): large business organisations whose activities span international boundaries

troposphere: the lowest layer of the Earth's **atmosphere** which varies in depth between 10 and 12 km over the poles to 17 km over the equator. It is the layer in which most weather features occur

trypanosomiasis: a group of debilitating, long-lasting diseases caused by infestation with microscopic single-celled organisms. In Africa, sleeping sickness among humans and ngana in cattle, both transmitted by tsetse flies, are common forms of trypanosomiasis. A common form in the Americas is the incurable Chagas' disease

tsunami: a seismic sea wave caused by a submarine earthquake, underwater volcanic explosion or massive slide of sea-bed sediments

vector: an organism that conveys a parasite from one host to another (e.g. *Anopheles* mosquito, which transmits malaria)

volatile organic compounds (VOCs): the term used for a class of atmospheric pollutants, also known as hydrocarbons, which can produce ozone in combination with NO_x and sunlight. There are many hundreds of VOCs (e.g. methane, propane, toluene, ethanol) emitted into the atmosphere from both natural and human sources

volatilisation: the collective term for evaporation and sublimation

zooplankton: see **plankton**

BIBLIOGRAPHY

Abrol, I.P., Yadav, J.S.P. and Massoud, F.I. 1988 Salt-affected soils and their management. *FAO Soils Bulletin* 39.

Abu-Zeid, M. 1989 Environmental impacts of the Aswan High Dam. *Water Resources Development* 5: 147–157.

Acheson, J.M. 1975 The lobster fiefs: economic and ecological effects of territoriality in the Maine lobster industry. *Human Ecology* 3: 183–207.

Adams, J. 1993 The emperor's old clothes: the curious comeback of cost–benefit analysis. *Environmental Values* 2: 247–260.

Adams, R. 1975 The Haicheng earthquake of 4 February 1975: the first successfully predicted major earthquake. *Earthquake Engineering and Structural Dynamics* 4: 423–437.

Adriaanse, A. 1993 *Environmental policy performance indicators*. The Hague, Sdu Uitgeverij Koninginnegracht.

Agnew, C. and Anderson, E. 1992 *Water resources in the arid realm*. London, Routledge.

Ågren, C. 1993 SO_2 emissions: the historical trend. *Acid News* 5: 1–4.

Agricola, G. 1556 *De re metallica*. Translated by Hoover, H.C. and Hoover, L.H. 1950, New York, Dover Publications.

Akiner, S., Cooke, R.U. and French, R.A. 1992 Salt damage to Islamic monuments in Uzbekistan. *The Geographical Journal* 158: 257–272.

Akleyev, A.V. and Lyubchansky, E.R. 1994 Environmental and medical effects of nuclear weapon production in the southern Urals. *Science of the Total Environment* 142: 1–8.

Alexander, D. 1993 *Natural disasters*. London, University College London Press.

Al-Saleh, M.A. 1992 Declining groundwater level of the Minjur Aquifer, Tebrak area, Saudi Arabia. *The Geographical Journal* 158: 215–222.

Alström, K. and Åkerman, A.B. 1992 Contemporary soil erosion rates on arable land in southern Sweden. *Geografiska Annaler* 74(A): 101–108.

Amin, M.A. 1977 Problems and effects of schistosomiasis in irrigation schemes in Sudan. In Worthington, E.B. (ed.), *Arid land irrigation in developing countries*. London, Pergamon: 407–411.

Anderson, A.B. and Ioris, E.M. 1992 Valuing the rain forest: economic strategies by small-scale forest extractivists in the Amazon estuary. *Human Ecology* 20: 337–369.

Archer, L.J. 1993 *Aircraft emissions and the environment: CO_x, SO_x, HO_x, & NO_x*. Oxford, Oxford Institute for Energy Studies.

Aubreville, A. 1949 *Climats, forêts et désertification de l'Afrique tropicale*. Paris, Société d'éditions géographiques maritimes et coloniales.

Bailey, R.S. and Steele, J.H. 1992 North Sea herring fluctuations. In Glantz, M.H. (ed.), *Climatic variability, climate change and fisheries*. Cambridge, Cambridge University Press: 213–230.

Bakir, F., Damlaji, S., Amin-Zaki, L., Murtadha, M., Khalidi, A., Al-Rawi, N., Tikriti, S., Dhakir, H., Clarkson, T., Smith, J. and Doherty, R. 1973 Methylmercury poisoning in Iraq. *Science* 181: 230–241.

Baltz, D.M. 1991 Introduced fishes in marine systems and inland seas. *Biological Conservation* 56: 151–177.

Banks, G. 1993 Mining multinationals and developing countries: theory and practice in Papua New Guinea. *Applied Geography* 13: 313–327.

Barker, J.R., Thurow, T.L. and Herlocker, D.J. 1990 Vegetation of pastoralist campsites within the coastal grassland of Somalia. *African Journal of Ecology* 28: 291–297.

Barlaz, M.A., Eleazer, W.E. and Whittle, D.J. 1993 Potential to use waste tires as a supplemental fuel in pulp and paper mill boilers, cement kilns and in road pavement. *Waste Management and Research* 11: 463–480.

Barrow, C.J. 1981 Health and resettlement consequences and opportunities created as a result of river impoundment in developing countries. *Water Supply and Management* 5: 135–150.

——1991 *Land degradation*. Cambridge, Cambridge University Press.

Bartone, C. 1990 Economic and policy issues in resource recovery from municipal solid wastes. *Resources, Conservation and Recycling* 4: 7–23.

Bates, T.S., Lamb, B.K., Guenther, A., Dignon, J and Stoiber, R.E. 1992 Sulphur emissions to the atmosphere from natural sources. *Journal of Atmospheric Chemistry* 14: 315–337.

Battarbee, R.W. 1994 Surface water acidification. In Roberts, N. (ed.), *The changing global environment*. Oxford, Blackwell: 213–241.

Bayfield, N.G., Barker, D.H. and Yah, K.C. 1992 Erosion of road cuttings and the use of bioengineering to improve slope stability in Peninsular Malaysia. *Singapore Journal of Tropical Geography* 13: 75–89.

BCAS (Bangladesh Centre for Advanced Studies) 1993 Bangladesh vulnerability profile to climate change and sea level rise. *Bangladesh Environmental Newsletter* 4(4): 3.

Beaumont, P., Blake, G.H. and Wagstaff, J.M. 1988 *The Middle East: a geographical study*, 2nd edn. London, David Fulton.

Been, V. 1994 Locally undesirable land uses in minority neighbourhoods: disproportionate siting or market dynamics? *The Yale Law Journal* 103: 1383–1422.

Beg, M.A.A. 1990 *Report on status of air pollution in Karachi, past, present and future*. Karachi, Pakistan Council of Scientific and Industrial Research.

Bennett, E.L. and Reynolds, C.J. 1993 The value of a mangrove area in Sarawak. *Biodiversity and Conservation* 2: 359–375.

Bennett, O. (ed.) 1991 *Greenwar: environment and conflict*. London, Panos.

Bergström, M. 1990 The release in war of dangerous forces from hydrological facilities. In Westing, A.H. (ed.), *Environmental hazards of war*. London, Sage: 38–47.

Berkes, F. (ed.) 1989 *Common property resources: ecology and community-based sustainable development*. London, Belhaven.

Bertrand, F. 1993 *Contribution à l'étude de l'environnement et de la dynamique des mangroves de Guinée*. Paris, ORSTOM.

Berz, G.A. 1991 Global warming and the insurance industry. *Nature and Resources* 27(1): 19–28.

Bhatti, N., Streets, D.G. and Foell, W.K. 1992 Acid rain in Asia. *Environmental Management* 16: 541–562.

Bird, E.F.C. 1985 *Coastline changes: a global review*. Chichester, Wiley.

Biswas, A.K. 1990 Watershed management. In Thanh, N.C. and Biswas, A.K. (eds), *Environmentally-sound water management*. Delhi, Oxford University Press: 155–175.

Björk, S. and Digerfeldt, G. 1991 Development and degradation, redevelopment and preservation of Jamaican wetlands. *Ambio* 20: 276–284.

Black, R. 1994 Environmental change in refugee-affected areas of the Third World: the role of policy and research. *Disasters* 18: 107–116.

Blaikie, P. 1985 *The political economy of soil erosion*. London, Longman.

——, Cannon, T., Davis, I. and Wisner, B. 1994 *At risk: natural hazards, people's vulnerability and disasters*. London, Routledge.

Blunden, J. 1985 *Mineral resources and their management*. London, Longman.

Boardman, J. 1990a Soil erosion on the South Downs: a review. In Boardman, J., Foster, I.D.L. and Dearing, J.A. (eds), *Soil erosion on agricultural land*. Chichester, Wiley: 87–105.

——1990b *Soil erosion in Britain: costs, attitudes and policies*. Social Audit Paper 1. University of Sussex, Brighton, Education Network for Environment and Development.

Bond, A.R. and Piepenburg, K. 1990 Land reclamation after surface mining in the USSR: economic, political, and legal issues. *Soviet Geography* 31: 332–365.

Boserüp, E. 1965 *The conditions of agricultural growth: the economics of agrarian change under population pressure*. London, Allen and Unwin.

Bosson, R. and Varon, B. 1977 *The mining industry and the developing countries*. New York, Oxford University Press.

Bowonder, B., Prasad, S.S.R. and Unni, N.V.M. 1988 Dynamics of fuelwood prices in India. *World Development* 16: 1213–1229.

Boyce, J.K. 1990 Birth of a megaproject: political economy of flood control in Bangladesh. *Environmental Management* 14: 419–428.

Bradshaw, A.D. and Chadwick, M.J. 1980 *The restoration of land*. Oxford, Blackwell Scientific.

Brandon, C. and Ramankutty, R. 1993 Toward an environmental strategy for Asia. *World Bank Discussion Paper* 224.

Briscoe, J. 1987 A role for water supply and sanitation in the child survival revolution. *Bulletin of the Pan American Health Organization* 21: 92–105.

Broad, R. 1994 The poor and the environment: friends or foe? *World Development* 22: 811–822.

Brookes, A. 1985 River channelization: traditional engineering methods, physical consequences, and alternative practices. *Progress in Physical Geography* 9: 44–73.

Brown, S. and Lugo, A.E. 1990 Tropical secondary forests. *Journal of Tropical Ecology* 6: 1–32.

Bryson, R.A. and Barreis, D.A. 1967 Possibility of major climatic modifications and their implications: northwest India, a case for study. *Bulletin of the American Meteorological Society* 48: 136–142.

Burt, T. 1993 From Westminster to Windrush: public policy in the drainage basin. *Geography* 78: 388–400.

——, Heathwaite, A.L. and Trudgill, S.T. 1993 *Nitrate: processes, patterns and management*. Chichester, Wiley.

Burton, I., Kates, R.W. and White, G.F. 1978 *The environment as hazard*. New York, Oxford University Press.

Caddy, J.F. and Gulland, J.A. 1983 Historical patterns of fish stocks. *Marine Pollution* 7: 267–278.

Carson, R. 1962 *Silent spring*. Boston, Houghton Mifflin.

Carter, F.W. 1993a Czechoslovakia. In Carter, F.W. and Turnock, D. (eds), *Environmental problems in Eastern Europe*. London, Routledge: 63–88.

——1993b Poland. In Carter, F.W. and Turnock, D. (eds), *Environmental problems in Eastern Europe*. London, Routledge: 107–134.

Carwardine, M. 1994 *On the trail of the whale*. Guilford, Thunder Bay.

Cavanagh, J.E., Clarke, J.H. and Price, R. 1993 Ocean energy systems. In Johansson, T.B., Kelly, H., Reddy, A.K.N. and Williams, R.H. (eds), *Renewable energy: sources for fuels and electricity*. New York, Island Press: 513–547.

Caviedes, C.N. and Fik, T.J. 1992 The Peru–Chile eastern Pacific fisheries and climatic oscillation. In Glantz, M.H. (ed.), *Climatic variability, climate change and fisheries*. Cambridge, Cambridge University Press: 355–375.

Chakela, Q. and Stocking, M. 1988 An improved methodology for erosion hazard mapping. Part II, application to Lesotho. *Geografiska Annaler* 70(A): 181–189.

Charlson, R.J. and Wigley, T.M.L. 1994 Sulfate aerosol and climatic change. *Scientific American* 270(2): 28–35.

Charney, J., Stone, P.H. and Quirk, W.J. 1975 Drought in the Sahara: a bio-geophysical feedback mechanism. *Science* 187: 434–435.

Chen, R.Z., Peng, G.Y. and Hunag, F.X. 1990 Effect of simulated acid rain on the growth and yield of soybean and peanut. *Journal of Ecology (China)* 9: 58–60.

Chester, D. 1993 *Volcanoes and society*. London, Edward Arnold.

Chien, N. 1985 Changes in river regime after the construction of upstream reservoirs. *Earth Surface Processes and Landforms* 10: 143–159.

Christensen, B. 1983 Mangroves – what are they worth? *Unasylva* 35: 2–15.

CIPEL (Commission Internationale pour la Protection des Eaux du Léman) 1992 *Rapports sur les Etudes et Recherches Enterprises dans le Bassin Lémanique: Campagne 1991*. Lausanne, CIPEL.

CIPRCP (Commission Internationale pour la Protection du Rhin Contre la Pollution) 1984 *Rapport Annual 1984*. Koblenz, CIPRCP.

CIRIA (Construction Industry Research and Information Association) 1989 *The engineering implications of rising ground water levels in the deep aquifer beneath London*. London, CIRIA.

Clark, R.B. 1992 *Marine pollution*, 3rd edn. Oxford, Clarendon Press.

Clark, W.C. 1989 Managing planet earth. *Scientific American* 261(3): 19–26.

Clayton, K. 1991 Scaling environmental problems. *Geography* 76: 2–15.

Cohen, M.N. 1977 *The food crisis in prehistory: overpopulation and the origins of agriculture*. New Haven, Yale University Press.

Cohn, J.P. 1990 Elephants: remarkable and endangered. *BioScience* 40: 10–14.

Colinvaux, P. 1993 *Ecology 2*. New York, Wiley.

Collins, C.O. and Scott, S.L. 1993 Air pollution in the Valley of Mexico. *Geographical Review* 83: 119–133.

Collins, N.M., Sayer, J.A. and Whitmore, T.C. (eds) 1991 *The conservation atlas of tropical forests: Asia and the Pacific*. London, Macmillan/IUCN.

Connell, J.H. 1978 Diversity in tropical rain forests and coral reefs. *Science* 199: 1302–1310.

Cooke, R.U. 1984 *Geomorphological hazards in Los Angeles*. London, Allen and Unwin.

——1989 Geomorphological contributions to acid rain research: studies of stone weathering. *The Geographical Journal* 155: 361–366.

—— and Doornkamp, J.C. 1990 *Geomorphology in environmental management*, 2nd edn. Oxford, Clarendon Press.

Coull, J.R. 1993 Will a blue revolution follow the green revolution? The modern upsurge of aquaculture. *Area* 25: 350–357.

Crowder, B.M. 1987 Economic costs of reservoir sedimentation: a regional approach to estimating cropland erosion damages. *Journal of Soil and Water Conservation* 42: 194–197.

Crowson, P. 1992 *Mineral resources: the infinitely finite*. Ottawa, The International Council on Metals and the Environment.

——1994 *Minerals handbook 1994–95*. New York, Stockton Press.

Csirke, J. 1988 Small shoaling pelagic fish stocks. In Gulland, J.A. (ed.), *Fish population dynamics*, 2nd edn. Chichester, Wiley: 277–284.

Curtis, D., Hubbard M. and Shepherd, A. (eds) 1988 *Preventing famine: policies and prospects for Africa*. London, Routledge.

Daly, H.E. 1987 The economic growth debate: what some economists have learned but many others have not. *Journal of Environmental Economics and Management* 14: 323–336.

——1993 The perils of free trade. *Scientific American* 269(5): 24–29.

Davis, M.B. 1976 Erosion rates and land use history in southern Michigan. *Environmental Conservation* 3: 139–148.

de Jong, J. and Wiggens, A.J. 1983 Polders and their environment in the Netherlands. In *Polders of the world, an international symposium: final report*. Wageningen, International Institute for Land Reclamation and Improvement: 221–241.

De Noni, G., Trujillo, G. and Viennot, M. 1986 L'érosion et la conservation des sols en Equateur. *Cahiers ORSTOM Série Pédologie* 22: 235–245.

Décamps, H. and Fortuné, M. 1991 Long-term ecological research and fluvial landscapes. In Risser, P.G. (ed.), *Long-term ecological research*. SCOPE Report 47. Chichester, Wiley: 135–151.

Deegan, J. 1987 Looking back at Love Canal. *Environmental Science and Technology* 21: 328–331.

Deichmann, U. and Eklundh, L. 1991 *Global digital datasets for land degradation studies: a GIS approach*. UNEP/GEMS GRID Case Study Series 4. Nairobi, UNEP.

Dejene, A. and Olivares, J. 1991 Integrating environmental issues into a strategy for sustainable agricultural development: the case of Mozambique. *World Bank Technical Paper* 146.

Diamond, J.M. 1984 Historic extinctions: a Rosetta Stone for understanding prehistoric extinctions.

In Martin, P.S and Klein, R.G. (eds), *Quaternary extinctions: a prehistoric revolution*. Tucson, AZ, University of Arizona Press: 824–862.

Dittrich, V., Hassan, S.O. and Ernst, G.H. 1985 Sudanese cotton and the whitefly: a case study of the emergence of a new primary pest. *Crop Protection* 4: 161–176.

Dixon, J.A., Talbot, L.M. and Le Moigne, J.-M. 1989 Dams and the environment. *World Bank Technical Paper* 110.

Dobson, M. 1980 'Marsh fever' – the geography of malaria in England. *Journal of Historical Geography* 6: 357–389.

DoE (Department of the Environment) 1992 *The UK environment*. London, HMSO.

Doolette, J.B. and Smyle, J.W. 1990 Soil and moisture conservation technologies: review of literature. In Doolette, J.B. and Magrath, W.B. (eds), Watershed development in Asia: strategies and technologies. *World Bank Technical Paper* 127.

Döös, B.R. 1994 Why is environmental protection so slow? *Global Environmental Change* 4: 179–184.

Doumenge, F. 1986 La révolution aquacôle. *Annales de Géographie* 530: 445–482, 527–586.

Dove, M.R. 1993 A revisionist view of tropical deforestation and development. *Environmental Conservation* 20: 17–24, 56.

Dovers, S.R. and Handmer, J.W. 1993 Contradictions in sustainability. *Environmental Conservation* 20: 217–222.

Down, C.G. and Stocks, J. 1977 *Environmental impact of mining*. London, Applied Science.

Drake, J.A., Mooney, H.A., di Castri, F., Groves, R.H., Kruger, F.J. and Williamson, M. (eds) 1989 *Biological invasions: a global perspective*. SCOPE Report 37. Chichester, Wiley.

Dresch, J. 1986 Degradation of natural ecosystems in the countries of the Maghreb as a result of human impact. In Glantz, M.H. (ed.), *Arid land development and the combat against desertification*. Moscow, UNEP: 65–67.

Drysdale, J. 1994 *Whatever happened to Somalia?* London, Haan.

Dudley, N. 1986 Acid rain and British pollution control policy. In Goldsmith, E. and Hildyard, N. (eds), *Green Britain or industrial wasteland?* Cambridge, Polity Press: 95–107.

Durham, W.H. 1979 *Scarcity and survival in Central America: ecological origins of the Soccer War*. Stanford, Stanford University Press.

Durning, A. 1991 Asking how much is enough. In Brown, L.R. (ed.), *State of the world 1991*. New York, W.W. Norton: 153–169.

Durning, A.B. and Brough, H.B. 1991 *Taking stock: animal farming and the environment*. Worldwatch Paper 103. Worldwatch Institute, Washington, DC.

D'Yakanov, K.N. and Reteyum, A.Y. 1965 The local climate of the Rybinsk reservoir. *Soviet Geography* 6: 40–53.

Earthquest 1991 Science capsule. Vol. 5(1). Washington, D.C. Office for Interdisciplinary Earth Studies.

Edwards, P. 1985 *Aquaculture a component of low cost sanitation technology*. Washington DC, World Bank.

Ehrenfeld, D.W. 1988 Why put a value on biodiversity? In Wilson, E.O. and Peter, F.M. (eds), *Biodiversity*. Washington DC, National Academy Press: 212–216.

Ehrlich, P.R. and Ehrlich, A.H. 1981 *Extinction: the causes and consequences of the disappearance of species*. New York, Random House.

El-Hinnawi, E. and Hashmi, M.H. 1987 *The state of the environment*. London, Butterworths.

Elkington, J. and Burke, T. 1987 *The green capitalists*. London, Victor Gollancz.

Ellis, D. 1989 *Environments at risk: case histories of impact assessment*. Berlin, Springer-Verlag.

El-Swaify, S.A., Dangler, E.W. and Armstrong, C.L. 1982 *Soil erosion by water in the Tropics*. Research Extension Series 024, Honolulu, College of Tropical Agriculture and Human Resources, University of Hawaii.

Elvingson, P. 1993 Younger stands now affected. *Acid News* 5: 8–9.

EMEP (European Monitoring and Evaluation Programme) 1993 *Calculated budgets for airborne acidifying components in Europe, 1985, 1987, 1988,*

1989, 1990, 1991 and 1992. EMEP Report 1/93. Oslo, EMEP.

Endler, J.A. 1982 Pleistocene forest refuges: fact or fantasy? In Prance, G.T. (ed.) *Biological diversification in the tropics.* New York, Columbia University Press: 641–657.

Environment Canada 1991 *A report on Canada's progress towards a national set of environmental indicators.* SOE Report 91-1. Ottawa, Environment Canada.

Evans, R. 1990 Assessment of soil erosion risk in England and Wales. *Soil Use and Management* 1: 127–131.

Ezcurra, E. 1990 The Basin of Mexico. In Turner, B.L., II, Clark, W.C., Kates, R.W., Richards, J.F., Mathews, J.T. and Meyer, W.B. (eds), *The earth as transformed by human action.* Cambridge, Cambridge University Press: 577–588.

FAO (Food and Agriculture Organization) 1983 *Fuelwood supplies in developing countries.* Rome, FAO.

——1991 *Environment and sustainability in fisheries.* Rome, FAO Committee on Fisheries.

——1993 *The state of food and agriculture.* Rome, FAO.

Farman, J.C., Gardiner, B.G. and Shanklin, J.D. 1985 Large losses in total ozone in Antarctica reveal seasonal CLO_x/NO_x interaction. *Nature* 315: 207–210.

Farrington, J. 1992 Transport, environment and energy. In Hoyle, B.S. and Knowles, R.D. (eds), *Modern transport geography.* London, Belhaven: 51–66.

Fearnside, P.M. 1989 The charcoal of Carajás: a threat to the forests of Brazil's eastern Amazon region. *Ambio* 18: 141–143.

——1990 Environmental destruction in the Brazilian Amazon. In Goodman, D. and Hall, A. (eds), *The future of Amazonia: destruction or sustainable development?* Basingstoke, Macmillan: 179–225.

Fernando, C.H. 1991 Impact of fish introductions in tropical Asia and America. *Canadian Journal of Fisheries and Aquatic Sciences* 48: 24–32.

Feshbach, M. and Friendy, A. 1992 *Ecocide in the USSR: health and nature under siege.* New York, Basic Books.

Fey, M.V., Manson, A.D. and Schutte, R. 1990 Acidification of the pedosphere. *South African Journal of Science* 86: 403–406.

Feyerabend, P.K. 1978 *Science in a free society.* London, NLB.

Fickett, A.P., Gellings, C.W. and Lovins, A.B. 1990 Efficient use of electricity. *Scientific American* 263(3): 29–36.

Fischer, G., Frohberg, K., Parry, M.L. and Rosenzweig, C. 1994 Climate change and world food supply, demand and trade: who benefits, who loses? *Global Environmental Change* 4: 7–23.

Fisher, H.I. and Baldwin, P.H. 1946 War and the birds of Midway Atoll. *Condor* 48: 3–15.

Fleischer, S., Andersson, G., Brodin, Y., Dickson, W., Herrmann, J. and Muniz, I. 1993 Acid water research in Sweden – knowledge for tomorrow? *Ambio* 22: 258–263.

Folke, C. and Jansson, A.M. 1992 The emergence of an ecological economics paradigm: examples from fisheries and aquaculture. In Svedin, U. and Aniansson, B. (eds), *Society and the environment: a Swedish perspective.* Dordrecht, Kluwer: 69–87.

Forman, D., Cook-Mozaffari, P., Darby, S., Davey, G., Stratton, I., Doll, R. and Pike, M. 1987 Cancer near nuclear installations. *Nature* 329: 499–505.

Forslund, J. 1986 Groundwater quality today and tomorrow. *World Health Statistical Quarterly* 39: 81–92.

Freeman, D.B. 1992 Prickly pear menace in eastern Australia. *The Geographical Review* 82: 411–429.

Fricke, H. and Hissman, K. 1990 Natural habitat of the coelacanths. *Nature* 346: 323–324.

Fritsch, J.M. and Sarrailh, J.M. 1986 Les transports solides dans l'écosytème forestier tropical humide guyanais: effects du défrichement et de l'aménagement de pâturages. *Cahiers ORSTOM Série Pédologie* 22: 209–222.

Fryrear, D.W. 1981 Long-term effect of erosion and cropping on soil productivity. In Péwé, T.L. (ed.), *Desert dust: origins, characteristics and effects on man.* Geological Society of America Special Paper 186: 253–259.

Fujita, M.S. and Tuttle, M.D. 1991 Flying foxes (Chiroptera: Pteropodidae): threatened animals of key ecological and economic importance. *Conservation Biology* 5: 455–463.

Gammelsrød, T. 1992 Improving shrimp production by Zambezi River regulation. *Ambio* 21: 145–147.

Gardner, J.S. 1993 Mountain hazards. In French, H.M. and Slaymaker, O. (eds), *Canada's cold environments*. Montreal, McGill-Queen's University Press: 247–267.

Geertz, C. 1963 *Agricultural involution: the process of change in Indonesia*. Berkeley, CA, University of California Press.

GEMS (Global Environment Monitoring System) 1988 *Assessment of freshwater quality*. Nairobi, UNEP/WHO.

George, D.J. 1992 Rising groundwater: a problem of development in some urban areas of the Middle East. In McCall, G.J.H., Laming, D.J.C. and Scott, S.C. (eds), *Geohazards: natural and man-made hazards*. London, Chapman & Hall: 171–182.

GESAMP (Group of Experts on the Scientific Aspects of Marine Pollution) 1990 *The state of the marine environment*. Oxford, Blackwell Scientific.

Gibbons, J.R.H. and Clunie, F.G.A.U. 1986 Sea level changes and Pacific prehistory. *Journal of Pacific History* 21: 58–82.

Gilland, B. 1993 Cereals, nitrogen and population: an assessment of the global trends. *Endeavour* 17: 84–87.

Glantz, M.H. and Orlovsky, N. 1983 Desertification: a review of the concept. *Desertification Control Bulletin* 9: 15–22.

——, Rubinstein, A.Z. and Zonn, I. 1993 Tragedy in the Aral Sea basin: looking back to plan ahead. *Global Environmental Change* 3: 174–198.

Gleick, P.H. (ed.) 1993 *Water in crisis: a guide to the world's fresh water resources*. New York, Oxford University Press.

Goldblat, J. 1975 The prohibition of environmental warfare. *Ambio* 4: 186–190.

——1990 The mitigation of environmental disruption by war: legal approaches. In Westing,

A.H. (ed.), *Environmental hazards of war*. London, Sage: 48–60.

Goldsmith, E. and Hildyard, N. 1984 *The social and environmental effects of large dams* Volume 1. Wadebridge, Cornwall, Wadebridge Ecological Centre.

Gómez-Pompa, A., Flores, J.S. and Sosa, V. 1987 The 'pet kot': a man-made tropical forest of the Maya. *Interciencia* 12: 10–15.

Goodland, R. 1986 Hydro and the environment: evaluating the tradeoffs. *Water Power and Dam Construction*, November: 25–29.

——1990 The World Bank's new environmental policy for dams and reservoirs. *Water Resources Development* 6: 226–239.

Goodland, R.J.A. and Irwin, H.S. 1974 An ecological discussion of the environmental impact of the highway construction program in the Amazon Basin. *Landscape Planning* 1: 123–254.

——, Daly, H.E. and Serafy, S. El 1993a The urgent need for rapid transformation to global environmental sustainability. *Environmental Conservation* 20: 297–309.

——, Juras, A. and Pachauri, R. 1993b Can hydro-reservoirs in tropical moist forests be environmentally sustainable? *Environmental Conservation* 20: 122–130.

Goodrich, J.A., Lykins, B.W. and Clark, R.M. 1991 Drinking water from agriculturally contaminated groundwater. *Journal of Environmental Quality* 20: 707–717.

Gorham, E. 1958 The influence and importance of daily weather conditions in the supply of chloride, sulphate and other ions to freshwaters from atmospheric precipitation. *Philosophical Transactions of the Royal Society, London* 241B: 147–178.

Goriup, P.D. 1989 Acidic air pollution and birds in Europe. *Oryx* 23: 82–86.

Goudie, A.S. 1993a *The nature of the environment*, 3rd edn. Oxford, Blackwell.

——1993b *The human impact on the natural environment*, 4th edn. Oxford, Blackwell.

——1993c Human influence on geomorphology. *Geomorphology* 7: 37–59.

—— and Middleton, N.J. 1992 The changing frequency of dust storms through time. *Climatic Change* 20: 197–225.

Grainger, A. 1993a *Controlling tropical deforestation.* London, Earthscan.

——1993b Rates of deforestation in the humid tropics: estimates and measurements. *The Geographical Journal* 159: 33–44.

Granéli, E. and Haraldson, C. 1993 Can increased leaching of trace metals from acidified areas influence phytoplankton growth in coastal waters? *Ambio* 22: 308–311.

Graveland, J., van der Wal, R., van Balen, J.H. and van Noordwijk, A.J. 1994 Poor reproduction in forest passerines from decline of snail abundance on acidified soils. *Nature* 368: 446–448.

Graves, H.S. 1918 Effect of the war on forests of France. *American Forestry* 24: 707–717.

Gray, R.E. and Bruhn, R.W. 1984 Coal mine subsidence: eastern United States. *Geological Society of America, Reviews in Engineering Geology* 6: 123–149.

Grenon, M. and Batisse, M. (eds) 1989 *Futures for the Mediterranean basin: the Blue Plan.* New York, Oxford University Press.

Grigg, D.B. 1992 *The transformation of agriculture in the West.* Oxford, Blackwell.

——1993 The role of livestock products in world food consumption. *Scottish Geographical Magazine* 109: 66–74.

Grossman, L.S. 1992 Pesticides, caution, and experimentation in Saint Vincent, Eastern Caribbean. *Human Ecology* 20: 315–336.

Grove, R. 1990 The origins of environmentalism. *Nature* 345: 11–14.

Grubb, M.J. and Meyer, N.I. 1993 Wind energy: resources, systems and regional strategies. In Johansson, T.B., Kelly, H., Reddy, A.K.N. and Williams, R.H. (eds), *Renewable energy: sources for fuels and electricity.* New York, Island Press: 157–212.

Gunn, J.M. and Keller, W. 1990 Biological recovery of an acid lake after reductions in industrial emissions of sulphur. *Nature* 345: 431–433.

Hägerstrand, T. and Lohm, U. 1990 Sweden. In Turner, B.L., II, Clark, W.C., Kates, R.W., Richards, J.F., Mathews, J.T. and Meyer, W.B. (eds), *The earth as transformed by human action.* Cambridge, Cambridge University Press: 605–622.

Hall, D.O. and Rosillo-Calle, F. 1991 *Biomass in developing countries.* Report to the Office of Technology Assessment, Washington, DC.

——, ——, Williams, R.H. and Woods, J. 1993 Biomass for energy: supply prospects. In Johansson, T.B., Kelly, H., Reddy, A.K.N. and Williams, R.H. (eds), *Renewable energy: sources for fuels and electricity.* New York, Island Press: 595–651.

Hall, D.R. 1993 Albania. In Carter, F.W. and Turnock, D. (eds), *Environmental problems in Eastern Europe.* London, Routledge: 7–37.

Hameed, S. and Dignon, J. 1992 Global emissions of nitrogen and sulphur oxides from fossil fuel combustion 1970–1986. *Journal of the Air and Waste Management Association* 42: 159–163.

Hanan, N.P., Prevost, Y., Diouf, A. and Diallo, O. 1991 Assessment of desertification around deep wells in the Sahel using satellite imagery. *Journal of Applied Ecology* 28: 173–186.

Hardin, G. 1968 The tragedy of the commons. *Science* 162: 1243–1248.

Hardoy, J.E., Mitlin, D. and Satterthwaite, D. 1992 *Environmental problems in third world cities.* London, Earthscan.

Harper, P.P. 1992 La Grande Rivière: a subarctic river and hydroelectric megaproject. In Calow, P. and Petts, G.E. (eds), *The rivers handbook.* Vol. 1. Oxford, Blackwell Scientific: 411–425.

Hartig, J.H. and Vallentyne, J.R. 1989 Use of an ecosystem approach to restore degraded areas of the Great Lakes. *Ambio* 18: 423–428.

Hassan, F.A. 1980 Prehistoric settlement along the Main Nile. In Williams, M.A.J. and Faure, H. (eds), *The Sahara and the Nile.* Rotterdam, Balkema: 421–450.

Hawksworth, D.L. 1990 The long-term effects of air pollutants on lichen communities in Europe and North America. In Woodwell, G.M. (ed.), *The earth in transition: patterns and processes of biotic impoverishment.* Cambridge, Cambridge University Press: 45–64.

Hays, J.D., Imbrie, J. and Shackleton, N.J. 1976 Variations in the earth's orbit: pacemaker of the ice ages. *Science* 235: 1156–1167.

Heath, J., Pollard, E. and Thomas, J.A. 1984 *Atlas of butterflies in Britain and Ireland.* Harmondsworth, Viking.

Hellawell, J.M. 1988 River regulation and nature conservation. *Regulated Rivers: Research and Management* 2: 425–443.

Hellden, U. 1988 Desertification monitoring: is the desert encroaching? *Desertification Control Bulletin* 17: 8–12.

——1991 Desertification – time for an assessment? *Ambio* 20: 372–383.

Henderson-Sellers, A. 1991 Global climatic change: the difficulties of assessing impacts. *Australian Geographical Studies* 29: 202–225.

——1994 Numerical modelling of global climates. In Roberts, N. (ed.), *The changing global environment.* Oxford, Blackwell: 99–124.

Hengeveld, R. 1989 *Dynamics of biological invasions.* London, Chapman & Hall.

Hilz, C. and Radka, M. 1991 Environmental negotiation and policy: the Basel Convention on transboundary movement of hazardous wastes and their disposal. *International Journal of Environment and Pollution* 1: 55–72.

Hinrichsen, D. and Láng, I. 1993 Hungary. In Carter, F.W. and Turnock, D. (eds), *Environmental problems in Eastern Europe.* London, Routledge: 89–106.

Hira, P.R. 1969 Transmission of schistosomiasis in Lake Kariba, Zambia. *Nature* 2124: 670–672.

Hobbelink, H. 1991 *Biotechnology and the future of world agriculture.* London, Zed Books.

Holdgate, M.W. 1991 Conservation in a world context. In Spellerberg, I.F., Goldsmith, F.B. and Morris, M.G. (eds), *The scientific management of temperate communities for conservation.* Oxford, Blackwell Scientific: 1–26.

Holmberg, J. 1991 *Poverty, environment and development: proposals for action.* Stockholm, Swedish International Development Agency.

Homer-Dixon, T.F., Boutwell, J.H. and Rathjens, G.W. 1993 Environmental change and violent conflict. *Scientific American* 268(2): 16–23.

Houghton, J.T., Jenkins, G.J. and Ephraums, J.J. (eds) 1990 *Climate change: the IPCC scientific assessment.* Cambridge, Cambridge University Press.

Howells, G. 1990 *Acid rain and acid waters.* London, Ellis Horwood.

Hudson, N.W. 1991 A study of the reasons for success or failure of soil conservation projects. *FAO Soils Bulletin* 64.

Hughes, T.P. 1994 Catastrophes, phase shifts, and large-scale degradation of a Caribbean coral reef. *Science* 265: 1547–1551.

Humphreys, D. 1993 *The phantom of full cost pricing.* London, Rio Tinto Zinc.

Hurni, H. 1993 Land degradation, famine, and land resource scenarios in Ethiopia. In Pimental D. (ed.), *World soil erosion and conservation.* Cambridge, Cambridge University Press: 27–61.

Hurst, P. 1992 Pesticide reduction programs in Denmark, the Netherlands, and Sweden. *International Environmental Affairs* 4: 234–253.

IAEA (International Atomic Energy Authority) 1991 *The International Chernobyl Project: summary brochure, assessment of radiological consequences and evaluation of protective measures.* Report by International Advisory Committee, IAEA, Vienna.

Innes, J.L. 1992 Forest decline. *Progress in Physical Geography* 16: 1–64.

Intergovernmental Panel on Climate Change (IPCC) 1992 *Climate change 1992: the supplementary report to the IPCC scientific assessment.* Report by working group I. Cambridge, Cambridge University Press.

Irish Peatland Conservation Council (IPCC) 1992 *Policy statement and action plan 1992–1997.* Dublin, IPCC.

Iraq, Government of, 1980 Desertification in the Greater Mussayeb Project, Iraq. In UNESCO/UNEP/UNDP *Case studies on desertification.* Natural Resources Research Series XVIII. Paris, UNESCO: 176–213.

IUCN (International Union for Conservation of Nature and Natural Resources) 1988 *Coral reefs of the world Volume 3: Central and Western Pacific.* Cambridge, IUCN.

——1992 *Angola: environmental status quo assessment.* Harare, IUCN Regional Office Southern Africa.

——1993 Fears for the tiger. *IUCN Bulletin* 2: 5.

——/UNEP 1986a *Review of the protected areas system in the Indo-Malayan realm.* Gland, IUCN.

——/——1986b *Review of the protected areas system in the Afrotropical realm.* Gland, IUCN.

——/——/WWF 1980 *World conservation strategy: living resource conservation for sustainable development.* Gland, IUCN.

——/——/——1991 *Caring for the earth: a strategy for sustainable living.* Gland, IUCN.

Ives, J.D. and Messerli, B. 1989 *The Himalayan dilemma: reconciling development and conservation.* London, Routledge.

Iwama, G.K. 1991 Interactions between aquaculture and the environment. *Critical Reviews in Environmental Control* 21: 177–216.

Jagtap, T.G., Chavan, V.S. and Untaawale, A.G. 1993 Mangrove ecosystems of India: a need for protection. *Ambio* 22: 252–254.

Jasani, B. 1975 Environmental modification - new weapons of war? *Ambio* 4: 191–198.

Jenkins, R. 1987 *Transnationals and uneven development.* London, Croom Helm.

Jeyaratnam, J. 1990 Acute pesticide poisoning: a major global health problem. *World Health Statistics Quarterly* 43: 139–144.

Jickells, T.D., Carpenter, R. and Liss, P.S. 1990 Marine environment. In Turner, B.L., II, Clark, W.C., Kates, R.W., Richards, J.F., Mathews, J.T. and Meyer, W.B. (eds), *The earth as transformed by human action.* Cambridge, Cambridge University Press: 313–334.

Jimenez, R.D. and Velasquez, A. 1989 Metropolitan Manila: a framework for its sustained development. *Environment and Urbanization* 1: 51–58.

Johansson, T.B., Kelly, H., Reddy, A.K.N. and Williams, R.H. 1993 Renewable fuels and electricity for a growing world economy: defining and achieving the potential. In Johansson, T.B., Kelly, H., Reddy, A.K.N. and Williams, R.H. (eds), *Renewable energy: sources for fuels and electricity.* New York, Island Press: 1–71.

Johns, A.D. 1992 Species conservation in managed tropical forests. In Whitmore, T.C. and Sayer, J.A. (eds), *Tropical deforestation and species extinction.* London, Chapman & Hall: 15–53.

Johnston, G.H. 1981 *Permafrost engineering, design and construction.* Toronto, Wiley.

Jones, D.K.C., Cooke, R.U. and Warren, A. 1986 Geomorphological investigation, for engineering purposes, of blowing sand and dust hazard. *Quarterly Journal of Engineering Geology* 19: 251–270.

Jutila, E. 1992 Restoration of salmonid rivers in Finland. In Boon, P.J., Calow, P. and Petts, G.E. (eds), *River conservation and management.* Chichester, Wiley: 353–362.

Kajak, Z. 1992 The River Vistula and its floodplain valley (Poland): its ecology and importance for conservation. In Boon, P.J., Calow, P. and Petts, G.E. (eds), *River conservation and management.* Chichester, Wiley: 35–49.

Kallend, A.S., Marsh, A.R.W., Pickles, J.H. and Proctor, M.V. 1983 Acidity of rain in Europe. *Atmospheric Environment* 17: 127–137.

Kates, R.W., Turner, B.L., II and Clark, W.C. 1990 The great transformation. In Turner, B.L., II, Clark, W.C., Kates, R.W., Richards, J.F., Mathews, J.T. and Meyer, W.B. (eds), *The earth as transformed by human action.* Cambridge, Cambridge University Press: 1–17.

Kerpelman, C. 1990 *Preliminary study on the identification of disaster-prone countries based on economic impact.* Geneva, Office of the United Nations Disaster Relief Organisation.

Khalaf, F.I. 1989 Desertification and aeolian processes in Kuwait. *Journal of Arid Environments* 12: 125–145.

Khalil, G.M. 1992 Cyclones and storm surges in Bangladesh: some mitigative measures. *Natural Hazards* 6: 11–24.

Khordagui, H.K. 1992 A conceptual approach to selection of a control measure for residual chlorine discharge in Kuwait Bay. *Environmental Management* 16: 309–316.

Kiersch, G.A. 1965 The Vaiont reservoir disaster. *Mineral Information Service* 18: 129–138.

Kim, S.S. 1984 *The quest for a just world*. Boulder, CO, Westview Press.

Kinlen, L.J. 1988 Evidence for an infective cause of childhood leukaemia: comparison of a Scottish new town with nuclear reprocessing sites in Britain. *The Lancet* 10 December: 1323–1327.

Kiss, A. 1985 The protection of the Rhine against pollution. *Natural Resources Journal* 25: 613–632.

Kivinen, E. and Pakarinen, P. 1981 Geographical distribution of peat resources and major peatland complex types in the world. *Annals, Academy Sciencia Fennicae* A132: 1–28.

Klein, B.C. 1989 Effects of forest fragmentation on dung and carrion beetle communities in central Amazonia. *Ecology* 70: 1715–1725.

Klein, D.R. 1971 Reaction of reindeer to obstructions and disturbances. *Science* 173: 393–398.

Knox, P. and Agnew, J. 1994 *The geography of the world economy*, 2nd edition. London, Edward Arnold.

Kobayashi, Y. 1981 Causes of fatalities in recent earthquakes in Japan. *Journal of Disaster Science* 3: 15–22.

Koike, K. 1985 Japan. In Bird, E.C.F. and Schwartz, M.L. (eds), *The world's coastline*. Stroudsburg, PA, Van Nostrand Reinhold: 843–855.

Kreimer, A., Lobo, T., Menezes, B., Munasinghe, M. and Parker, R. 1993 Towards a sustainable urban environment: the Rio de Janeiro study. *World Bank Discussion Paper* 195.

Krimgold, F. 1992 Modern urban infrastructure: the Armenian case. In Kreimer, A. and Munasinghe, M. (eds), 1992 Environmental management and urban vulnerability. *World Bank Discussion Paper* 168: 263–265.

Kummer, D.M. 1991 *Deforestation in the postwar Philippines*. Chicago, University of Chicago Press.

Lal, R. 1993 Soil erosion and conservation in West Africa. In Pimental D. (ed.), *World soil erosion and conservation*. Cambridge, Cambridge University Press: 7–25.

Lambert, J.H., Jennings, J.N., Smith, C.T., Green, C. and Hutchinson, J.N. 1970 *The making of the Broads: a reconsideration of their origin in the light of new evidence*. Royal Geographical Society Research Series 3. London, Royal Geographical Society.

Lamprey, H.F. 1975 *Report on the desert encroachment reconnaisance in northern Sudan, October 21–November 10, 1975*. Khartoum, National Council for Research, Ministry of Agriculture, Food and Resources.

Landsberg, H.E. 1981 *The urban climate*. New York, Academic Press.

Langford, T.E.L. 1990 *Ecological effects of thermal discharges*. London, Elsevier Applied Science.

Larson, W.E., Pierce, F.J. and Dowdy, R.H. 1983 The threat of soil erosion to long-term crop production. *Science* 219: 458–465.

Lee, K.N. 1989 The Columbia River basin: experimenting with sustainability. *Environment* 31: 6–11, 30–33.

Leprun, J.-C., da Silveira, C.O. and Sobral Filho, R.M. 1986 Efficacité des practiques culturales antiérosives testées sous différents climats brésiliens. *Cahiers ORSTOM Série Pédologie* 22: 223–233.

Lerner, D.N. and Tellam, J.H. 1993 The protection of urban groundwater from pollution. In Currie, J.C. and Pepper, A.T. (eds), *Water and the environment*. Chichester, Ellis Horwood: 322–335.

Levins, R., Awerbuch, T., Brinkman, U., Eckardt, I., Epstein, P., Makhoul, N., Albuquerque de Possas, C., Puccia, C., Spielman, A. and Wilson, M.E. 1994 The emergence of new diseases. *American Scientist* 82: 52–60.

Lewis, L.A. and Berry, L. 1988 *African environments and resources*. London, Allen and Unwin.

Lisk, D.J. 1991 Environmental effects of landfills. *Science of the Total Environment* 100: 415–468.

Lloyd, G.O. and Butlin, R.N. 1992 Corrosion. In Radojevic, M. and Harrison, R.M. (eds), *Atmospheric acidity: sources, consequences and abatement.* London, Elsevier Applied Science: 405–434.

Lockeretz, W. 1978 The lessons of the dust bowl. *American Scientist* 66: 560–569.

Lonergan, S.C. 1993 Impoverishment, population, and environmental degradation: the case for equity. *Environmental Conservation* 20: 328–334.

Lorius, C., Jouzel, J., Raynaud, D., Hansen, J. and Le Treut, H. 1990 The ice-core record: climate sensitivity and future greenhouse warming. *Nature* 347: 139–145.

Lottermoser, B.G. and Morteani, G. 1993 Sewage sludge: toxic substances, fertilizers, or secondary metal resources? *Episodes* 16: 329–333.

Lovelock, J.E. 1989 *The ages of Gaia.* Oxford, Oxford University Press.

Lowe, M.D. 1991 *Shaping cities: the environmental and human dimensions.* Worldwatch Paper 105. Worldwatch Institute, Washington, DC.

——1994 Reinventing transport. In Brown, L.R. (ed.), *State of the world 1994.* New York, W.W. Norton: 81–98.

Lugo, A.E. 1988 Estimating reductions in the diversity of tropical forest species. In Wilson, E.O. and Peter, F.M. (eds), *Biodiversity.* Washington DC, National Academy Press: 58–70.

MacArthur, R.H. and Wilson, E.O. 1967 *The theory of island biogeography.* Princeton, Princeton University Press.

McCabe, J.T. 1990 Turkana pastoralism: a case against the tragedy of the commons. *Human Ecology* 18: 81–103.

McCauley, J.F., Breed, C.S., Grolier, M.J. and Mackinnon, D.J. 1981 The US dust storm of February 1977. In Péwé, T.L. (ed.), *Desert dust: origins, characteristics and effects on man.* Geological Society of America Special Paper 186: 123–147.

McFarland, M. 1989 Chlorofluorocarbons and ozone. *Environmental Science and Technology* 23: 1203–1208.

McNeely, J. 1988 *Economics and biological diversity.* Gland, IUCN.

——1994 Lessons from the past: forests and biodiversity. *Biodiversity and Conservation* 3: 3–20.

Maddox, J. 1990 Clouds and global warming. *Nature* 247: 329.

Magrath, W. and Arens, P. 1989 The costs of soil erosion in Java: a natural resource accounting approach. *World Bank Environment Department Working Paper* 18.

Maitland, P.S. 1991 Conservation of fish species. In Spellerberg, I.F., Goldsmith, F.B. and Morris, M.G. (eds), *The scientific management of temperate communities for conservation.* Oxford, Blackwell Scientific: 129–148.

Maki, A.W. 1991 The Exxon oil spill: initial environmental impact assessment. *Environmental Science and Technology* 25: 24–29.

Malm, O., Pfeiffer, W.C., Souza, C.M.M. and Reuther, R. 1990 Mercury pollution due to gold mining in the Madeira River basin, Brazil. *Ambio* 19: 11–15.

Malthus, T. R. 1798 *An essay on the principle of population.* London, Johnson.

Mann, R.H.K. 1988 Fish and fisheries of regulated rivers in the UK. *Regulated Rivers: Research and Management* 2: 411–424.

Manner, H.I., Thaman, R.R. and Hassall, D.C. 1984 Phosphate mining induced changes on Nauru Island. *Ecology* 65: 1454–1465.

Maragos, J.E. 1993 Impact of coastal construction on coral reefs in the US-affiliated Pacific Islands. *Coastal Management* 21: 235–269.

Marples, D.R. 1992 Post-Soviet Belarus and the impact of Chernobyl. *Post-Soviet Geography* 33: 419–431.

Marsh, G.P. 1874 *The earth as modified by human actions.* New York, Sampson Low.

Martin, P.S. and Klein, R.G. 1984 *Pleistocene extinctions.* Tucson, AZ, University of Arizona Press.

Mather, A.S. 1990 *Global forest resources.* London, Belhaven.

May, R.M. 1978 Human reproduction reconsidered. *Nature* 272: 491–495.

Meade, R.B. 1991 Reservoirs and earthquakes. *Engineering Geology* 30: 245–262.

Meadows, D.H., Meadows, D.L., Randers, J. and Behrens, W.W., III 1972 *The limits to growth: a report to the Club of Rome's project on the predicament of mankind*. New York, Potomac Associates.

Meadows, P.S. and Campbell, J.I. 1988 *An introduction to marine science*. London, Blackie & Son.

Mee, L.D. 1992 The Black Sea crisis: a need for concerted international action. *Ambio* 21: 278–286.

Meierding, T.C. 1993 Marble tombstone weathering and air pollution in North America. *Annals of the Association of American Geographers* 83: 568–588.

Mellquist, P. 1992 River management – objectives and applications. In Boon, P.J., Calow, P. and Petts, G.E. (eds), *River conservation and management*. Chichester, Wiley: 1–8.

Meybeck, M. 1982 Carbon, nitrogen and phosphorus transport by world rivers. *Science (New York)* 282: 401–450.

——, Chapman, D. and Helmer, R. 1989 *Global freshwater quality: a first assessment*. Oxford, Blackwell.

Meynell, P.-J. 1993 Developing collaboration to protect Saudi Arabia's wetlands. *IUCN Wetlands Programme Newsletter* 7: 11–12.

Middleton, N.J. 1985 Effect of drought on dust production in the Sahel. *Nature* 316: 431–434.

——1991 *Desertification*. Oxford, Oxford University Press.

Milliman, J.D. 1990 Fluvial sedimentation in coastal seas: flux and fate. *Nature and Resources* 26(4): 12–22.

Mittermeir, R., Konstant, B., Nicoll, M. and Langrand, O. 1992 *Lemurs of Madagascar: an action plan for their conservation 1993–1999*. Gland, IUCN.

Molina, M.J. and Rowland, F.S. 1974 Stratospheric sink chlorofluoromethanes: chlorine atom catalyzed destruction of ozone. *Nature* 249: 810–814.

Monteiro, C.A. de F. 1989 Environmental quality in the great national metropolis and its industrial portuary appendix. In Gerasimov, I.P. (ed.), *Environmental problems in cities of developing countries*. Moscow, UNEP: 26–39.

Moore, D. and Driver, A. 1989 The conservation role of water supply reservoirs. *Regulated Rivers: Research and Management* 4: 203–212.

Moore, N.W., Hooper, M.D. and Davis, B.N.K. 1967 Hedges, I. Introduction and reconnaissance studies. *Journal of Applied Ecology* 4: 201–220.

Moran, E.F. 1981 *Developing the Amazon*. Bloomington, Indiana University Press.

Moreira, J.R. and Poole, A.D. 1993 Hydropower and its constraints. In Johansson, T.B., Kelly, H., Reddy, A.K.N. and Williams, R.H. (eds), *Renewable energy: sources for fuels and electricity*. New York, Island Press: 73–119.

Mortimore, M. 1987 Shifting sands and human sorrow: social response to drought and desertification. *Desertification Control Bulletin* 14: 1–14.

Murray, I. 1994 Time and tide rip into the frontier of old England. *The Times*, 23 March: 7.

Murray, J.E., Johnson, M.S. and Clarke, B. 1988 The extinction of *Partula* on Moorea. *Pacific Science* 42: 150–153.

Musannif, B. 1992 Integrated approach to energy efficient housing refurbishment. *Energy Management* July/August: 22–23.

Myers, N. 1979 *The sinking ark: a new look at the problem of disappearing species*. Oxford, Pergamon.

Mylona, S. 1993 *Trends of sulphur dioxide emissions, air concentrations and depositions of sulphur in Europe since 1880*. EMEP/MSC-W Report 2/93. Oslo, EMEP.

Nelson, P.M. (ed.) 1987 *Transportation noise reference handbook*. London, Butterworth.

Neumann, A.C. and Macintyre, I. 1985 Reef response to sea level rise: keep-up, catch-up or give-up. *Proceedings of the 5th International Coral Reef Congress, Tahiti* 3: 105–110.

Newcombe, K. 1984 An economic justification for rural afforestation: the case of Ethiopia. *World Bank Energy Department Paper* 16.

Nicol, S. and de la Mare, W. 1993 Ecosystem management and the Antarctic krill. *American Scientist* 81: 36–47.

Nilsson, K. 1991 Emission standards for waste incineration. *Waste Management and Research* 9: 224–227.

Nixon, S.W. 1993 Nutrients and coastal waters: too much of a good thing? *Oceanus* 36(2): 38–47.

Noble, A.G. 1980 Noise pollution in selected Chinese and American cities. *GeoJournal* 4: 573–575.

Nortcliff, S. and Gregory, P.J. 1992 Factors affecting losses of soil and agricultural land in tropical countries. In McCall, G.J.H., Laming, D.J.C. and Scott, S.C. (eds), *Geohazards: natural and man-made hazards*. London, Chapman & Hall: 183–190.

Norton, D.A. 1991 *Trilepidea adamsii*: an obituary for a species. *Conservation Biology* 5: 52–57.

Nriagu, J. 1981 *Cadmium in the environment: health effects*. New York, Wiley.

Nunn, P.D. 1990 Recent environmental changes on Pacific islands. *The Geographical Journal* 156: 125–140.

——1994 *Oceanic islands*. Oxford, Blackwell.

Obeng, L. 1978 Environmental impacts of four African impoundments. In Gunnerson, C.G. and Kalbermatten, J.M. (eds), *Environmental impacts of international civil engineering projects and practices*. New York, American Society of Civil Engineers.

Odén, S. 1968 *The acidification of air precipitation and its consequences in the natural environment*. Energy Committee Bulletin 1. Stockholm, Swedish Natural Sciences Research Council.

Odum, E.P. 1979 The value of wetlands: a hierarchial approach. In Greeson, P.E., Clark, J.R. and Clark, J.E. (eds), *Wetland functions and values: the state of our understanding*. Minneapolis, MN, American Water Resources Association Technical Publication: 16–25.

OECD (Organisation for Economic Co-operation and Development) 1991a *The state of the environment*. Paris, OECD.

——1991b *Environmental indicators: a preliminary set*. Paris, OECD.

——1993 *OECD environmental data compendium*. Paris, OECD.

O'Hara, S.L., Street-Perrot, F.A. and Burt, T.P. 1993 Accelerated soil erosion around a Mexican highland lake caused by prehispanic agriculture. *Nature* 362: 48–51.

Ohkita, T. 1984 Health effects on individuals and health services of the Hiroshima and Nagasaki bombs. In WHO, *Effects of nuclear war on health and health service*. Geneva, WHO.

Oldeman, L.R., Hakkeling, R.T.A. and Sombroek, W.G. 1990 *World map of the status of human-induced soil degradation. An explanatory note*. Wageningen, ISRIC/UNEP.

Oliver, F.W. 1945 Dust storms in Egypt and their relation to the war period, as noted in Maryut, 1939–45. *The Geographical Journal* 106: 26–49.

Olson, S.L. and James, H.F. 1984 The role of Polynesians in the extinction of the avifauna of the Hawaiian Islands. In Martin, P.S and Klein, R.G. (eds), *Quaternary extinctions: a prehistoric revolution*. Tucson, AZ, University of Arizona Press: 768–780

Olsson, L. 1993 On the causes of famine – drought, desertification and market failure in the Sudan. *Ambio* 22: 395–403.

Orians, G.H. and Pfeiffer, E.W. 1970 Ecological effects of the war in Vietnam. *Science* 168: 544–54.

Otzen, U. 1993 Reflections on the principles of sustainable agricultural development. *Environmental Conservation* 20: 310–316.

Pakistan 1990 *Towards sustainable development: the Pakistan National Conservation Strategy*. Islamabad, National Conservation Strategy Secretariat.

Palmerini, C.G. 1993 Geothermal energy. In Johansson, T.B., Kelly, H., Reddy, A.K.N. and Williams, R.H. (eds), *Renewable energy: sources for fuels and electricity*. New York, Island Press: 549–591.

Pape, R. 1993 Air pollution in the Taiga. *Acid News* 2: 13–14.

Paskett, C.J. and Philoctete, C.-E. 1990 Soil conservation in Haiti. *Journal of Soil and Water Conservation* 45: 457–459.

Patrick, S.T., Timberlid, J.A. and Stevenson, A.C. 1990 The significance of land-use and land management change in the acidification of lakes in Scotland and Norway: an assessment utilizing documentary sources and pollen analysis. *Philosophical Transactions of the Royal Society, London* 327(1240): 363–367.

Pearce, D. 1993 *Economic values and the natural world*. London, Earthscan.

—— and Mäler, K. 1991 Environmental economics and the developing world. *Ambio* 20: 52–54.

—— and Turner, R.K. 1992 Packaging waste and the polluter pays principle: a taxation solution. *Journal of Environmental Planning and Management* 35: 5–15.

——, Barde, J.-P. and Lambert, J. 1984 Estimating the cost of noise pollution in France. *Ambio* 13: 27–28.

Pennell, C.R. 1994 The geography of piracy: northern Morocco in the mid-nineteenth century. *Journal of Historical Geography* 20: 272–282.

Perkins, J.S. and Thomas, D.S.G. 1993 Environmental responses and sensitivity to permanent cattle ranching, semi-arid western central Botswana. In Thomas, D.S.G. and Allison, R.J. (eds), *Landscape sensitivity*. Chichester, Wiley: 273–286.

Pernetta, J.C. 1989 Projected climate change and sea-level rise: a relative impact rating for the countries of the Pacific Basin. In Pernetta, J.C. and Hughes, P.J. (eds), *Implications of expected climate changes in the South Pacific: an overview*. UNEP Regional Seas Reports and Studies 128. Nairobi, UNEP: 14–24.

——1992 Impacts of climate change and sea-level rise on small island states: national and international responses. *Global Environmental Change* 2: 19–31.

—— and Sestini, G. 1989 *The Maldives and the impacts of expected climate changes*. UNEP Regional Seas Reports and Studies 104. Nairobi, UNEP.

Peters, C.M., Gentry, A.H. and Mendelsohn, R.O. 1989 Valuation of an Amazonian rain forest. *Nature* 339: 655–656.

Peters, R.L. and Lovejoy, T.E. 1990 Terrestrial fauna. In Turner, B.L., II, Clark, W.C., Kates, R.W., Richards, J.F., Mathews, J.T. and Meyer, W.B. (eds), *The earth as transformed by human*

action. Cambridge, Cambridge University Press: 353–369.

Petts, G.E. 1984 *Impounded rivers – perspectives for ecological management*. Chichester, Wiley.

——1988 Regulated rivers in the United Kingdom. *Regulated Rivers: Research and Management* 2: 201–220.

Phantumvanit, D. and Liengcharernsit, W. 1989 Coming to terms with Bangkok's environmental problems. *Environment and Urbanization* 1: 31–39.

Pimental, D. 1991 Diversification of biological control strategies in agriculture. *Crop Protection* 10: 243–253.

—— and Levitan, L. 1988 Pesticides: where do they go? *The Journal of Pesticide Reform* 7(4): 2–5.

Plucknett, D.L. and Smith, N.J.H. 1986 Sustaining agricultural yields. *BioScience* 36: 40–45.

Poore, D. and Sayer, J. 1987 *The management of tropical moist forest lands: ecological guidelines*. Gland, IUCN.

Potter, G.L., Ellsaesser, H.W., MacCracken, M.C. and Luther, F.M. 1975 Possible climatic impact of tropical deforestation. *Nature* 258: 697–698.

Prance, G.T. 1990 Flora. In Turner, B.L., II, Clark, W.C., Kates, R.W., Richards, J.F., Mathews, J.T. and Meyer, W.B. (eds), *The earth as transformed by human action*. Cambridge, Cambridge University Press: 387–391.

Price, L.W. 1972 *The periglacial environment, permafrost, and man*. Commission on Geographical Resources Paper 14. Washington, DC, Association of American Geographers.

Pryde, P.R. 1972 *Conservation in the Soviet Union*. Cambridge, Cambridge University Press.

——1991 *Environmental management in the Soviet Union*. Cambridge, Cambridge University Press.

Punning, J.-M. 1993 Environmental problems in Estonia. In Punning, J.-M. and Hult, J. (eds), *Human impact on environment*. Tallinn, Estonian Academy of Sciences: 7–23.

Rabinovitch, J. 1992 Curitiba: towards sustainable urban development. *Environment and Urbanization* 4: 62–73.

Rabinowitz, A. 1993 Estimating the indochinese tiger *Panthera tigris corbetti* population in Thailand. *Biological Conservation* 65: 213–217.

Rackham, O. 1986 *The history of the countryside*. London, Dent.

Rapin, F., Blanc, P. and Corvi, C. 1989 Influence des apports sur le stock de phosphore dans le lac Léman et sur son eutrophisation. *Revue des Sciences de L'eau* 2: 721–737.

Rasid, H. and Pramanik, M.A.H. 1990 Visual interpretation of satellite imagery for monitoring floods in Bangladesh. *Environmental Management* 14: 815–821.

Readman, J.W., Fowler, S.W., Villeneuve, J.-P., Cattini, C., Oregioni, B. and Mee, L.D. 1992 Oil and combustion product contamination of the Gulf marine environment following the war. *Nature* 358: 662–665.

Reid, W.V. 1992 How many species will there be? In Whitmore, T.C. and Sayer, J.A. (eds), *Tropical deforestation and species extinction*. London, Chapman & Hall: 55–73.

Reinking, R.F, Mathews, L.A. and St-Amand, P. 1975 Dust storms due to desiccation of Owens Lake. *International Conference on Environmental Sensing and Assessment*. Las Vegas, IEEE.

Renberg, I., Korsman, T. and Birks, H.J.B. 1993 Prehistoric increases in the pH of acid-sensitive Swedish lakes caused by land-use change. *Nature* 362: 824–827.

Renner, M. 1991 Assessing the military's war on the environment. In *State of the world 1991*. New York, W.W. Norton.

Repetto, R. and Gillis, M. (eds) 1989 *Public policies and the misuse of forest resources*. New York, Cambridge University Press.

Richards, J.F. 1990 Agricultural impacts in tropical wetlands: rice paddies for mangroves in south and southeast Asia. In Williams, M. (ed.), *Wetlands: a threatened landscape*. Institute of British Geographers Special Publication 25. Oxford, Blackwell: 217–233.

Richardson, M.L. 1993 The assessment of hazards and risks to the environment caused by war damage to industrial installations in Croatia.

Paper presented at the International Conference on the Effects of War on the Environment, Zagreb, 15–17 April.

Riebsame, W.E. 1990 The United States Great Plains. In Turner, B.L., II, Clark, W.C., Kates, R.W., Richards, J.F., Mathews, J.T. and Meyer, W.B. (eds), *The earth as transformed by human action*. Cambridge, Cambridge University Press: 561–575.

Robb, G.A. 1994 Environmental consequences of coal mine closure. *The Geographical Journal* 160: 33–40.

Roberts, L. 1988 Conservationists in Panda-monium. *Science* 241: 529–531.

Rosenberry, P., Knutson, R. and Harmon, L. 1980 Predicting effects of soil depletion from erosion. *Journal of Soil and Water Conservation* 35: 123–134.

Ruckelshaus, W.D. 1989 Toward a sustainable world. *Scientific American* 261(3): 114–120C.

Ruel, S. 1993 The scourge of land mines. *UN Focus*, October, 4pp.

Rukang, F. 1989 Environment and cancer in Shanghai. *Journal of Environmental Science (China)* 1: 1–9.

Runnels, D.D., Shepherd, T.A. and Angino, E.E. 1992 Metals in water. *Environmental Science and Technology* 26: 2316–2323.

Ryding, S.-O. and Rast, W. (eds) 1989 *The control of eutrophication of lakes and reservoirs*. UNESCO Man and the Biosphere Series, Vol. 1. Paris, Parthenon.

Salvat, B. 1992 Coral reefs – a challenging ecosystem for human societies. *Global Environmental Change* 2: 12–18.

Sandford, S. 1977 *Dealing with drought and livestock in Botswana*. London, Overseas Development Institute.

Sapozhnikova, S.A. 1973 *Map diagram of the number of days with dust storms in the hot zone of the USSR and adjacent territories*. Report HT-23-0027. Charlottesville, VA, US Army Foreign and Technology Center.

Sather, J.M. and Smith, R.D. 1984 *An overview of major wetland functions and values*. Washington, DC, Fish and Wildlife Service, FWS/OBS-84/18.

Savchenko, V.K. 1991 The Chernobyl catastrophe and the biosphere. *Nature and Resources* 27(1): 37–46.

Savidge, J.A. 1987 Extinction of an island forest avifauna by an introduced snake. *Ecology* 68: 660–668.

Sayer, J.A., Harcourt, C.S. and Collins, N.M. (eds) 1992 *The conservation atlas of tropical forests: Africa*. London, Macmillan/IUCN.

Schlesinger, W.H. 1991 *Biogeochemistry: an analysis of global change*. San Diego, Academic Press.

Schneider, S.H. 1989a *Global warming: are we entering the greenhouse century?* San Francisco, Sierra Club Books.

——1989b The changing climate. *Scientific American* 261(3): 38–47.

Schteingart, M. 1989 The environmental problems associated with urban development in Mexico City. *Environment and Urbanization* 1: 40–50.

SCOPE (Scientific Committee on Problems of the Environment) 1989 *Environmental consequences of nuclear war. Vol. II: ecological and agricultural effects*, 2nd edn. SCOPE Report 28. Chichester, Wiley.

Sen, A. 1981 *Famines and poverty*. Oxford, Oxford University Press.

Sevaldrud, I.H., Muniz, I.P. and Kalvenes, S. 1980 Loss of fish populations in southern Norway: dynamics and magnitude of the problem. In Drablos, D. and Tollan, A. (eds), *Ecological impact of acid precipitation*. Norway, SNSF-Project: 350–351.

Shahgedanova, M. and Burt, T.P. 1994 New data on air pollution in the former Soviet Union. *Global Environmental Change* 4: 201–227.

Shaw, D.G. 1992 The Exxon Valdez oil spill: ecological and social consequences. *Environmental Conservation* 19: 253–258.

Sheail, J. 1988 River regulation in the United Kingdom: an historical perspective. *Regulated Rivers: Research and Management* 2: 221–232.

Sheeline, L. 1993 Pacific fruit bats in trade: are CITES controls working? *Traffic USA* 12(1): 1–4.

Shields, L.M. and Wells, P.V. 1962 Effects of nuclear testing on desert vegetation. *Science* 135: 38–40.

Shiklomanov, I.A. 1993 World fresh water resources. In Gleik, P.H. (ed.), *Water in crisis: a guide to the world's fresh water resources*. New York, Oxford University Press: 13–24.

Sibley, C.G. and Monroe, B.L. 1990 *Distribution and taxonomy of birds of the world*. Yale, Yale University Press.

sigma 1994 *Natural catastrophes and major losses in 1993: insured damage drops significantly*. Zurich, Swiss Reinsurance Company.

Sipilä, K. 1993 New power-production technologies: various options for biomass cogeneration. *Bioresource Technology* 46: 5–12.

SIPRI 1990 *World armaments and disarmament*, SIPRI Yearbook 1990. Oxford, Oxford University Press.

Skaf, R. 1988 A story of a disaster: why locust plagues are still possible. *Disasters* 12: 122–127.

Slaymaker, O. and French, H.M. 1993 Cold environments and global change. In French, H.M. and Slaymaker, O. (eds), *Canada's cold environments*. Montreal, McGill-Queen's University Press: 313–334.

Smith, K. 1992 *Environmental hazards: assessing risk and reducing disaster*. London, Routledge.

Smith, N.J.H., Alvim, P., Homma, A., Falesi, I and Serráo, A. 1991 Environmental impacts of resource exploitation in Amazonia. *Global Environmental Change* 1: 313–320.

Smith, R.A. 1852 On the air and rain of Manchester. *Memoirs and Proceedings of the Manchester Literary and Philosophical Society* 2: 207–217.

Soboleva, O.V. and Mamadaliev, U.A. 1976 The influence of Nurek Reservoir on local earthquake activity. *Engineering Geology* 10: 293–305.

Sorensen, J.C. and McCreary, S.T. 1990 *Coasts: institutional arrangements for managing coastal resources and environments*. Washington, DC, US Department of the Interior and US Agency for International Development.

Soutar, A. 1967 The accumulation of fish debris in certain California coastal sediments. *Californian Cooperative Ocean Fisheries Investment Report* 11: 136–139.

Spencer, J.W. and Kirby, K.J. 1992 An inventory of ancient woodland for England and Wales. *Biological Conservation* 62: 77–93.

Stanley, D.J. and Warne, A.G. 1993 Nile delta: recent geological evolution and human impact. *Science* 260: 628–634.

Stock, R.F. 1976 Cholera in Africa: diffusion of the disease 1970–75, with special reference to West Africa. *African Environment Special Report* 3, London, International African Institute.

Stocking, M.A. 1987 Measuring land degradation. In Blaikie, P.M. and Brookfield, H.C. *Land degradation and society*. London, Routledge: 49–63.

Stolzenburg, W. 1992 The mussels' message. *Nature Conservancy* 42 (Nov./Dec.): 16–23.

Svidén, O. 1993 Clean fuel and engine systems for twenty-first-century road vehicles. In Giannopoulos, G. and Gillespie, A. (eds), *Transport and communications innovation in Europe*. London, Belhaven: 122–148.

Swinnerton, C.J. 1984 Protection of groundwater in relation to waste disposal in Wessex Water Authority. *Quarterly Journal of Engineering Geology* 17: 3–8.

Terborgh, J. 1992 Why American songbirds are vanishing. *Scientific American* 266(5): 56–62.

Teufel, D. 1989 Die zukunft des autoverkehrs. Heidelberg, Bericht 17 Umwelt- und Prognose Institut (unpublished).

Thanh, N.C. and Tam, D.M. 1990 Water systems and the environment. In Thanh, N.C. and Biswas, A.K. (eds), *Environmentally-sound water management*. Delhi, Oxford University Press: 1–29.

Thiel, H. and Schriever, G. 1990 Deep-sea mining, environmental impact and the DISCOL Project. *Ambio* 19: 145–150.

Thomas, D.S.G. and Middleton, N.J. 1994 *Desertification: exploding the myth*. Chichester, Wiley.

Thomas, J.A. 1991 Rare species conservation: case studies of European butterflies. In Spellerberg, I.F., Goldsmith, F.B. and Morris, M.G. (eds), *The scientific management of temperate communities for conservation*. Oxford, Blackwell Scientific: 149–197.

Thompson, D.R., Becker, P.H. and Furness, R.W. 1993 Long-term changes in mercury concentrations in herring gulls *Larus argentatus* and common terns *Sterna hirundo* from the German North Sea coast. *Journal of Applied Ecology* 30: 316–320.

Thrupp, L.A. 1991 Sterilization of workers from pesticide exposure: causes and consequences of DBCP-induced damage in Costa Rica and beyond. *International Journal of Health Services* 21: 731–739.

TMG (Tokyo Metropolitan Government) 1985 *Protecting Tokyo's environment*. Tokyo, TMG.

Tolba, M.K. 1992 *Saving our planet*. London, Chapman & Hall.

—— and El-Kholy, O.A. (eds) 1992 *The world environment 1972–1992*. London, Chapman & Hall.

Tolmazin, D. 1985 Changing coastal oceanography of the Black Sea. I: northwestern shelf. *Progress in Oceanography* 15: 217–276.

Tolouie, E., West, J.R. and Billam, J. 1993 Sedimentation and desiltation in the Sefid-Rud reservoir, Iran. In McManus, J. and Duck, R.W. (eds), *Geomorphology and sedimentology of lakes and reservoirs*. Chichester, Wiley: 125–138.

Transnet 1990 *Energy, transport and the environment*. London Transnet.

Tsehai, F.M. 1991 Refugees in Ethiopia: some reflections on the ecological impact. *IDOC Internazionale* 22(2): 2–6.

Tucker, C.J., Dregne, H.E. and Newcombe, W.W. 1991 Expansion and contraction of the Sahara Desert from 1980 to 1990. *Science* 253: 299–301.

Turner, B.L., II, Moss, R.H. and Skole, D.L. 1993 Relating land use and global land-cover change: a proposal for an IGBP-HDP core project. *International Geosphere–Biosphere Programme Report* 24.

Turnock, D. 1993 Romania. In Carter, F.W. and Turnock, D. (eds) *Environmental problems in Eastern Europe*. London, Routledge: 135–163.

UN (United Nations) 1989a Prospects of world urbanization. *Population Studies* 112.

——1989b *Study on the economic and social consequences of the arms race and military expenditures*. Disarmament Study Series 19. New York, UN.

UNDP (UN Development Programme) 1993 *Human development report.* New York, UNDP.

UNECLA (UN Economic Commission for Latin America and the Caribbean) 1990 *The water resources of Latin America and the Caribbean: planning, hazards, and pollution.* Santiago, UNECLA.

——1991 *Sustainable development: changing production patterns, social equity and the environment.* Santiago, UNECLA.

UNEP (UN Environment Programme) 1984 *Socio-economic activities that may have an impact on the marine and coastal environment of the East African region.* UNEP Regional Seas Reports and Studies 41. Nairobi, UNEP.

——1987 *Environmental data report.* Oxford, Blackwell.

——1989a *State of the Mediterranean marine environment.* Athens, UNEP.

——1989b Sustainable water development. *Water Resources Development* 5: 223–251.

——1990 *Green energy: biomass fuels and the environment.* Nairobi, UNEP.

——1991 *Environmental data report 1991/92.* Oxford, Blackwell.

——1992 *World atlas of desertification.* Sevenoaks, Edward Arnold.

——1993 *Environmental data report 1993/94.* Oxford, Blackwell.

——IE/PAC, 1993 *Cleaner production worldwide.* Paris, IE/PAC.

——/UNESCO/UN-DIESA 1985 *Coastal erosion in west and central Africa.* UNEP Regional Seas Reports and Studies 67. Nairobi, UNEP.

——/WHO 1988 *Assessment of freshwater quality.* Nairobi, UNEP.

——/——1992 *Urban air pollution in megacities of the world.* Oxford, Blackwell.

UNHCR (United Nations High Commissioner for Refugees) 1994 *Environmental issues in Benaco refugee camp.* Press release, 21 June.

USEPA (US Environmental Protection Agency) 1983 *Chesapeake Bay Program: findings and recommendations.* Philadelphia, EPA.

——1989 *The solid waste dilemma: an agenda for action.* Washington, DC, EPA Office of Solid Waste.

——1990 *National water quality inventory.* 1988 Report to Congress, Office of Water. EPA 440-4-90-003. Washington, DC.

Valencia, R., Balslev, H. and Paz y Miño, G. 1994 High tree alpha-diversity in Amazonian Ecuador. *Biodiversity and Conservation* 3: 21–28.

Viet Nam 1985 *Viet Nam: National conservation strategy.* Ho Chi Minh City, Committee for Rational Utilisation of National Resources and Environmental Protection Programme 52-02.

Viner, A. 1992 Biodiversity, fisheries, and the future of Lake Victoria. *IUCN Wetlands Programme Newsletter* 6: 3–6.

Voight, B. 1990 The 1985 Nevado del Rúiz volcano catastrophe: anatomy and retrospection. *Journal of Volcanology and Geothermal Research* 42: 151–188.

von Below, M.A. 1993 Sustainable mining development hampered by low mineral prices. *Resources Policy* 19: 177–181.

Wali, A. 1988 *Kilowatts and crisis: a study of development and social change in Panama.* Boulder, CO, Westview.

Walker, H.J. 1990 The coastal zone. In Turner, B.L., II, Clark, W.C., Kates, R.W., Richards, J.F., Mathews, J.T. and Meyer, W.B. (eds), *The earth as transformed by human action.* Cambridge, Cambridge University Press: 271–294.

Walsh, R. and Reading, A. 1991 Historical changes in tropical cyclone frequency within the Caribbean since 1500. *Würzburger Geographische Arbeiten* 80: 198–240.

Wang, W.C., Yung, Y.L., Lacis, A.A., Mo, T. and Hansen, J.E. 1976 Greenhouse effect due to manmade perturbations of other gases. *Science* 194: 685–690.

Warren, A. and Agnew, C. 1988 *An assessment of desertification and land degradation in arid and semi-arid areas.* Drylands paper 2. London, International Institute for Environment and Development.

—— and Khogali, M. 1992 *Assessment of desertification and drought in the Sudano-Sahelian region 1985-1991.* New York, United Nations Sudano-Sahelian Office.

WCED (World Commission on Environment and Development) 1987 *Our common future.* Oxford, Oxford University Press.

——1992 *Our common future reconvened.* London, WCED.

WCMC (World Conservation Monitoring Centre) 1992 *Global biodiversity: status of the Earth's living resources.* London, Chapman & Hall.

Weber, P. 1994 Safeguarding oceans. In Brown, L.R. (ed.), *State of the world 1994.* New York, W.W. Norton: 41–60.

WEC (World Energy Council) 1992 *1992 survey of energy resources.* London, WEC.

——1993 *Energy for tomorrow's world – the realities, the real opinions and the agenda for achievement.* London, Kogan Page.

Weinberg, B. 1991 *War on the land: ecology and politics in Central America.* London, Zed Books.

Weinberg, C.J. and Williams, R.H. 1990 Energy from the sun. *Scientific American* 263(3): 98–106.

Weiss, H., Courty, M.-A., Wetterstrom, W., Guichard, F., Senior, L., Meadow, R. and Curnow, A. 1993 The genesis and collapse of third millenium north Mesopotamian civilisation. *Science* 261: 995–1004.

Wellner, F.-W. and Kürsten, M. 1992 International perspective on mineral resources. *Episodes* 15: 182–194.

Westing, A.H. 1980 *Warfare in a fragile world: military impact on the human environment.* London, Taylor and Francis.

——(ed.) 1984 *Herbicides in war: the long-term ecological and human consequences.* London, Taylor and Francis.

—— and Pfeiffer, E.W. 1972 The cratering of Indochina. *Scientific American* 226(5): 21–9.

Whitelegg, J. 1992 Transport and the environment. *Geography* 77: 91–93.

——1993 *Transport for a sustainable future: the case for Europe.* London, Belhaven.

——1994 Transportation: for a sustainable policy. *Acid News* 1 (February): 8–9.

Whitlow, R.J. 1990 Mining and its environmental impacts in Zimbabwe. *Geographical Journal of Zimbabwe* 21: 50–80.

Whittaker, R.H. and Likens, G.E. 1973 Carbon in the biota. In Woodwell, G.M. and Pecan, E.V. (eds), *Carbon and the biosphere.* Washington, DC, US Department of Commerce: 281–302.

WHO (World Health Organization) 1972 Health hazards of the human environment. Geneva, WHO.

——1992a *Our planet, our health.* Geneva, WHO.

——1992b Cholera in the Americas. *Weekly Epidemiological Record* 67: 33–40.

Wijkman, A. and Timberlake, L. 1984 *Natural disasters: acts of God or acts of man?* London, Earthscan.

Williams, E.H. and Bunkley-Williams, L. 1990 The worldwide coral reef bleaching cycle and related sources of coral mortality. *Atoll Research Bulletin* 335: 1–71.

Williams, J. 1994 The great flood. *Weatherwise* 47: 18–22.

Williams, M. 1990a Understanding wetlands. In Williams, M. (ed.), *Wetlands: a threatened landscape.* Institute of British Geographers Special Publication 25. Oxford, Blackwell: 1–41.

——1990b Agricultural impacts in temperate wetlands. In Williams, M. (ed.), *Wetlands: a threatened landscape.* Institute of British Geographers Special Publication 25. Oxford, Blackwell: 181–216.

Williams, M.A.J., Dunkerley, D.L., De Deckker, P., Kershaw, A.P. and Stokes, T. 1993 *Quaternary environments.* London, Edward Arnold.

Willson, B. 1902 *Lost England: the story of our submerged coasts.* London, George Newnes.

Wilson, E.O. 1989 Threats to biodiversity. *Scientific American* 261(3): 60–66.

Wilson, J.S. 1858 The general and gradual desiccation of the earth and atmosphere. *Report of the Proceedings of the British Association for the Advancement of Science:* 155–156.

Wirawan, N. 1993 The hazard of fire. In Brookfield, H. and Byron, Y. (eds), *South-East Asia's environmental future*. Tokyo, UN University Press: 242–260.

Wischmeier, W.H. 1976 Use and misuse of the Universal Soil Loss Equation. *Journal of Soil and Water Conservation* 31: 5–9.

Wood, L.B. 1982 *The restoration of the tidal Thames*. London, Hilger.

Woodruff, N.P. and Siddoway, F.H. 1965 A wind erosion equation. *Proceedings of the Soil Science Society of America* 29: 602–608.

Woodwell, G.M., Wurster, C.F. and Isaacson, P.A. 1967 DDT residues in an east coast estuary: a case of biological concentration of a persistent insecticide. *Science* 156: 821–824.

World Bank 1985 *Desertification in the Sahelian and Sudanian zones of West Africa*. Washington, DC, World Bank.

——1989 *Philippines: environmental and natural resource management study*. Washington, DC, World Bank.

——1992a *World development report 1992: development and the environment*. New York, Oxford University Press.

——1992b Strategy for African mining. *World Bank Technical Paper* 181.

Worster, D. 1979 *Dust bowl*. New York, Oxford University Press.

Worthington, E.B. 1978 Some ecological problems concerning engineering and tropical diseases. *Progress in Water Technology* 11: 5–11.

WRI (World Resources Institute) 1986 *World resources 1986*. New York, Basic Books.

——1987 *World resources 1987*. New York, Basic Books.

——1990 *World resources 1990–1991*. New York, Oxford University Press.

——1992 *World resources 1992–1993*. New York, Oxford University Press.

——1994 *World resources 1994–1995*. New York, Oxford University Press.

Wright, D.H. 1990 Human impacts on energy flow through natural ecosystems, and implications for species endangerment. *Ambio* 19: 89–194.

Yakowitz, H. 1993 Waste management: what now? what next? an overview of policies and practices in the OECD area. *Resources, Conservation and Recycling* 8: 131–178.

Yhdego, M. 1991 Scavenging solid wastes in Dar es Salaam, Tanzania. *Waste Management and Research* 9: 259–265.

Young, J.E. 1992 *Mining the earth*. Worldwatch Paper 109. Worldwatch Institute, Washington, DC.

Zaika, B.E. 1990 Change in macrobenthic populations in the Black Sea with depth (50–200m). *Proceedings of the Academy of Sciences of the Ukrainian SSR, Series B*, 11: 68–71 (In Russian).

INDEX

Aberfan 243–4, 246
Abu Simbel 175
acid mine drainage 244, 245
acid rain 60, 90, 114–26, 144, 203, 245
  combatting effects 123–6, 246, 247
  effects 118–23, 245
  geography 115–18
  nature 114–15
acid shock 119
Aegean 141
aerial photography 72
Africa 48, 64–5, 88, 97, 104, 150, 153–4, 168, 263, 277
  see also specific countries
Agadez 296
Agent Orange 152
Agricola, Georgius 240
agriculture see food production
agronomic measures 82
Akosombo Dam 172, 175, 176
Åland Convention 262
Alaska 145–6, 187, 295
Alaska Highway 200
Albania 168
albedo change 100
aldrin 64, 89
alewife 202
algae 119, 281
algal blooms 70, 145, 147, 164, 277
Algeria 79
alkalinisation 119
Altamount Pass 228
Alzheimers disease 122
Amazon Basin 32–3, 35, 38, 42, 89, 246
Amazonian peacock bass 163
Ambukloo Dam 179
Amu Darya River 162–3
anchoveta, Peruvian 129–30

Andes Mountains 56, 273–4
Angkor Wat 284
Angola 261
Antarctic 93, 130
  krill 130–1
  Treaty 262
  see also Southern Ocean
Appalachian Mountains 241
aquaculture 69–70
Arabian Oryx (Oryx leucoryx) 98
Aral Sea 51, 62, 162–3
Arctic 291
  bowhead whale 131
  circle 4
  Ocean 108
Armenia 195
Armero 273
Asia 48, 88, 150, 168, 171
  see also South Asia; South East Asia; specific countries
Asia Minor 90
'Asian Tigers' 239
asthma 122, 162
Aswan High Dam 171–2, 175
Ataturk Dam 182
Atitlan Lake 162
Atlantic sperm whale 131
atmospheric bomb testing 73
Auglaize River 90
Australia 6, 15, 48, 77, 87, 201, 238, 250, 287, 291
  biological control 65–6
Austria 234
Azores 58
Azov, Sea of 138–9
Aztecs 15

Baikal Lake 165
Balbina Dam 33, 174
Bali 142

Baltic Sea 234
Banc d'Arguin, Parc National du 153–4
Bangalore 34
Bangkok 109, 186
Bangladesh 88, 110, 275–6, 277, 280–1
  Flood Control Action Plan 275, 281
barsha 281
Basel Convention on the Control of Transboundary Movements of Hazardous Wastes and their Disposal 217
Basques 131
Bayano Hydroelectric Complex 172–3
behavioural control 65
Beijing 192, 194, 209
Beira corridor 260
Belarus 167, 236
Belém–Brasilia Highway 202
Belgium 253
Benaco 259
benthic invertebrates 119
Berlin 218
Bhopal 195
bilharzia 164
Bingham Canyon copper mine 243, 247
bioaccumulation 63–4, 89, 157, 174
biochemical oxygen demand (BOD) 59, 156
biodiversity 36, 85
biogeochemical cycles 6–8
bioleaching 246, 249
biological control 65
biological invasions 201
biomass 1–2, 206–7
  as energy source 229–30
  biomes 3–6

biotechnology 65, 66–9
bison, European 257
Black River wetlands 249
Black Sea 138–9
blanket bogs 169
blue baby syndrome 60, 158
Blue Revolution 69
blue whale 131
Bogota 274
Bohemia 116
bois de prune blanc (*Drypetes caustica*) 91
Bolivia 271
bollworm, American 63
Bombay 178
bonna 281
boreal forest 4
Borneo 152
Boston 6
Botswana 48, 97, 168
Bougainville 246
Boxer Rebellion 257
Brahmaputra
  Delta 152
  River 36, 281
Brakopondo reservoir 174
Brandt line 23
Brazil 82, 185, 226, 239, 294
  biomass fuel 206, 229
  dams 171, 174
  deforestation 32–3, 37, 38, 89, 201, 185, 241–2
  diseases 202, 246
  mining pollution 245
  sustainable city 196–7
Breckland 81
Britain *see* UK
British Phosphate Commission 250
Brittany 230
Brokopondo 177
bronchitis 162
Bronze Age 236
Brooklyn 188
brown tree snake (*Boiga irregularis*) 92
brown trout 119
BST 68
buffalo 90
buffering capacity 118
Bulgaria 234
Buriganga River 280
Burundi 88
Büyük Menderes River 141

cacti 90
Cactoblastus 65–6
cadmium pollution 122, 244–5
caesium–134 234
caesium–137 73, 234–5
Cahora Bassa Dam 172, 173, 179

Cairo 13, 194
Calcutta 70
Calha Norte programme 33
California 64, 133, 162, 216, 227–8, 244, 270
Cambodia 284
Campeche, Bay of 256
campesino 263
Canada 12, 13, 98, 120, 124, 126, 174, 178, 187, 194, 219, 226, 234, 245, 246, 292
  effects of global warming 108
  Fraser River blockage 203
Canal del Dique 149–50
Canary Islands 16
cancers 64, 100, 158, 276
Captain Cook 92
Caracas 195
carbon cycle 7–8
carbon dioxide, atmospheric 8, 36, 103–4, 107, 111, 115, 206, 230, *see also* global warming; greenhouse effect
carbon–14 235
Caribbean 16, 80–1, 110, 147, 150
carp 69
carrying capacity 51
Caspian Sea 143, 144, 179, 202
catalytic converters 123, 205–6
catchment management plans 162
Central Africa 257, *see also* specific countries
Central America 168, 263, 284, *see also* specific countries
Central Andes 56
Central Asia 43, 62, 64, 163, *see also* specific countries
centre-pivot irrigation 53
cetaceans 90, *see also* whales
chlorofluorocarbons (CFCs) 103, 104–5, 111
Chagas' Disease (American trypanosomiasis) 202
Chalk River 234
Chao Phraya River 186
chaparral 6
Chengchow 255
Chengdu 195
Chernobyl 234–6
Chesapeake Bay 154, 167
Chickamauga River 90
Chile 6, 91, 247
China 13, 124, 125, 142, 192, 194, 203, 207, 239
  aquaculture 69
  dams 171, 175, 177, 179, 180, 255
  earthquakes 271–2
  giant panda 88–9
china clay 240–1, 246
chloride pollution 248–9

cholera 145, 157, 159, 203, 276–7
Churchill 194
cities *see* urbanisation; specific cities
class struggles 84
cleaner production 221–3
climatic change 90, 99–113, 276
  around reservoirs 178–9
  greenhouse gases 36, 103–5
  human impacts on 99–100
  nuclear war 259
  past 100–3
  see also global warming; ozone depletion
Club of Rome 237
coasts 109–10, 120, 134, 136, 141–54
  habitat destruction on 146–52
  management and conservation 153–4
  pollution 144–6
  physical changes 141–4
  see also coral reefs; mangroves
coelacanths 97
coliform organisms 186
Colombia 149–50, 273–4
Colon 173
Colorado 252
Colorado beetle 62
Colorado River 142, 171
Columbia River 171
Commission for the Conservation of Antarctic Marine Living Resources (CCAMLR) 130–1
commodity concentration of exports 23
Comoros 97, 142
computer-controlled spark ignition 206
coniferous forest 4
Convention for the Protection of the Black Sea 139
Convention for the Regulation of Whaling 132
Convention on Biodiversity 69
Convention on International Trade in Endangered Species (CITES) 97
Convention on Long–Range Transboundary Air Pollution (CLRTAP) 124
Convention on the Protection of the Rhine against Chlorides 248
Cook Islands 270
Copper Basin 241
Copper Belt 246
coral bleaching 150
coral reefs 77, 111, 147–50, 242
Cordero coal mine 243, 247
Cornwall 241, 246
corrosion 122
cost–benefit analysis 286–7

Costa Rica 64, 153, 262
Côte d'Ivoire 31, 151, 171
cotton plantations 62, 63, 64,
    162–63
Country Environmental Profiles 292
Cracow 159
cradle to grave approach 213, 286
credit, access to 25
Cretaceous period 85
critical loads 118, 124, 165
Croatia 257
crop
    cloning 68
    failures 68
    yields 53, 58, 60, 67, 68
    see also food production
cross-breeding 67
Crown-of-Thorns starfish
    (*Acanthaster planci*) 150
cultural eutrophication 164
cultural imperialism 24
Cumberland River 202
Cumbria 233, 234
Curitiba 196–7
cyclones, tropical 34, 110, 147,
    274–6, *see also* hurricanes
Czechoslovakia 252

Dakar 50, 200
dams 142, 160, 171–83
    benefits 171–3
    environmental impacts 173–80
    political impacts 180–3
    see also specific dams
Danjiangkou Dam 179, 180
Danube River 138, 158
Dar es Salaam 218
Darwin, Charles 16
DDT 62, 63, 89, 277
debt crisis 23
debt-for-nature swaps 38, 293–4
defoliation 121
deforestation 16, 28–42, 50, 79,
    104, 144, 165, 179, 185, 200, 201,
    242
    by refugees 259–60
    causes 29–34
    consequences 34–7
    during war time 253–4
    rates 28–9
    see also mangroves
Delhi 192, 195, 296
demersal fish species 127
denitrification 60
Denmark 60–1, 64, 128, 228
deoxygenation 244
desalination 145
desert 5–6
desertification 43–55
    areas affected by 45–6

causes 46–51
definition 43–5
Great Plains 53
Sahel 53–5
theoretical problems 51–3
desuphurisation 123
detergents 138, 144, 165
Detroit River 202
Devon 241, 246
Dhaka 280
diarrhoea 157, 164
dibromochloropropane 64
Didcot 103
dieldrin 64, 89
diesel engines 206
dilute and disperse philosophy 214,
    233
dinosaurs 85
diseases 37, 52, 59, 63, 65, 67, 69, 76,
    89, 93, 107, 120, 122, 145, 147,
    185, 284
    epidemics 276–7
    exotic 246
    irrigation 61–2, 175–6
    near nuclear installations 233
    polluted water 157, 186, 195–6,
    244–5
    transport 202–3
    see also specific diseases
diversity 3
Djibouti 232
DNA 67
dodo (*Raphus cucullatus*) 90
domestication 56–7
Dominican Republic 81
Dounreay 233
drainage 166–8, 277
drinking water standards 61, 158,
    186
drought 45, 46, 49, 52, 53, 54, 79–80,
    100, 120, 271
drylands 44
Dubrovnik 257
dung beetles 89
Dust Bowl 43, 79
dust storms 45, 75, 77, 108, 162, 257
Dutch elm disease 203

Earth Summit 69, 286
earthquakes 178, 195, 271–2, 273
East Africa 51
East Asia 69
East Kalimantan 37
ecocide 255
ecoimperialism 38
ecosystem 1
ecotourists 133
Ecuador 24, 33, 80, 153
Eder Dam 255
Egypt 61, 62, 63, 168, 171–2, 179

El Niño 129, 130, 150
El Salvador 232, 255, 263, 272
Elbe River 146
electronic fuel injection 206
elephants 90, 97, 260–1
emphysema 162
endemic species 66
energy 111, 224–36
    efficiency and conservation 123,
    225–6
    nuclear 232–6
    renewable 226–32
    sources 224–5
    use projections 111–12
England 38, 58, 61, 81, 116, 141, 154,
    231, 240–1, 253, 277, *see also* UK;
    specific locations
enteric fermentation 104
environmental activists 17, 247
environmental control 65
environmental economics 250, 286–7
environmental hazards 268
Environmental Impact Assessment
    181, 247
environmental law 16, 95–7, 224,
    247–8
Environmental Modification
    Convention 253
environmental services 282
EPIC 73
epidemics 276–7
epiphytes 4
equilibrium
    dynamic 10–11
    thermal 187, 200
Erie Lake 90, 202
erodibility 71, 72
erosion *see* soil erosion
erosivity 71, 72
Estonia 241
ethanol 206
Ethiopia 50, 81, 259
Ethiopian Highlands 81, 172
Euphrates Dam 182
Europe 103, 115–16, 117, 121, 126,
    143–4, 166, 206, 221, *see also*
    specific countries
European Monitoring and
    Evaluation Programme (EMEP)
    117
eutrophication 76, 138, 164, 244, 284
ex situ conservation 97–8
Exclusive Economic Zones 127
exhaust recirculation systems 206
experimental erosion plots 72
exploitation and dependency
    global scale 22–4
    national scale 24–5
Exxon 23
Exxon Valdez 145–6, 295

Faial 58
Falklands Islands 257
Falu copper mine 245
famine 52–3, 197, 259, 271
feedback 11–12, 35, 100, 108, 289
Felmersham Gravel Pits 249
fenitrothion 65
Fens 81
fertilisers 59–61, 75, 144, 167, 221, 292
Ferussac snail (*Euglandina rosea*) 66, 92
fin whale 131
Finland 161, 234
Finland, Gulf of 144
fire 5, 6, 51, 89, 114, 193
fish farming 69
floods 36, 76, 143–4, 178, 195, 255, 277–81
Florida 275
flue gas desulphurisation 123
fluidised bed technology 123
foxes 194, 257
    flying 97
food chain 8–9
food production 52–3, 56–70, 79–84, 107, 154, 155
    agricultural innovations 19
    agricultural change 56–9
    agricultural pests 37, 52, 59, 62–6, 120
    agricultural pollution 60–1, 63–4, 158
    alternatives to pesticides 65–6
    aquaculture 69–70
    biotechnology 66–9
    fertiliser use 59–61
    irrigation 61–2
    pesticides 62–5
    see also oceans
Ford Motor Company 23
forest cover changes 29
fossil fuels 7, 8, 103–4, 111, 115, 116, 123, 171, 189, 203, 224–5, 226
fragmentation 37
Frains Lake 79
France 62, 123, 179, 231, 232, 234, 253, 262
    noise and house prices 207
    pollution of Rhine 248
    tidal power 230
    urban sanitation 195–6
Franklin River Project 180
Fraser River 203
free-range animal products 59
French Guiana 79
fruit bats 97
fuelwood crisis 33–4, 50, 151, *see also* biomass
full cost pricing 250, 286–7
fungicides 62

Gadani 146
Gaia hypothesis 10
Gambia 88
Ganges River 36, 281
Ganges–Brahmaputra
    Delta 152, 276
    floodplains 291
garbage 185, 193–4, 218, *see also* waste
garimpeiros 246
Garonne River 179
Gatun Lake 163
General Agreement on Tariffs and Trade (GATT) 24
general circulation models (GCMs) 36, 106–7, 259
General Motors 23
genetic and sterile male techniques 65
genetic engineering 57, 67
Geneva Conventions 261
geological timescale 11, 13
geomorphological change 108, 187, 200, 242–4, *see also* specific processes
geothermal energy 231–2
Germany 117, 124, 128, 215, 220, 234, 248
Gezira project 61
Ghana 40, 151, 172, 175, 178
giant African snail (*Achatina fulica*) 66, 92
giant panda (*Ailuropoda melanoleuca*) 88–9
glacial maxima 34, 103
glass recycling 221
Global Assessment of Human-Induced Soil Degradation (GLASOD) 73
Global Environmental Monitoring System (GEMS) 157
global warming 36, 105–12, 144
    predicting impacts 106–7
    impacts 107–10
    responses 110–12
    see also greenhouse effect; sea–level change
goats 91
Gobi Desert 49
Göksu Delta 169
gold mining 245, 246
gorillas 257
Grand Comore 142
Grande Carajás Programme 33, 241–42
grasslands 5, 88, 95
Great Barrier Reef 150
Great Depression 53
Great Lakes 165, 166, 202
    Water Quality Agreement 165

great whales 131
Greater Mussayeb Project 51
Greece 23
green GNP 289
Green Revolution 58
green winged orchid (*Orchis morio*) 88
greenhouse effect 103–5, 119
    greenhouse gases 12, 99, 103–5
    see also global warming
Greenland bowhead whale 131
Greenland National Park 93
Greenpeace 133
grey whales 133
Gross Domestic Product (GDP) 289
Gross National Product (GNP) 23, 270, 271, 289
groundwater 7, 10, 59, 161, 177, 227, 235, 247, 252, 257, 262
    agricultural effects on 60–1, 64, 158
    landfill effects on 214–16
    urban effects on 187–9
Group of Experts on the Scientific Aspects of Marine Pollution (GESAMP) 134, 147
Guam 92, 97, 150
Guanabarra Bay 185
Guatemala 64, 162
Guatemala City 195
Guayaquil, Gulf of 153
Guelaya Peninsula 262
Guinea 151
Guinea-Bissau 151
gypsum 122

habitat destruction and modification 58, 69, 87–90, 146–7
    threat to species 87–90
    acid rain 118–22
    coral reefs 147–50
    deforestation 34–7
    desertification 47–51
    global warming 107–10
    lakes 162–5, 245
    mangroves 150–2
    mining 33, 240–4
    rivers 157–60
    war 253–61
    wetlands 166–9, 277
habitat protection 37–9, 93–5, 169–70, *see also* protected areas
Haicheng 272
Haiti 81
Hall–Héroult process 219
halocarbons 104, 111
Han River 179
Haracleia 141
Hararghe 259

Hartisheikh 259
Hawaii 68, 78, 92, 108, 109, 147–9,
    273
Haworth 228
hazards 265–81
    areas 269–71
    classifications 265–268
    examples 271–81
    responses 268–9
    urban environments 195–6
    see also earthquakes; disease;
        floods; cyclones; volcanoes
health 122, 175–6, *see also* disease
heath fritillary (*Mellicta athalia*) 95
heavy metals 119, 122, 138, 146, 155,
    157, 189, 244
hedgerows 58
Hell's Gate 203
herbicides 62, 253
Hercynia 90
herring
    North Atlantic 127
    North Sea 127–8, 129
Highway BR-364 33, 201
highway construction 78
Himalayas 36
Hispaniola 81
Hitachi 23
Ho Chi Min trail 253
Högvadsån River 126
Holocene 56, 103
hominids 10, 224
Homo habilis 237
Homo sapiens 10, 56, 237
Honduras 259, 263
Hong Kong 88
Honshu Island 178
Hoover Dam 171, 178
Horn of Africa 259
house sparrow 202
Huang Ho River 56, 142, 255
Huang–Huai–Hai plains 291
Huayuankow dike 255
Hull 154
human rights 25, 37
Humberside 141, 142, 154
Humboldt, Alexander von 16
humpback whale 131
Hungary 158
hunger 107, *see also* famine
hunting 38, 90–1, 95–7
hurricanes 110, 147, 275, *see also*
    cyclones
hydroelectricity 227, *see also* dams
hydrological cycle 6–7, 10, 36

IBM 23
ice ages 101
icebergs 108
ice-core data 105

Iceland 132, 231, 273
ICI 23
incineration 216
India 16, 88, 116, 124–5, 171, 178,
    180, 185, 223, 239, 271, 296
    deforestation 33–4
    tigers 93, 95
    urban pollution 186, 195
    see also specific locations
Indian Ocean 142, 242
indigenous peoples 37
Indonesia 142, 153, 171, 277
Indus River 56
Industrial Revolution 19, 157, 237,
    295
insecticides 62
insurance
    industry 110, 269, 275, 280
    strategies 53–5
integrated pest management 65
International Boundary and Water
    Commission 182
International Commission on Large
    Dams 171
International Council on Metals and
    the Environment 247
International Soil Reference Center
    74
International Tropical Timber
    Agreement (ITTA) 40
International Tropical Timber
    Organization (ITTO) 40
International Union for
    Conservation of Nature and
    Natural Resources (IUCN) 86,
    292
International Whaling Commission
    (IWC) 132–4
iodine-131 234
Iran 169, 177, 178, 185
Iraq 51, 64, 182
Ireland 23, 169
Irish Sea 138, 233
Irkutsk 201
Iron Age 236
Irrawaddy
    Delta 152
    River 152
irrigation 43, 50–1, 61, 100, 154, 162–3
island states, small 109, 111, 270
Israel 188, 262
itai-itai disease 244–5
Italy 90, 91, 178, 231, 273
ITT 23
ivory 90, 287
Ixtoc 1 136, 256

Jaba River 246
Jakarta 186
Jamaica 147, 249

Jamuna River 186
Japan 23, 24, 69, 115, 116, 141, 152,
    158, 178, 216, 239, 244, 262, 271,
    273
    effects of earthquakes 272
    effects of nuclear war 258
    mining pollution 244–5
    Tokyo subsidence 187–8
    whaling 131, 132–3
    see also specific locations
Java 24, 76, 153
Johannesburg 245
Jonglei Canal 168
Jordan River 262
Juan Fernández islands 91

K-strategists 86
Kainji Dam 175
Kakadu National Park 287
Kalabsha 175
Kalahari Desert 48
Kampala 194, 207
Kaneohe Bay 147–9
kaolin 240
Karachay Lake 253
Karachi 184, 204
Karaj River 185
Kariba Dam 175, 176
Kawerong River 246
Kazakhstan 80, 252
Kenya 48, 232, 287
keystone species 86
Khartoum 50
Khorezm oasis 43
Kiribati 109
Konya Dam 178
Kora-Bogaz-Gol 142–3
Korea 116
Krasnodar 177
Kuala Lumpur–Ipoh highway 200
Kuban reservoir 177
Kuban River 177
Kurobe Dam 178
Kuroshio current 130
Kuta Beach 142
Kuwait 145, 243, 244, 256, 262
Kuznetsk Basin 241
Kwae Yai River 174

Labrador-Ungava plateau 12
Lac Léman (Lake Geneva) 164–5
Lagos 109, 142
La Grande 2 174
lahars 272
lake charr 202
Lake District 114
lakes 162–5
    acid deposition 118–19, 125–6, 245
    degradation 162–5
    management 165

lakes *contd*
  rehabilitation 246
  see also specific lakes; irrigation;
    urbanisation
land mines 257
land tenure 25, 84
landfill 213–16, 246
landslides 12, 35, 178, 195, 244, *see
  also* mass movement
La Rance estuary 230
Lar River 185
large blue butterfly (*Maculinea
  arion*) 95
laterisation 35
Latin America 64, 80, 187, 239, *see
  also* specific countries
Laysan finch 257
Laysan rail 257
leaching 120
lead pollution 122, 192, 204–5
lead-free petrol 204–5
lean-burn engines 123, 206
lemurs 91
Lesotho 73
levées 160, 280
Liaoning Province 272
Liberia 88
lichens 90, 120
life expectancy 64
Line Islands 109
lions 90
liquefied petroleum gas (LPG) 206
Little Ice Age 103, 105
livestock farming 59, *see also*
  overgrazing
llianas 4
locally undesirable land uses
  (LULUs) 216
locust control programmes 64–5
loess 77, 142
logging 28, 39–40, 165, *see also*
  deforestation
London 16, 109, 117, 143, 159–60,
  184, 188–9, 193, 224
London Dumping Convention 136,
  139
Long Beach 244
Long Island 63, 64
Los Angeles 193, 204, 244
  county 270
  Department of Water 162
Louvre 123
Love Canal 215
Lower Mesopotamia 43
Luz International 232
Luzon Island 273

Maas River 143
macrobenthic species 138
Madagascar 91

Madeira 16
Madeira River 245
Maghreb 79
Magna Carta 161
malaria 63, 157, 163, 176, 246, 277,
  284
Malawi 97, 260
Malaysia 152, 200
Maldives 110
mallee 6
Malta 90, 154, 227
Malthus, Thomas 18
mammoth 90
Manaus 174
manganese 33, 122, 242
mangroves 33, 69, 110, 111, 122,
  150–2, 168, 179, 185, 260, 276
Manic 3 Dam 178
Manila Metro 158, 185–6
Manitoba 13, 194
Maoris 87
Maputo 260
maquiladoras 24
maquis 6
Maraba 202
marabou storks 194
Maracaibo Lake 244
marsh ague 277
Marshall Islands 109
Masai Mara Reserve 287
mass extinction events 85
mass movement 71, *see also*
  landslides
Matsushita Electronics 23
Maúa 260
Mauritania 45, 153–4
Mauritius 90, 91
maximum sustainable yield 164
Mayan civilisation 34, 284
mechanical methods 82
Mediterranean Sea 139–40, 146, 169
megacities 184, 191
Meghna River 281
Mekong Delta 253
Melbourne 77
memory effect 126
mercury pollution 64, 122, 146, 159,
  174, 245
Mersey Estuary 231
Merseyside 217
Merthyr Vale 244
MesoAmerica 56
Mesolithic 28
Mesopotamia 13, 61, 107
methane 12, 103, 104, 108, 214
methanol 206
methylation 174
Mexico 34, 80, 171, 182, 239
  Basin of 291
  Gulf of 136, 167

Mexico City 168, 184, 192, 194, 272
Michigan state 79
Middle East 48, 189, 200, *see also*
  specific countries
Midway islands 257
migration 25, 26, 32–3, 175, 246,
  259–60
Milankovitch 101
Miletus, Gulf of 141
military activity 33, 78, 147, *see also*
  war
mineral production 237
minimum tillage 82
mining 33, 78, 154, 155, 233,
  237–50
  economic aspects 237–39
  environmental impacts 240–6
  rehabilitation and damage
    reduction 246–9
  true costs 249–50
Minjur aquifer 61
Minnesota Mining and
  Manufacturing (3M) 218
Miskito coast 259
Mississippi River 159, 167, 278–80
Missouri River 280
mixed farming 58
Möhne Dam 255
Mojave Desert 229
Mongolia 5, 16, 49, 79, 98
Mono Lake 162
monoculture 63
Montreal Protocol 105, 111
Montserrat 270
Moorea 66, 92
Moravia 252
Morocco 262
morphoclimatic regions 3
Mount Hope Bay 144–5
Mount Pinatubo 273
Mozambique 172, 173, 179
  effects of civil war 260–1
mulching 82
Myanmar (Burma) 26, 152
Myrmica spp. 95
Mysis relicta 120

Nam Choan Dam 174, 180
Namib Desert 5
Namibia 97, 218
Nanjing 207
Narmada Dam 180
Narrangansett Bay 144–5
Nasser Lake 175, 179
National Conservation Strategies
  292
National Environmental Action
  Plans 292
National Environmental Policy Plan
  (NEPP) 292–3

NATO 295
natterjack toad 119
Nauru 240, 250
Negev Desert 43
Negril wetlands 249
Neolithic 20, 28
Neolithic Revolution 56, 295
Nepal 93
Netherlands 59, 64, 98, 128, 215, 248
  coastal protection 143
  sustainable development
      programme 292–3
  wetland drainage 166–7
Nevada Desert 258
Nevado del Rúiz 273–4
Neves Corvo mine 238, 247
New York 184, 188, 193, 194
New York State Canal 202
New Zealand 4, 23, 86–7, 231
New Zealand mistletoe (*Trilepidea
  adamsii*) 86–7
Niagara Falls 215
Niamey 50
Nicaragua 232, 259, 271
nickel pollution 122, 245
Niger 48, 50, 52, 296
Niger River 200
Nigeria 54, 75, 76, 171, 175
Nile
  Delta 181
  Perch 164
  River 13, 56, 61, 62, 142, 168,
    171–2, 179
nitrates 59–61, 158
nitrogen oxides (NOx) 59, 103, 115,
  116, 123, 189, 203, 205
nitrogen cycle 10, 115
noise 207–8
non-governmental organisations 17,
  292
Norfolk Broads 249
North America 5, 36, 48, 73, 89, 90,
  103, 114, 115, 116, 118, 122, 125,
  126, 165, *see also* specific
  countries
North American Free Trade
  Agreement (NAFTA) 24
North Atlantic 58, 127
North Sea 136, 143, 146
North/South divide 24, 111, 293–5
Norway 69, 117–18, 119, 128, 132,
  203, 234
Nouakchott 45
nuclear power 232–6
nuclear war 258–9
nuclear waste *see* radioactive waste
nuclear weapons 252
nuclear winter 259
Nurek Dam 178
nutrients 3, 7

Oceania 239
oceans 111, 127–40
  fisheries 127–31
  pollution 134–40
  whaling 131–4
  see also specific oceans and seas
Ogallala aquifer 61
Ogasawara Islands 133
oil shale 241
oil spills 136, 137, 138, 145–6, 255–6,
  295
Okavango 168
Oman 98, 243
Omori-Minami 125
Onchoceriasis (river blindness) 157,
  175
Ontario 234, 245
OPEC oil crisis 33
open-field farming 58
Operation Tiger 93, 95
orchids 88, 90
Organisation for Economic
  Co-operation and Development
  (OECD) 217, 292
organochlorine compounds 64, 89,
  146
Orinoco Basin 14
Oslo Convention 136, 139
Otto, Nicolaus 218
Ouagadougou 50
overcultivation 49, 260
overgrazing 46, 47–9, 51–2
Owens Lake 162
ownership 21–2
oxidation 7
ozone
  depletion 100, 204
  pollution 189, 204–5

Pa Mong Dam 177
Pacific Ocean 109, 110, 147, 242
Pacific rim 239
Pakistan 50–1, 64, 146
Palabora mine 238
palaeoenvironmental indicators 13,
  100–3
Palenque 284
pampa 5
Panama 163, 172–3
  Canal 163
Panama City 173
pandas 88
pandemic 145
Pangaea 10
Papua New Guinea 246
Pará state 33, 241–2
Paraná state 33
Paris 196, 123
passenger pigeon 90
pathogens 62, 76, 155, 157, 276

Patzcuaro, Lake Basin 80
peanuts 119
peat 169, 249
Pelada gold mine 246
pelagic fish species 127
Pennsylvania 234
perception 269
Père David's deer 257
peregrine falcons (*Falco peregrinus*)
  89
permafrost 4, 107
  disturbance 187, 200
Permian period 85
Persia 90
Peru 33, 41, 91
pesticides 24, 62–5, 135, 138, 139,
  157, 292
pests 37, 52, 59, 62–6, 120
pH scale 114
Philae Island 175
Philippines 147, 158, 232
  deforestation 31–2, 151–2, 179
  Manila 185, 185–6
  natural hazards 271, 273
Philips 23
phosphates 59, 60
phosphorus 164
photosynthesis 8
photovoltaic (PV) power 228
phytoplankton 128
  blooms 145
pigs 59
piospheres 52
plaice, North Sea 127
Plain of Reeds 253
plantation agriculture 16
Plantation de Dautieng 253
Pleistocene 90, 101
plutonium–239 235
Po Delta 244
Poland 116, 124, 158–9, 257
polar bears 194, 257
polio 157
political acceptance of
  environmental issues 17
polluter pays principle 219, 248, 292
pollution 60–1, 63–4, 75–6, 89–90
  air 107–10, 118–23, 189–93, 203–7
  coastal 144–6
  fresh water 119–20, 138, 154–60,
    185, 245, 246, 248–9
  marine 134–40
  mining 244–6
  see also noise; pesticides; thermal;
    specific types
polychlorinated biphenyls (PCBs)
  213, 257
polymetallic (manganese) nodules
  242
Polynesians 92

population, human 18–19, 57, 84, 141, 289
Port Howard 257
Portugal 23, 58, 128, 158, 238, 247
potash 60
poverty 20, 32, 52, 84, 262–3, 285
Powder River Basin 243, 247
prairie 5
prickly pear 65–6
primary production 2
primary/secondary forest 34
Prince William Sound 145–6, 295
prior informed consent 217
Pripyat River 167
producers 9
Production Association Mayak 252–3
productivity 2–3
progressive desiccation 43
protected areas 39, 93–5, 153–4, 161, 165, 174, 200, 249
Przewalski's horse (*Equus caballus przewalski*) 98
Pteropodidae 97

Qatar 104
Quaternary 103
Quebec 174
Quito Basin 80

r–strategists 86
radioactive waste 137–8, 139, 233–4, 252–3
radionuclides 73, 138, 156, 234–6
rain-making 253
rainbow trout 119
raised bogs 169
Ramsar Convention 154, 169
raptors 89
refugees 25, 26, 259–60
resettlement 32–3, 175
resilience 12
resistance breeding 65
resources 15–16
  as cause of conflict 246, 262–3
  commonly owned 286
  military consumption 251–2
  renewable and non–renewable 225
  sustainable use 285–6
  valuation 286–7
respiration 8
return period 269
Réunion 91
Rhanthambore National Park 95
Rhine
  Delta 277
  River 146, 159, 248, 292
rhino 90
Richter scale 272

rills and gullies 72, 75
Rio de Janeiro 185, 195, 286
Rio Grande 182
Rio Grande do Sul state 33
risk assessment 269
rivers 157–62
  blockage 203
  pollution 119–20, 157–60, 186, 248–9
  management 160–2
  see also specific rivers; irrigation; dams; urbanisation
Riyadh 186
Riyadh River 186
road networks 30
Rocky Mountain Arsenal 252
Rocky Mountains 73
Romania 64, 234
Rondônia state 33, 201
Rosario Coral Reef National Park 149–50
Ruhr Valley 255
Russia 93, 95–6, 98, 125, 165, 169, 177, 178–9, 241
Rwanda 88, 257
Rybinsk reservoir 178–9

Sabah 37
Safeway 133
Sahara Desert 43, 45, 47, 90
Sahel 45, 48, 53–5, 100
Saiga antelope (*Saiga tatarica*) 95–6
Saint Louis 200
Saint Petersburg 144
salinisation 43, 50–1, 110, 177, 188, 247–8
salmon
  Atlantic 69, 126, 161, 202
  Pacific 203
salmonids 119
saltation 71
Samreboi 40
San Francisco 195
San Salvador 272
sand dunes 5, 14, 46
sandalo tree (*Santalum fernandezianum*) 91
Sanmen Gorge Project 175
Sanmenxia reservoir 177
Sanriku 272
Sao Paulo 184, 185
Sarawak Mangroves Forest Reserve 152
sardine
  California 130
  Japanese 130
satellite imagery 29, 33, 45, 275
Saudi Arabia 61, 154, 256
savanna 5, 88
Saxony 116

Scandinavia 114, 118, 169, 229, 234
Scheldt River 143
Schistosomiasis (bilharzia) 61, 157, 164, 176
science 17
scorched earth policy 253
Scotland 119, 169
sea lamprey 202
sea-level change 107–11, 144, 150, 276
sediment delivery ratios 73
Sefid-Rud reservoir 177, 178
Sellafield 138, 233
Semipalatinsk 252
Senegal 48, 50, 200
sensitivity 12
Seoul 184, 192
Severn Estuary 231
sewage 136, 145, 148, 155, 157, 158, 159, 186
  reuse 70, 220–1
Shanghai 192
shelter belts 82
shrimp farming 153, 283–4
Siberia 80, 93, 96, 116, 165
Siemens 23
Sierra Leone 88, 151
siltation 36, 43, 157, 158, 185
  reservoirs 76–7
Sites of Special Scientific Interest 161, 200, 249
skin cancers 100
slash and burn agriculture 33
slope instability 178, 200, 244, 272
smallpox 93, 203
smog 90, 99, 117
Smokey Mountain 185
snails 66, 92, 121
Soccer War 263
socio-economic system, as part of ecosystem 282–6
sociocultural organisation 20–1
soil
  compaction 35, 48
  conservation 82–4
  degradation 35, 48, 49
  formation 78
  liquefaction 272
  management techniques 82
  productivity 52, 75
soil erosion 35, 49, 71–84, 114, 155, 200
  accelerated 77–82
  effects 74–7
  factors affecting 71
  techniques for measurement 71–4
Solar One 229
solar power 228–9
solar thermal electric generator 228–9

Somalia 259
songbirds 89
Sorpe Dam 255
South Africa 5, 6, 25, 97, 118, 172, 238, 245, 261
South America 5, 43, 65, 203, 277, *see also* specific countries
South Asia 69, 70, 276, *see also* specific countries
South East Asia 116, *see also* specific countries
Southeastern Anatolian Project 182
southern Africa 43, *see also* specific countries
Southern Ocean 130–1
Soviet Union, former 53, 93, 157, 227
 Aral Sea 162–3
 Chernobyl 234–6
 Lake Baikal 165
 mining damage 241–2
 nuclear contamination 252–3
 pesticide use 64
 urban air pollution 190–1
 Virgin Lands Scheme 79–80
 see also specific countries
Spain 124, 230
spatial scales 12–13
species–area relationship 89
species, threatened 85–98
 conservation efforts 92–8
 definition and theory 86
 extinction rates 85–6
 island species 91–2
 threats to 86–92, 257
sperm whale 131
Spitzbergen Treaty 262
splitmouth (*Lagochila lacera*) 90
Sri Lanka 88
St Helena 91
stability 12
stent 241
steppe 5, 79
Stewart Island 4
Stone Age 90, 236
strontium–90 235
structural inequalities 22–7
sturgeon 179
Subir civilisation 107
subsidence 143, 187–8, 200, 215, 232, 244
succession 258
Sudan 46, 50, 52–3, 61, 168
Sudbury 124, 125–6, 245, 246
Sudd Swamps 168
sugar-cane 68
suicide 64
Sulawesi 153
sulphur dioxide 115–18, 120, 122–3, 124, 192–3, 203, 206, 245, 246, 247

Sumatra 153
Sumerian civilisation 43
Sundarbans 110, 152, 276
Superfund 216
Surinam 174, 177
suspended sediment loads 72
suspension 71
sustainable development 27, 282–97
 and economic growth 288–90
 national plans for 291–3
Sweden 64, 78, 95, 114, 117–19, 124, 126, 234, 245
Switzerland 62, 164–5, 220, 234, 248
synergy 108, 212
synthetic organic compounds 155
Syr Darya River 162–3
Syria 90, 182
systems 9

taiga 4
Taiwan 142, 283
Tajikistan 178
tambalocque tree (*Sideroxylon sessiliflorum*) 90
Tanzania 218, 259
Tasmania 180
Tebrak 61
Techa River 253
technology 19–20, 289
tectonic movements 10, *see also* earthquakes
Tehran 185
Tel Aviv 188
telle 79
temperate forest 4
temperate grassland 5
Tennessee River 202
Tennessee Valley Authority 176
Texas Panhandle 53
Thailand 26, 93, 152, 171, 174, 180, 283
Thames
 Barrier 143
 River 159–60
 Water Authority 161
thermal pollution 156
Thessaly 90
30% Club 124
Three Gorges Dam 175, 180
Three Mile Island 234
thresholds 12
Thung Yai Wildlife Sanctuary 174
Thymus praecox 95
tidal power 144, 230–1
tigers
 Asian 90, 93, 95, 257
 sabre-toothed 90
Tigris–Euphrates Rivers 56, 61, 182
tilapia 69
time lags 12, 278

timescales 10–12
tissue culture 57, 67
Tokelau 109
Tokyo 109, 124, 125, 187–8, 193, 206
Tonga tribe 175
Toronto 219
Toyama prefecture 244–5
trachoma 157
trade 24, 238–9
 in threatened species 90–1, 97
 terms of 23, 84
tragedy of the commons 21
transport 197–211
 access 30, 31, 33
 congestion 185
 impacts on atmosphere 192, 203–7
 impacts on biosphere 201–3
 impacts on land 200–1
 noise effects 207–8
 policy 208–10
Transnational corporations (TNCs) 23, 295
Trans-Siberian railway 200
Transvaal highveld 118
trophic level 9
tropical forest management 37–42
Tropical Forestry Action Plan (TFAP) 41
tropical rain forest 4–5, 29, 111, *see also* deforestation
trypanosomiasis 175
tsunamis 272
Tucuruí Dam 33, 174, 176
tundra 4, 12, 13
Tunisia 74, 75, 79
turbidity problems 156
Turkana 51
Turkey 141, 169, 182
Turkish Society for the Protection of Nature 169
Turkmenistan 143
Tuvalu 109
2,4,5-T 64
2,4-D 63, 64
typhoid 157
typhoons 275, *see also* cyclones
tyre disposal 216

Uganda 194, 257
UK 136, 179, 203, 204, 207, 228
 acid rain 114, 116, 117, 118, 119, 124
 Clean Air Act 193
 coastal erosion 141, 154
 deforestation 38
 mining impacts 241, 246
 nuclear waste 137–8, 233–4
 river management 160–1
 road programme 200
 soil erosion 81–2

UK *contd*
  Thames pollution 159–60
  threatened species 87, 88, 95
  waste arisings 213
  see also England; Scotland; Wales;
    specific locations
Umbria 91
UN Conference on Desertification
  (UNCOD) 43, 45, 52
UN Convention to Combat
  Desertification 52
UN Disaster Relief Organisation
  (UNDRO) 270–1
UN Economic Commission for
  Europe 124
UN Environment Programme
  (UNEP) 44, 45, 74, 139, 157,
  191, 293
UN Framework Convention on
  Climate Change 111
UN International Decade for
  Natural Disaster Reduction 265
UNCED 291
unequal commodity exchange 23
Unilever 23
Union Carbide 23
Universal Soil Loss Equation
  (USLE) 73
Urals 252
urban heat islands 99
urbanisation 154, 184–96
  effects on atmosphere 189–93
  effects on groundwater 187–9
  effects on surface water 186–7
  garbage 193–4
  hazards 195–6
  rates 184
US Agency for International
  Development 292
US Environmental Protection
  Agency 157, 217
US Soil Conservation Service 53
USA 4, 62, 68, 124, 137, 171, 170,
  178, 182, 195, 220, 270, 275, 295
  air pollution 204–5
  coastal eutrophication 144–5
  energy issues 225, 227–8, 229, 231,
    234
  flooding 278–80
  Great Plains 43, 53, 79
  groundwater problems 61
  incineration sites 216
  landfill leakage 215
  military 252, 253–5, 262

mining impacts 239, 241, 243, 244,
  247
river pollution 157
soil erosion 73, 75, 77, 78, 79
trade ban 97
wetlands 166–7, 169
whales 133, 134
see also specific locations
Utah 243
Uzbekistan 43, 61

Vaiont Dam 178
Vakhish river 178
value 21–2
Vanuatu 270
vectors 157, 276, 277
veldt 5
Venezuela 171, 244
Vibrio 276–7
Victoria Lake 164
Viet Nam 88, 177
  deforestation 30–1
  War 30, 253–5, 257, 277
viral hepatitis 64
Virgin Lands Scheme 79–80
Virunga volcanoes 257
Vistula River 158–9
volatile organic compounds (VOCs)
  165, 189, 205
volatilisation 60
volcanic eruptions 100, 114, 272–4
Volga River 178, 179
Volta Lake 178
vulnerability to disasters 26, 55,
  270–1, 281

Wadi Hanifa 186
waldsterben 124
Wales 38, 81, 116, 231, 243–4
war 30, 52, 168, 182, 243, 251–64
  cost of preparation 251–3
  direct impacts 253–9
  environment as cause 262–3
  indirect impacts 259–61
  limiting effects 261–2
waste 212–23
  types 212–13
  disposal 136–8, 193–4, 213–16
  international movement 217
  hazardous 213, 214, 215–16, 252
  prevention 221–3
  radioactive 137–8, 139, 233–4, 252–3
  reuse, recovery, recycling 185,
    194, 217–21

water erosion *see* soil erosion
water hyacinth 164, 177
waterlogging 50, 61, 177
weathering of building stone 61,
  122
Welland Canal 202
welwitschia 5
West Africa 43, 50, 151, 200, *see also*
  specific countries
Western Sahara 262
wetlands 104, 144, 153–4, 165–70, 249
  destruction 166–9, 277
  natural functions 165–6
  protection 169–70
  see also mangroves; specific
    wetlands
Wetlands Reserve Program 169
whale watching 133–4
whales 131–4
Whaling Convention 132
willingness to pay 287
Wilmington 244
wind energy 227–8
wind erosion *see* soil erosion
wind erosion equation 73
Windscale 234, *see also* Sellafield
Witwatersrand 245
World Bank 33, 281, 292, 297
World Commission on Environment
  and Development 286, 288–9,
  293
World Conservation Strategy 286
World Health Organization (WHO)
  191
World Heritage Convention 154

Xochimilco 168

Yangtze River 175
Yanomani indians 246
Year of the Great Stink 159
York, Vale of 81
Yucatan Peninsula 34
Yugoslavia 234

Zabbalean 194
Zaire 257
Zambezi River 173, 175, 179
Zambia 176, 246
zebra 90
zebra mussel 202
Zimbabwe 97, 171
zooplankton 119
Zuiderzee 143